Universal Algebra

Mathematics and Its Applications

Volume 6

P.M. Cohn, F.R.S.
Bedford College, University of London,
London, England

Universal Algebra

D. REIDEL PUBLISHING COMPANY
Dordrecht: Holland / Boston: U.S.A. / London: England

Library of Congress Cataloging in Publication Data

CIP

Cohn, Paul Moritz.
Universal algebra.

(Mathematics and its applications ; v. 6)
1. Algebra, Universal. I. Title. II. Series: Mathematics and its applications (Dordrecht) ; v. 6)
QA251.C55 1981 512 80-29568
ISBN 90-277-1213-1
ISBN 90-277-1254-9 (pbk.)

Published by D. Reidel Publishing Company,
P.O. Box 17, 3300 AA Dordrecht, Holland

Sold and distributed in the U.S.A. and Canada
by Kluwer Boston Inc.,
190 Old Derby Street, Hingham, MA 02043, U.S.A.

In all other countries, sold and distributed
by Kluwer Academic Publishers Group,
P.O. Box 322, 3300 AH Dordrecht, Holland

D. Reidel Publishing Company is a member of the Kluwer Group

Printed in The Netherlands

To Deirdre, and her Parents

Notes to the Reader

The central part of the book is Chapter II, and this may be read directly after I.3, referring back to other parts of Chapter I when necessary. Much of the material in Chapters V and VI can be read after Chapter II, and most of Chapter VII can be read after Chapter IV. Any exceptions to these rules are usually indicated by a reference to the relevant chapter and section.

All theorems, propositions, and lemmas are numbered in a single series; thus 'Theorem IV.3.5' refers to item 5 of Section 3 of Chapter IV. Cross-references within a single chapter omit the roman numeral. The end of a proof is indicated thus: ∎

The Bibliography includes, besides the works referred to in the text, a few papers of general interest on universal algebra, but it is not intended to be exhaustive in any direction. Practically all the items listed have appeared since 1900, and this fact is utilized by referring to an item by the author's name and the last two digits of the year of publication, with primes to distinguish papers published in the same year.

Contents

CHAPTER VII. APPLICATIONS

CHAPTER VIII. CATEGORY THEORY AND UNIVERSAL ALGEBRA

CHAPTER IX. MODEL THEORY AND UNIVERSAL ALGEBRA

CHAPTER X. MISCELLANEOUS FURTHER RESULTS

CHAPTER XI. ALGEBRA AND LANGUAGE THEORY

Preface to the Revised Edition

The present book was conceived as an introduction for the user of universal algebra, rather than a handbook for the specialist, but when the first edition appeared in 1965, there were practically no other books entirely devoted to the subject, whether introductory or specialized. Today the specialist in the field is well provided for, but there is still a demand for an introduction to the subject to suit the user, and this seemed to justify a reissue of the book.

Naturally some changes have had to be made; in particular, I have corrected all errors that have been brought to my notice. Besides errors, some obscurities in the text have been removed and the references brought up to date. I should like to express my thanks to a number of correspondents for their help, in particular C. G. d'Ambly, W. Felscher, P. Goralčik, P. J. Higgins, H.-J. Hoehnke, J. R. Isbell, A. H. Kruse, E. J. Peake, D. Suter, J. S. Wilson. But I owe a special debt to G. M. Bergman, who has provided me with extensive comments, particularly on Chapter VII and the supplementary chapters. I have also consulted reviews of the first edition, as well as the Italian and Russian translations.

In addition there are four new chapters. Chapter VIII deals with category theory, in so far as it affects our subject. The construction of monads (triples) is described, with free algebras as an illustration, and Lawvere's definition of algebraic theories is outlined. Chapter IX presents the various notions of algebraic closure developed in model theory, particularly the existential closure and A. Robinson's infinite forcing, and its applications in algebra. Chapter X contains a number of isolated remarks related to the main text, and the final chapter is an article on algebraic language theory which appeared in 1975 in the Bulletin of the London Mathematical Society; I am grateful to the Society for permission to include it here. Although not directly concerned with our topic, it describes the links with automata theory, itself of considerable relevance in universal algebra.

Several friends have read the new chapters and have provided helpful comments; in addition to G. M. Bergman they are W. A. Hodges and M. Y. Prest, and I should like to thank them here. I am also grateful to the publisher, D. Reidel and Co., for accepting the book in their Mathematical Series, and to their staff, for their efficiency in seeing it through the press.

Bedford College, London P. M. Cohn
December, 1980

Preface

Universal algebra is the study of features common to familiar algebraic systems such as groups, rings, lattices, etc. Such a study places the algebraic notions in their proper setting; it often reveals connexions between seemingly different concepts and helps to systematize one's thoughts. The actual ideas involved are quite simple and follow as natural generalizations from a few special instances. However, one must bear in mind that this approach does not usually solve the whole problem for us, but only tidies up a mass of rather trivial detail, allowing us to concentrate our powers on the hard core of the problem.

The object of this book is to provide a simple account of the basic results of universal algebra. The book is not intended to be exhaustive, or even to achieve maximum generality in places where this would have meant a loss of clarity. Enough background has been included to make the text suitable for beginning graduate students who have some knowledge of groups and rings, and, in the case of Chapter V, of the basic notions of topology. Only the final section of the book (VII.7) requires a somewhat greater acquaintance with representation theory.

The discussion centres on the notion of an algebraic structure, defined roughly as a set with a number of finitary operations. The fact that the operations are finitary may be regarded as characteristic of algebra, and its consequences are traced out in Chapter II. Those consequences, even more basic, that are independent of finitarity are treated separately in Chapter I. This chapter also provides the necessary background in set theory, as seen through the eyes of an algebraist.

One of the main tools for the study of general algebras is the notion of a *free algebra.* It is of particular importance for classes like groups and rings which are defined entirely by laws—i.e., *varieties* of algebras—and this has perhaps tended to obscure the fact that free algebras exist in many classes of algebras which are not varieties. To emphasize the distinction, free algebras are developed as far as possible without reference to varieties in Chapter III, while properties peculiar to varieties are treated separately in Chapter IV.

These two chapters present the only contact the book makes with homological algebra, and a word should perhaps be said about the connexion. The central part of homological algebra is the theory of abelian categories; this is highly developed, but is too restrictive for our purpose and does not concern us here. The general theory of categories, though at an earlier stage of development, has by now enough tools at its disposal to yield the main theorems on the existence of free algebras, but in an account devoted exclusively to algebra these results are much more easily proved directly; in particular the hypotheses under which the theorems are obtained here are usually easier to verify (in the case of algebras) than the corresponding hypotheses found in general category theory. For this reason we have borrowed little beyond the bare definitions of category and functor. These of course are indispensable in any satisfactory account of free algebras, and they allow us to state our results concisely without taking us too far from our central topic.

The notion of an algebraic structure as formulated in Chapter II is too narrow even in many algebraic contexts and has to be replaced by that of a *relational structure*, i.e., a set with a number of finitary relations defined on it. Besides algebraic structures themselves, this also includes structures with operations that are many-valued or not everywhere defined. In recent years, relational structures satisfying a given system of axioms, or *models*, have been the subject of intensive study and many results of remarkable power and beauty have been obtained. With the apparatus of universal algebra all set up, this seemed an excellent opportunity for giving at least a brief introduction to the subject, and this forms the content of Chapters V-VI.

The final chapter on applications is not in any way intended to be systematic; the aim was to include results which could be established by using the earlier chapters and which in turn illuminate the general theory, and which, moreover, are either important in another context (such as the development of the natural numbers in VII.1 or the representation theory of Lie and Jordan algebras in VII.5–7), or interesting in their own right (e.g., Malcev's embedding theorem for semigroups, VII.3).

Although the beginnings of our subject can be found in the last century (A. N.

Whitehead's treatise with the same title appeared in 1898), universal algebra as understood today only goes back to the 1930's, when it emerged as a natural development of the abstract approach to algebra initiated by Emmy Noether. As with other fields, there is now a large and still growing annual output of papers on universal algebra, but a curiously large portion of the subject is still only passed on by oral tradition. The author was fortunate to make acquaintance with this tradition in a series of most lucid and stimulating lectures by Professor Philip Hall in Cambridge 1947–1951, which have exercised a much greater influence on this book than the occasional reference may suggest. In other references an easily accessible work has often been cited in preference to the original source, and no attempt has been made to include remarks of an historical character; although such an attempt would certainly have been well worth while, it would have delayed publication unduly. For the same reason the bibliography contains, apart from papers bearing directly on the text, only a selection of writings on universal algebra. This was all the more feasible since a very full bibliography is available in *Mathematical Reviews;* besides, a comprehensive bibliography on universal algebra is available in G. Grätzer [79].

The book is based on a course of lectures which I gave at Yale University in 1961–1962. I am grateful to the audience there for having been such good listeners, and to the many friends who have performed the same office since then. In particular, D. E. Cohen and P. J. Higgins read parts of the manuscript and made many useful suggestions; J. L. MacDonald helped with the proofreading; A. J. Bowtell and F. E. J. Linton checked through the whole text and brought a number of inaccuracies to my attention. To all of them I should like to express my warmest thanks. I am also grateful to Messrs. Harper and Row for their willingness to carry out my wishes and to their editor, Mr. John Cronquist, for his help in preparing the manuscript for the press.

Queen Mary College, London
January, 1965

P. M. Cohn

Chapter I

Sets and Mappings

1. THE AXIOMS OF SET THEORY

The typical feature of a mathematical theory is that it deals with collections or sets of objects, where certain relations exist between the objects of these sets, or between different sets, while the nature of the objects is entirely immaterial. A simplification can be achieved by considering only objects which are themselves sets. At first sight this appears to lead to a vicious circle, but the difficulty may be resolved by beginning with the empty set. On the other hand it is necessary to restrict the sets which may appear as members of other sets, if one wants to avoid the contradictions arising from the consideration of 'the set of all those sets which are not members of themselves' (Russell's paradox). One therefore introduces a different term, such as 'class', for general collections of objects, and distinguishes those classes which are themselves members of other classes by calling them *sets*. Without entering fully into the question of axiomatics here (for which the reader may be referred to more detailed accounts such as Bourbaki [54], Gödel [40], Kelley [55], and Wang & McNaughton [53]), we shall give a list of axioms, which in the main express the conditions under which a class is to be regarded as a set.

Formally speaking, set theory consists of objects called *classes*, between which a binary relation can hold:

(1) $$A \in B.$$

We express (1) by saying 'A is a member (or element) of B', or 'A belongs to B', and we define a *set* to be a class which is a member of some class. Thus A is a set if and only if it stands in the relation (1) to some class B. To express the negation of (1) we write '$A \notin B$'. We denote classes by capital letters, except that classes which occur as members of other classes in a given context will often be denoted by lower-case letters.

Two objects are usually said to be equal if they have the same attributes, i.e. if the same statements are true of both. In set theory it is more convenient to frame the definition of equality more narrowly and add an axiom which in effect limits the statements one can make in the theory.

Definition

Two classes A and B are said to be *equal*, $A = B$, if they have the same members. The negation of the statement '$A = B$' is written '$A \neq B$'.

To obtain the usual interpretation of '=', we now add the following axiom:

A.1. If $A = B$ and $P(X)$ is any sentence about classes,[1] then $P(A)$ holds if and only if $P(B)$ holds.

For example, if $A = B$, then $A \in C$ if and only if $B \in C$. The axiom A.1 may be taken as limiting the kind of statement we are willing to discuss. Thus e.g., from the above definition of equality, the class of featherless bipeds is equal to the class of men (Russell), but any speculation whether a featherless biped is more likely to develop plumage than a man is outside the realm of set theory, which by axiom A.1 does not allow us to discriminate between the two descriptions.

In order to have a theory which conforms to our intuitive notions we require, for every meaningful statement in the theory which contains a variable X, a class whose members are precisely the sets X for which the statement holds; that is,

A.2. If $P(X)$ is any sentence about classes[2] then there is a class whose members are precisely the sets X for which $P(X)$ holds.

The class defined in A.2 is denoted by $\{X \mid P(X)\}$, so that for any set A,

$$A \in \{X \mid P(X)\} \text{ if and only if } P(A) \text{ holds.}$$

[1] To make the meaning quite precise one has to specify what sentences are allowed. In fact $P(X)$ may be any sentence involving set variables, class variables, the logical signs, and quantifiers acting on the set variables (cf. Chapter V).

[2] See previous footnote.

We note that the axiom A.2 may be made more explicit by replacing it by a small number of specific instances of the form A.2, from which the general form may be deduced (cf. Gödel [40]). The class $\{X \mid P(X)\}$ will not in general be a set. When it is a set, the sentence $P(X)$ is said to be *collectivising in* X. Thus e.g., if A is a given set, the sentence '$X \in A$' is collectivising in X, since the class of all X such that $X \in A$ is just the set A itself. On the other hand, '$X \notin X$' cannot be collectivising, if Russell's paradox, mentioned earlier, is to be avoided. In the presence of A.2, the class A of all sets X such that $X \notin X$ is well-defined, and the argument leading to Russell's paradox merely shows that there are classes which are not sets.[3]

By means of A.2 we can define the familiar operations on sets, although we cannot at this stage assert that the resulting classes are sets, but have to postulate special axioms to this effect.

The *empty class* is defined by the equation

$$\emptyset = \{X \mid X \neq X\}.$$

The *total class*[4] is defined as

$$T = \{X \mid X = X\}.$$

If A and B are any classes, the *singleton* consisting of A, and the *pair* consisting of A and B, are defined as

$$\{A\} = \{X \mid X = A\}, \qquad \{A,B\} = \{X \mid X = A \text{ or } X = B\},$$

when these are sets, and are not defined otherwise. Later we shall see that they are defined whenever A and B are sets.

If A is any class, then the *union* of A is defined as

$$\bigcup A = \{X \mid X \in Y \text{ and } Y \in A, \text{ for some } Y\}$$

and the *intersection* of A is given by

$$\bigcap A = \{X \mid X \in Y \text{ for all } Y \text{ such that } Y \in A\}.$$

If A and B are sets, then the *ordered pair* or *couple* consisting of A and B (in that order) is

$$(A,B) = \{\{A\}, \{A,B\}\}.$$

[3] It is possible to develop set theory without using classes if one restricts attention to properties which are collectivising (cf. Bourbaki [54]).

[4] This is sometimes called the *universal* class, but we shall not use this name, to avoid confusion with universal sets, which will be defined later (I.1; cf. also VI.2).

When $C = \{A,B\}$, we shall instead of '$\bigcup C$', '$\bigcap C$' write '$A \cup B$', '$A \cap B$' respectively; the set A is said to be *disjoint* from B in case $A \cap B = \emptyset$.

Given classes A and B, the class A is said to be a *subclass* of B, in symbols: $A \subseteq B$, if

$$X \in B \text{ for all } X \text{ such that } X \in A.$$

A subclass which is also a set is called a *subset*. If $A \subseteq B$ and $A \neq B$, A is said to be a *proper* subclass and we write '$A \subset B$'. We also write '$B \supseteq A$' or '$B \supset A$' in place of '$A \subseteq B$' or '$A \subset B$', respectively; further, the negation of any of these relations is indicated by the same symbol with a line drawn through it.

If A, B are any classes, then the class

$$A \backslash B = \{X \mid X \in A \text{ and } X \notin B\}$$

is called the *complement* of B in A.

If A is any class, then

$$\mathscr{B}(A) = \{X \mid X \subseteq A\}$$

is called the *Boolean* of A (after George Boole, 1815–1864; see also V.2).

If A and B are classes, then

$$A \times B = \{Z \mid Z = (X,Y) \text{ where } X \in A \text{ and } Y \in B\}$$

is called the *Cartesian product* of A and B.

A *function* from A to B is a subclass F of $A \times B$ such that for each $X \in A$ there exists just one $Y \in B$ for which $(X, Y) \in F$. The class A is called the *domain* of F and the class of *values* of F,

$$\{Y \mid Y \in B \text{ and } (X,Y) \in F \text{ for some } X \in A\}$$

is called the *range* of F. A somewhat different notation is often used when it is intended to focus attention on the range. If A is a class and I any set, then the values of a function from I to A are called a *family*[5] *of elements* of A, *indexed* or *coordinated* by I. If x_i is the element of A corresponding to $i \in I$, then the family is denoted by $(x_i)_{i \in I}$ and x_i is called its *i-coordinate*, i the *index*, and I the *index set*. Every set can be indexed, e.g. by itself; this means that we describe the set A by a function from A to itself, say by the identity function, in which every element of A corresponds to itself: $A = (a)_{a \in A}$. Thus in dealing with a set there is no loss of generality in taking it to be indexed.

[5] In using this definition the reader should bear in mind that a function is often identified with its range in practice.

We now come to the main group of axioms. They state essentially that the empty class is a set and that all reasonable constructions, applied to sets, again yield sets.

A.3. $\emptyset$ is a set.
A.4. Any subclass of a set is a set.
A.5. If A, B are sets, then so is $\{A,B\}$.
A.6. If A is a set, then so is the Boolean $\mathscr{B}(A)$.
A.7. If A is a set, then so is its union $\bigcup A$.
A.8. If F is a function whose domain is a set, then its range is a set.

We note some immediate consequences of these axioms.

(i) If A, B are sets, then so is the couple (A,B), and if (A',B') is another couple of sets, then $(A,B) = (A',B')$ if and only if $A = A'$ and $B = B'$.

For $\{A,B\}$ and $\{A\} = \{A,A\}$ are sets by A.5, hence so is $(A,B) = \{\{A\}, \{A,B\}\}$, and given (A,B) we can reconstruct (and distinguish) A and B by examining first the members of (A,B) and then their members.

(ii) If $P(X)$ is any sentence about classes[6] and A is a set, then $\{X \mid X \in A$ and $P(X)\}$ is a subclass of A, and hence, by A.4, a subset. This is often written $\{X \in A \mid P(X)\}$. In particular, the complement of any class in a set A is a subset of A.

(iii) If A, B are sets, then so is their Cartesian product $A \times B$. For if $X \in A$ and $Y \in B$ then $\{X\}, \{X,Y\} \subseteq A \cup B$; hence $\{X\}, \{X,Y\} \in \mathscr{B}(A \cup B)$, and so $(X,Y) \in \mathscr{B}\mathscr{B}(A \cup B)$. Therefore

$$A \times B = \{Z \in \mathscr{B}\mathscr{B}(A \cup B) \mid Z = (X,Y) \text{ for some } X \in A \text{ and } Y \in B\},$$

which shows $A \times B$ to be a set.

(iv) Any function whose domain is a set is itself a set. For let F be a function whose domain is a set A, say. By A.8, its range is a set B, and since $F \subseteq A \times B$, it follows from (iii) and A.4 that F is a set.

If I is a set and A is a class, then any function from I to A is a set, so we may form the class whose members are all the functions from I to A. This class is denoted by A^I and is called a *Cartesian power*. Since each function from I to A is a subset of $I \times A$, i.e. an element of $\mathscr{B}(I \times A)$, it follows that $A^I \subseteq \mathscr{B}(I \times A)$. In particular, if both I and A are sets, then so is A^I.

(v) If $A \neq \emptyset$, then $\bigcap A$ is a set. For if $a \in A$, then $\bigcap A$ is a subclass of a.

[6] Cf. footnote 1, page 2.

The axioms given so far allow us to construct arbitrarily large finite sets,[7] but no infinite sets, whose existence has to be postulated separately. This is usually done by means of an *axiom of infinity*, asserting the existence of an infinite set (once this notion has been defined). We shall adopt an alternative axiom, which asserts the existence of universal sets, now to be defined (cf. Sonner [62], Gabriel [62]). This will turn out to be considerably stronger than the axiom of infinity; it will also eliminate the need, in most problems, to consider classes which are not sets.

Definition

A set U is said to be *universal*, or a *universe*, if it satisfies the following conditions:

(i) If $X \in U$, then $X \subseteq U$.
(ii) If $X \in U$, then $\mathscr{B}(X) \in U$.
(iii) If $X, Y \in U$, then $\{X, Y\} \in U$.
(iv) If $F = (F_i)_{i \in I}$, where $F_i \in U$ and $I \in U$, then $\bigcup F \in U$.

We now add as axiom of infinity:

A.9. Every set is a member of some universe.

This axiom in effect does away with the need for considering classes. E.g., instead of the class of all sets we may consider the class of all sets in a given universe U, and this is again a set. In the sequel we shall therefore reserve the term 'class' to refer to a set which is not necessarily a member of the universe under consideration.

To illustrate A.9, we shall define the natural numbers and show that they form a set. Let U be a universe such that $\emptyset \in U$; for the moment we shall call a subset V of U *numeral*, if $\emptyset \in V$ and if $X \cup \{X\} \in V$ whenever $X \in V$. For example, U itself is a numeral set: if $X \in U$, then $\{X\} \in U$, hence $\{X,\{X\}\} \in U$, and indexing the set $\{X,\{X\}\}$, we see that $X \cup \{X\} \in U$. Let N be the intersection of all numeral subsets of U; then N itself is again numeral. Writing 'x'' for '$x \cup \{x\}$', we have the following properties for N:

N.1. $\emptyset \in N$.
N.2. If $x \in N$, then $x' \in N$.
N.3. Any subset of N satisfying N.1–2 coincides with N.

[7] Of course, the existence of finite sets may be deduced from experience; it is only for the infinite sets that any axioms are required at all.

Here N.3 follows from the fact that N is the intersection of all subsets of U satisfying N.1 and N.2. Of course N.3 is just the principle of induction. The natural numbers are now defined as the elements of N, the first few being

$$0 = \emptyset, \quad 1 = \{0\}, \quad 2 = \{0,1\}, \quad 3 = \{0,1,2\}, \cdots.$$

By a *positive integer* we mean a natural number $\neq 0$. A class is said to be *finite* if it can be indexed by a natural number; otherwise it is *infinite*. With this definition we can show that the set of natural numbers is infinite. First we note the following properties:

(a) If $n \in N$, then $n' \nsubseteq 0$. For $n' = n \cup \{n\} \neq \emptyset$, while $0 = \emptyset$.

(b) If $m,n \in N$ and $m \in n$, then $m \subseteq n$. This clearly holds if $n = 0$, because then $m \notin n$ for all $m \in N$. Now let M be the set of all $n \in N$ such that $m \subseteq n$ for all $m \in N$ satisfying $m \in n$. If $n \in M$ and $m \in n'$, then either $m = n$ or $m \in n$, whence $m \subseteq n$ in either case, and $n \subseteq n'$, so $m \subseteq n'$. Thus if $n \in M$, then $n' \in M$: and since M also contains 0, it must coincide with N.

(c) If $m' \subseteq n'$, then $m \subseteq n$. For if $m \cup \{m\} \subseteq n \cup \{n\}$, then $m \in n \cup \{n\}$; hence either $m \in n$, whence $m \subseteq n$ by (b), or $m = n$.

By applying (c) with m and n interchanged we obtain

(d) If $m' = n'$, then $m = n$.

We note that (a) and (d), together with N.1–3, constitute the usual Peano axioms for the natural numbers (see Chapter VII). Henceforth we also write '$n + 1$' instead of 'n'', and we shall assume that the elementary properties of natural numbers are known. We confine ourselves to proving

Theorem 1.1

The set N of natural numbers is infinite.

To prove this theorem we must show that if n is any natural number, there exists no function from n to N with N as range. This certainly holds for $n = 0$, because any function with domain $0 = \emptyset$ must have empty range, whereas $N \neq \emptyset$. Now let M be the set of numbers n such that no function from n to N with range N exists; then M satisfies N.1 by what has been shown. Let $n \in M$ and assume that $n' \notin M$; this means that there is a function f from n' to N with range N. Since every element $\neq 0$ of N has the form x', and since x' determines x uniquely, by (d), we may for any $y \in N$ such that $y \neq 0$ denote the unique element x of N satisfying

$x' = y$ by $y - 1$. Moreover, denote by if the unique element j such that $(i,j) \in f$. With these notations we may define a function g from n to N by the rule

$$kg = \begin{cases} kf & \text{if } kf \subseteq nf \\ kf - 1 & \text{otherwise} \end{cases} \qquad (k \in n).$$

Since $0 \subseteq nf$, this defines indeed a function from n to N; its range is easily seen to be N, which contradicts the hypothesis that $n \in M$. Therefore if $n \in M$, then $n' \in M$, i.e. M satisfies N.2 as well as N.1, and by N.3, $M = N$. ∎

In the construction which led to N, U was any universe containing $\emptyset$; now any nonempty universe must contain $\emptyset$, by definition, and so we have

Corollary 1.2

Any non-empty universe is infinite. ∎

Now A.9 ensures that there are universes with infinite sets as members. In the sequel when we operate in some universe it is always this kind we have in mind.

There are certain set constructions which, although they may not lead to contradictions, do give rise to some pathological situations which have no counterpart in the intuitive interpretation. E.g. we cannot on the basis of the above axioms decide whether a set can be a member of itself: $X \in X$, and thus we cannot tell in general whether the sets X and $\{X\}$ are equal or not. Since a situation where $X \in X$ never arises in practice, it is best to exclude it explicitly. This is done by the *axiom of foundation*:

A.10. Every nonempty class A has an element X which is disjoint from A.

With the help of this axiom one sees easily that $X \notin X$ for all sets X. More generally, there are no infinite descending chains $X_0, X_1, X_2, \cdots$ of sets (not necessarily distinct) such that

$$(2) \qquad X_{n+1} \in X_n \qquad (n \in N).$$

For if we had a family of sets satisfying (2), then the class whose members are $X_0, X_1, \cdots$ would contradict A.10.

Sometimes the following extension of the notation introduced is useful. Let $\mathscr{A}$ be a class of sets, indexed in some way, say $\mathscr{A} = (A_i)_{i \in I}$. Then in place of '$\bigcup \mathscr{A}$', '$\bigcap \mathscr{A}$' we shall also write '$\bigcup A_i$', '$\bigcap A_i$'. In particular, if $\mathscr{A}$ is finite, say $\mathscr{A} = \{A_0, \cdots, A_n\}$, then we write

$$\bigcup \mathscr{A} = \bigcup A_i = A_0 \cup \cdots \cup A_n,$$

$$\bigcap \mathscr{A} = \bigcap A_i = A_0 \cap \cdots \cap A_n.$$

Further, we define the *Cartesian product* of the family of sets $(A_i)_{i\in I}$ as

$$\prod A_i = \{x \in (\bigcup \mathscr{A})^I \mid x = (x_i)_{i\in I}, \text{ where } x_i \in A_i \text{ for all } i \in I\}.$$

This definition is not, in the case of two factors, the same as that given earlier for a Cartesian product of two factors, but the difference is of no consequence for us: we are free to interpret Cartesian products in either way, when there are two (or any finite number of) factors (cf. Exercise 2).

Clearly, $\prod A_i$ is again a set, but even when each A_i is nonempty there is in general nothing to show that the product is nonempty. This will, however, follow from the axiom of choice, introduced below in I.4.

Unlike the union and intersection, where the indexing of $\mathscr{A}$ was used only for convenience, the product depends essentially on the indexing as well as on the class of coordinates, since e.g., the product is affected if a coordinate appears more than once. To take an extreme case, if $A_i = A$ for all $i \in I$, then $\bigcup A_i = \bigcap A_i = A$, whereas $\prod A_i = A^I$.

When I is finite, $I = n + 1$ say, we write again

$$\prod A_i = A_0 \times A_1 \times \cdots \times A_n.$$

EXERCISES

1. Show that the total class T is not a set.

2. Define n-tuples $(x_1,\cdots,x_n)$ either as families indexed by $\{1,\cdots,n\}$ or by generalizing the definition of couples; and for each definition, show that for any sets $x_1,\cdots,x_n, y_1,\cdots,y_n,(x_1,\cdots,x_n) = (y_1,\cdots,y_n)$ if and only if $x_i = y_i$ $(i = 1,\cdots,n)$.

3. Show that $n \notin n$, for any natural number n, without using A.10.

4. How many distinct ways are there of indexing a set of n elements (i) by itself, (ii) by another set, of m elements?

2. CORRESPONDENCES

Let A and B be sets; then a *correspondence from A to B* is a subset of the Cartesian product $A \times B$. Thus a correspondence is a set of pairs (x,y), where $x \in A$, $y \in B$. We shall usually denote correspondences by Greek capitals. If Φ is a correspondence from A to B and $A' \subseteq A$, we define

$$A'\Phi = \{y \in B \mid (x,y) \in \Phi \text{ for some } x \in A'\}.$$

In case $B = A$, Φ is called a correspondence *in* A. Every set has the correspondence

$$\Delta_A = \{(x,x) \mid x \in A\},$$

which is called the *diagonal* in A.

To describe the different types of correspondence we introduce two operations, inversion and composition. Every correspondence Φ has the *inverse*

$$\Phi^{-1} = \{(x,y) \mid (y,x) \in \Phi\}.$$

Clearly, if $\Phi \subseteq A \times B$, then $\Phi^{-1} \subseteq B \times A$. A correspondence Φ in A is said to be *symmetric* if $\Phi^{-1} = \Phi$, *antisymmetric* if $\Phi \cap \Phi^{-1} \subseteq \Delta_A$, and *reflexive* if $\Phi \supseteq \Delta_A$.

For any correspondences Φ and Ψ we define a composition

$$\Phi \circ \Psi = \{(x,y) \mid (x,z) \in \Phi \text{ and } (z,y) \in \Psi \text{ for some } z\}.$$

This is indeed a set, for if $\Phi \subseteq A \times B$ and $\Psi \subseteq C \times D$, then $\Phi \circ \Psi \subseteq A \times D$; in particular we note that $\Phi \circ \Psi = \emptyset$ unless $B \cap C \neq \emptyset$.

The following laws are easily verified:

(1) $$\Phi \circ (\Psi \circ \Theta) = (\Phi \circ \Psi) \circ \Theta,$$

(2) $$(\Phi \circ \Psi)^{-1} = \Psi^{-1} \circ \Phi^{-1},$$

(3) $$(\Phi^{-1})^{-1} = \Phi.$$

They are valid for any correspondences Φ,Ψ,Θ. Further, if Φ is a correspondence from A to B, then

(4) $$\Phi \circ \Delta_B = \Delta_A \circ \Phi = \Phi.$$

With the help of these operations we can define several important types of correspondence which will frequently occur in the sequel.

A correspondence Φ in A is said to be *transitive*, if

$$\Phi \circ \Phi \subseteq \Phi.$$

A transitive reflexive correspondence Φ in A is called a *preordering* of A. Clearly Φ^{-1} is then also a preordering of A; it is said to be *opposite* to Φ. An antisymmetric preordering of A is also called an *ordering* of A, or sometimes a *partial ordering*, to distinguish it from a *total* ordering, which is required to satisfy in addition $\Phi \cup \Phi^{-1} = A^2$. By an *ordered set* we shall understand a set together with an ordering defined on it.

If Φ is any correspondence from A to B, then $\Phi \circ \Phi^{-1}$ is a correspondence in A and $\Phi^{-1} \circ \Phi$ a correspondence in B. If we have

(5) $$\Phi \circ \Phi^{-1} \supseteq \Delta_A \quad \text{and}$$

(6) $$\Phi^{-1} \circ \Phi \subseteq \Delta_B,$$

then Φ is a function from A to B, as defined in I.1. The following result is easily proved:

Proposition 2.1

If f and g are functions, then so is $f \circ g$. ▮

Note that $f \circ g$ may well be the empty set, namely if the range of f is disjoint from the domain of g. For functions we often write 'fg' instead of '$f \circ g$', and for a given x in the domain of f, we denote the unique element y such that $(x,y) \in f$ by xf.

A correspondence Φ from A to B is said to be a *bijection*, or one-one correspondence, if Φ is a function from A to B and Φ^{-1} is a function from B to A. Two sets are said to be *equipotent* if there is a bijection between them. The fact that A and B are equipotent is sometimes expressed by writing '$A \leftrightarrow B$'.

By an *n-ary relation* or *n-place relation* (where n is a positive integer) we mean a set A together with a subset Φ of A^n; we frequently refer to Φ as a relation in A. Thus, for example, a binary relation ($n = 2$) is merely a correspondence in a specified set, while a unary relation ($n = 1$) is a subset of a specified set.

EXERCISES

1. If A, B are sets, show that there is a bijection between the set of all couples $(x, y)(x \in A, y \in B)$ and the set of functions f from 2 to $A \cup B$ such that $0f \in A$, $1f \in B$.

2. Give a proof of Proposition 2.1, using (5) and (6).

3. Show that a finite set is not equipotent to a proper subset of itself. (Hint: If f is a bijection of S with a proper subset T of S and $a \in S$, $a \notin T$, show that the elements $a, af, aff, \cdots$ are all distinct, and use Theorem 1.1.)

4. Show that any class equipotent to a set is itself a set.

5. If X is a member of a universe U and Y is a set equipotent to X, does it follow that $Y \in U$?

3. MAPPINGS AND QUOTIENT SETS

Definition

A *mapping* is a triple (A,B,f) consisting of a set A, a second set B, and a function f from A to B. The set A is called the *source* and B the *target* of the mapping.

The mapping (A,B,f) is more usually denoted by $f{:}A \to B$, or sometimes $A \xrightarrow{f} B$. The latter notation is used especially in diagrams, to illustrate the composition of mappings. Unlike functions, which can always be composed, two mappings $f{:}A \to B$ and $g{:}C \to D$ are composable only if $B = C$, and the composite is then defined as $fg{:}A \to D$.

Given mappings $f{:}A \to B$, $g{:}B \to C$, and $h{:}A \to C$, if the equation $fg = h$ holds, this may be expressed by saying that the diagram of mappings

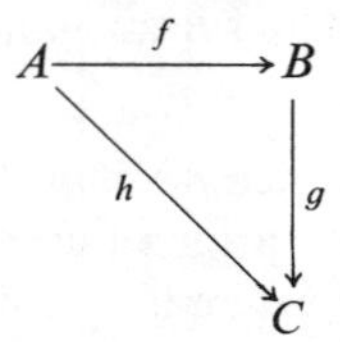

is *commutative*. More generally, by a commutative diagram one understands a network of arrows between sets, representing mappings, such that any two paths (going along the arrows) from one set to another define the same mapping between these sets. In practice most diagrams are made up of triangles as above, or squares,

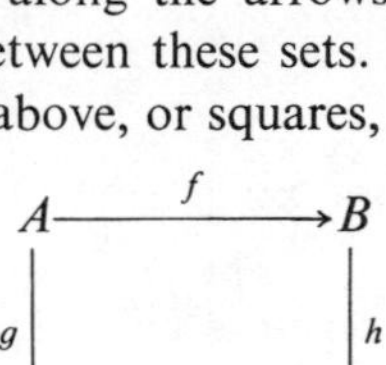

where commutativity means that $fh = gk$.

Given a mapping $f:A \to B$, then for each $x \in A$, the unique element xf of B is called the *image* of x under f. The set Af of all images of elements of A is also called the *image* of A; it is just the range of f, regarded as a function. If $Af = B$, i.e. $f^{-1} \circ f = \Delta_B$, f is said to be *surjective*, or *onto* B, or a *surjection*. If $f \circ f^{-1} = \Delta_A$, f is said to be *injective*, or one-one, or an *injection*. A mapping which is both surjective and injective is said to be *bijective*; this is merely a bijection between two given sets, as defined in I.2. We note that a correspondence Φ from A to B defines a bijection between A and B if and only if

$$\Phi \circ \Phi^{-1} = \Delta_A, \qquad \Phi^{-1} \circ \Phi = \Delta_B.$$

A bijection of a set A with itself is also called a *permutation* of A.

The following are important examples of mappings which will occur frequently in the sequel.

(i) In any set A the diagonal defines a mapping $\Delta_A : A \to A$, which is called the *identity mapping* and is denoted by 1 or 1_A.

(ii) If A is any set and B is a set containing A as subset, then the diagonal on A defines a mapping $\Delta_A : A \to B$, which is called the *inclusion mapping* from A to B.

(iii) Given a Cartesian product $P = \prod A_i$, then for any fixed element $i \in I$, the function which assigns to each $x \in P$ its i-coordinate defines a mapping $\varepsilon_i : P \to A_i$, called the *projection* of P on the factor A_i.

(iv) Given a mapping $f:A \to B$ and a subset A' of A, denote by i the inclusion mapping $A' \to A$; then the mapping $if:A' \to B$ is called the *restriction* of f to A', and is denoted by $f \mid A'$. Similarly, if B' is a subset of B and $Af \subseteq B'$, then f may be *cut down* to a mapping $f' : A \to B'$ by restricting the target. We shall not use any special notation for this mapping.

(v) Given two mappings $f:A \to B$ and $f' : A' \to B$, where A' is a subset of A, if $f' = f \mid A'$, then f is said to be an *extension* of f'. Thus e.g. if $f:A \to B$ is any mapping and $A' \subseteq A$, then f is an extension of $f \mid A'$, but of course, in general, f is not determined uniquely by $f \mid A'$.

(vi) Given a set A and a natural number n, any mapping $\alpha : A^n \to A$ is called an *n-ary operation* on A. For $n = 0$, we have a noughtary operation, essentially an element of A, while for $n = 1$ we just have the mappings of A into itself. Using the natural bijection between $A^n \times A$ and A^{n+1} given by

$$((x_1, \cdots, x_n), y) \leftrightarrow (x_1, \cdots, x_n, y)$$

we see that an n-ary operation may be considered as a special case of an $(n + 1)$-ary relation. The mappings from A^n to A for arbitrary finite n are sometimes called *finitary* operations, to distinguish them from infinitary operations, i.e. mappings $A^I \to A$, where I is infinite.

Let α be an n-ary operation in A and B a subset of A. Then B^n is also a subset of A^n, and so α defines a mapping of B^n into A, namely the restriction of α to B^n. If the image of B^n under this mapping is contained in B, we say that B is *closed with respect to* α or that B *admits the operation* α. The restriction can then be cut down to a mapping of B^n into B, i.e. an n-ary operation on B. This operation will be denoted by $\alpha|B$ and called the *restriction* of α to B.

(vii) Let $f: I \to A$ be a mapping and write 'a_i' instead of 'if' ($i \in I$). Then $(a_i)_{i \in I}$ is just a family of elements of A, as defined previously. This point of view of regarding a mapping is adopted when we want to stress the image set.

(viii) If A is a given set, then any subset B of A determines a mapping $\chi_B: A \to 2$ defined by

$$x\chi_B = \begin{cases} 0 & \text{if } x \notin B, \\ 1 & \text{if } x \in B. \end{cases}$$

χ_B is the *characteristic function* of the subset B. As is easily verified, the mapping

$$B \to \chi_B$$

from $\mathscr{B}(A)$ to 2^A is a bijection.

(ix) Given any mapping $f: A \to B$, we define a mapping

$$f^*: \mathscr{B}(B) \to \mathscr{B}(A)$$

by the rule

$$Yf^* = \{x \in A \mid xf \in Y\} \qquad \text{for any } Y \in \mathscr{B}(B).$$

The set Yf^* is said to be obtained by *pulling* Y *back along* f, and f^* is also referred to as the *pullback* mapping.

An *equivalence* on a set A is a correspondence Φ in A which is reflexive, symmetric, and transitive. Thus we have

(a) $\Phi \supseteq \Delta_A$,
(b) $\Phi^{-1} = \Phi$,
(c) $\Phi \circ \Phi \subseteq \Phi$.

We shall denote equivalences by lower-case German letters. If $\mathfrak{q}$ is an equivalence on A, then for each $x \in A$ we define a subset $x^{\mathfrak{q}}$ of A, the $\mathfrak{q}$-*class* of x, by[8]

$$x^{\mathfrak{q}} = \{y \in A \mid (x,y) \in \mathfrak{q}\}.$$

[8] This is not to be confused with the notation for Cartesian powers.

Instead of '$(x,y) \in \mathfrak{q}$', we shall often write '$x \equiv y \pmod{\mathfrak{q}}$'. From the properties of $\mathfrak{q}$ it follows easily that $x \in x^{\mathfrak{q}}$, and in fact $x \equiv y \pmod{\mathfrak{q}}$ if and only if $x^{\mathfrak{q}} = y^{\mathfrak{q}}$. In particular, the $\mathfrak{q}$-classes form a *partition* of A, i.e. a decomposition of A into pairwise disjoint nonempty sets, the *classes* of the partition. The $\mathfrak{q}$-classes themselves are members of the Boolean $\mathscr{B}(A)$; the subset of $\mathscr{B}(A)$ consisting of all $\mathfrak{q}$-classes will be denoted by $A/\mathfrak{q}$ and called the *quotient set* of A by $\mathfrak{q}$. If with each $x \in A$ we associate $x^{\mathfrak{q}}$, we obtain a mapping from A to $A/\mathfrak{q}$, called the *natural mapping* or *identification* associated with $\mathfrak{q}$ and denoted by nat $\mathfrak{q}$. Clearly this mapping is surjective, by definition.

We can now state the basic decomposition theorem for mappings:

Theorem 3.1

Let $f: A \to B$ be any mapping and put $\mathfrak{q} = f \circ f^{-1}$. Then $\mathfrak{q}$ is an equivalence on A, Af is a subset of B, and there is a decomposition of f,

(1) $$f = \varepsilon f' \mu,$$

where $\varepsilon =$ nat $\mathfrak{q}: A \to A/\mathfrak{q}$ is a surjection, $f': A/\mathfrak{q} \to Af$ is a bijection, and $\mu: Af \to B$, the inclusion mapping, is an injection.

The situation may be illustrated by the commutative diagram

$$\begin{array}{ccc} A & \xrightarrow{\ f\ } & B \\ {\scriptstyle \varepsilon}\downarrow & & \uparrow{\scriptstyle \mu} \\ A/\mathfrak{q} & \xrightarrow[\ f'\ ]{} & Af \end{array}$$

Proof:

By definition, $f \circ f^{-1} \supseteq \Delta_A$; further,

$$(f \circ f^{-1})^{-1} = (f^{-1})^{-1} \circ f^{-1} = f \circ f^{-1},$$

and since $f^{-1} \circ f \subseteq \Delta_B$, we have

$$f \circ f^{-1} \circ f \circ f^{-1} \subseteq f \circ \Delta_B \circ f^{-1} = f \circ f^{-1};$$

this shows $\mathfrak{q} = f \circ f^{-1}$ to be an equivalence on A. Clearly Af is a subset of B, and from the definition of $\mathfrak{q}$, $(x,y) \in \mathfrak{q}$ holds if and only if $(x,z) \in f$ and $(y,z) \in f$ for some $z \in B$, i.e. if and only if $xf = yf$; and so the mapping $f': A/\mathfrak{q} \to Af$ defined by $x^{\mathfrak{q}} f' = xf$ is a bijection. Now (1) follows by taking ε to be nat $\mathfrak{q}$ and μ to be the inclusion $Af \to B$. ∎

The equivalence $\mathfrak{q} = f \circ f^{-1}$ is called the *kernel* of f and is denoted by $\ker f$.

Corollary 3.2

Let $A = \bigcup A_i$ be a partition of A; then there is just one equivalence which gives rise to this partition.

To see this we define a mapping $f: A \to I$ by assigning to each $x \in A$ the unique index $i \in I$ such that $x \in A_i$. The equivalence classes of the kernel of f are just the subsets A_i, and clearly $\ker f$ is the only such equivalence. ∎

As a further application of Theorem 3.1 we have the factor theorem:

Theorem 3.3

Given a mapping $f: A \to B$ and an equivalence $\mathfrak{q}$ on A, if $\mathfrak{q} \subseteq \ker f$, then there exists a unique mapping $\bar{f}: A/\mathfrak{q} \to B$ such that the diagram

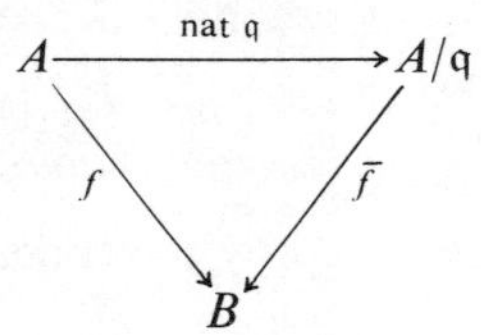

is commutative.

This means that $f = (\text{nat } \mathfrak{q})\bar{f}$, where nat $\mathfrak{q}$ is the natural mapping $A \to A/\mathfrak{q}$. The conclusion of the theorem may be expressed by saying that f can be factored (uniquely) by nat $\mathfrak{q}$. The theorem is sometimes briefly (though incorrectly) expressed by saying that any mapping may be factored by an equivalence which is contained in its kernel.

To prove the theorem, we note that if a mapping $\bar{f}$ to satisfy the conditions exists, then for every $x \in A$ we must have

$$(x^{\mathfrak{q}})\bar{f} = xf. \tag{2}$$

Thus there can be at most one mapping $\bar{f}$; on the other hand, putting $\mathfrak{k} = \ker f$, we have $\mathfrak{q} \subseteq \mathfrak{k}$ by hypothesis, hence $x^{\mathfrak{q}} = y^{\mathfrak{q}}$ implies $x^{\mathfrak{k}} = y^{\mathfrak{k}}$, i.e. $x^{\mathfrak{q}} = y^{\mathfrak{q}}$ implies $xf = yf$. Therefore xf depends only on $x^{\mathfrak{q}}$ and not on x itself. But this means that $\bar{f}$ as defined by (2) is single-valued, and so is the required mapping. ∎

The quotient sets of a given set A are to some extent dual to the subsets of A, but this duality is not complete. Thus, whereas the relation: *B is a* *subset of A* is transitive, the relation: *B is a quotient set of A* is not. In fact the set $\mathscr{B}(A)$ of subsets of A is an ordered set with respect to the relation $\subseteq$. If $\mathscr{Q}(A)$ denotes the set of quotients of A, then $\mathscr{Q}(A)$ may be ordered by putting $X \succ Y$, where $X, Y \in \mathscr{Q}(A)$, whenever the kernel of the natural mapping $A \to X$ is contained in the kernel of the natural mapping $A \to Y$. Applying the factor theorem (with $A/\mathfrak{q} = X$, $B = Y$), we see that in this case there is a surjection $X \to Y$ which has some claim to be called 'natural'. In this way a substitute for the missing transitivity is obtained. To express this formally, it is simpler to consider instead of $\mathscr{Q}(A)$ the set $\mathscr{C}(A)$ of all equivalences on A. By Corollary 3.2, there is a bijection between $\mathscr{C}(A)$ and $\mathscr{Q}(A)$; and clearly the ordering $\succ$ just defined on $\mathscr{Q}(A)$ corresponds under this bijection to the ordering by inclusion of $\mathscr{C}(A)$. We now have

Theorem 3.4

Let $\mathfrak{q},\mathfrak{r}$ be equivalences on the set A such that $\mathfrak{q} \subseteq \mathfrak{r}$. Then there is a unique mapping $\theta: A/\mathfrak{q} \to A/\mathfrak{r}$ such that (nat $\mathfrak{q}$) $\theta =$ nat $\mathfrak{r}$. *If* ker θ *is denoted by $\mathfrak{r}/\mathfrak{q}$, then $\mathfrak{r}/\mathfrak{q}$ is an equivalence on $A/\mathfrak{q}$ and θ induces a bijection*

$$\theta' : (A/\mathfrak{q})/(\mathfrak{r}/\mathfrak{q}) \to A/\mathfrak{r},$$

such that the diagram

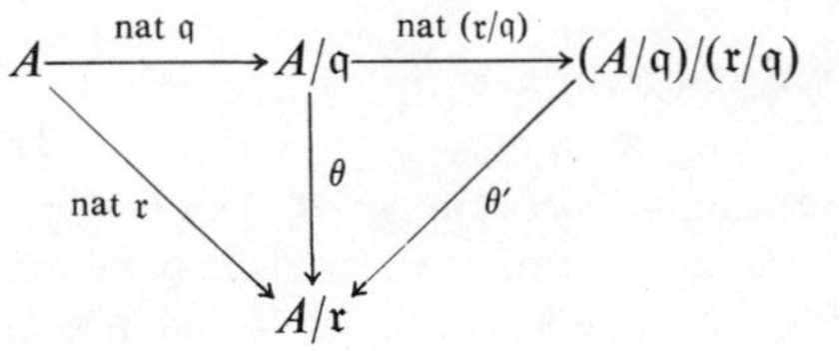

is commutative.

To obtain θ we apply Theorem 3.3, with $B = A/\mathfrak{r}$, $f =$ nat $\mathfrak{r}$; since nat $\mathfrak{q}$ and nat $\mathfrak{r}$ are surjective, it follows that θ is surjective, and if we now use the decomposition theorem 3.1, we obtain the remaining assertion. ∎

EXERCISES

1. If $\Phi_i (i \in I)$ and Ψ are any correspondences, show that

$$(\bigcup \Phi_i) \circ \Psi = \bigcup (\Phi_i \circ \Psi),$$

but in general,

$$(\bigcap \Phi_i) \circ \Psi \neq \bigcap (\Phi_i \circ \Psi).$$

2. If $\mathfrak{q}_i (i \in I)$ is any family of equivalences on A, then $\bigcap \mathfrak{q}_i$ is again an equivalence on A.

3. If p_n is the number of equivalences on a set of n elements, obtain the following recursion formula for p_n:

$$p_{n+1} = \sum \binom{n}{i} p_i \qquad (p_0 = 1),$$

where $\binom{n}{i} = \dfrac{n!}{i!\,(n-i)!}$. Prove also the formula

$$\sum \frac{p_n x^n}{n!} = \exp[(\exp x) - 1].$$

4. Show that every injective mapping from a finite set to itself is surjective. (Use Exercise 2.3).

5. In Theorem 3.1, if $f = \varepsilon_1 f_1' \mu_1$ is a second decomposition, where ε_1 is surjective, f_1' bijective, and μ_1 injective, show that there exist bijections α, β such that $\varepsilon_1 = \varepsilon\alpha$, $\mu_1 = \beta\mu$, and $f' = \alpha f_1' \beta$.

6. If Φ is a preordering on A and $\mathfrak{q} = \Phi \cap \Phi^{-1}$, verify that $\mathfrak{q}$ is an equivalence on A and show that $A/\mathfrak{q}$ may be ordered in a natural way in terms of the given preordering on A.

7. Let $\Phi \subseteq A \times B$ and write $A' = B\Phi^{-1}, B' = A\Phi, \mathfrak{a}_1 = \Phi \circ \Phi^{-1}, \mathfrak{b}_1 = \Phi^{-1} \circ \Phi$,

$$\mathfrak{a}_n = \mathfrak{a}_{n-1} \circ \mathfrak{a}_1, \quad \mathfrak{b}_n = \mathfrak{b}_{n-1} \circ \mathfrak{b}_1 \quad \text{and} \quad \mathfrak{a} = \bigcup \mathfrak{a}_n, \quad \mathfrak{b} = \bigcup \mathfrak{b}_n.$$

Show that $\mathfrak{a}$, $\mathfrak{b}$ are equivalences on A', B' respectively, and that Φ induces a bijection $A'/\mathfrak{a} \to B'/\mathfrak{b}$ in a natural way. (This may be regarded as the analogue of Theorem 3.1 for correspondences and is proved in the same way.)

8. Given any sets A, X, Y, show that a mapping $f: X \to Y$ induces a mapping $f^\circ: A^Y \to A^X$ and a mapping $f_\circ: X^A \to Y^A$. Express the projection operators of a direct power A^X as f°, for suitable f.

9. Given $f: A \to B$, define $f_*: \mathscr{B}(A) \to \mathscr{B}(B)$ by the rule

$$Xf_* = \{xf \mid x \in X\}$$

and f^* as in example (ix) of mappings (the pullback mapping). Under what conditions is $f^* f_* = 1$, or $f_* f^* = 1$?

4. ORDERED SETS

Let A be an ordered set, i.e. a set with an ordering defined on it. We now adopt the usual notation '$\leqslant$' for the ordering, so that the axioms read:

O.1. If $x \leqslant y$ and $y \leqslant z$, then $x \leqslant z$.
O.2. $x \leqslant x$.
O.3. If $x \leqslant y$ and $y \leqslant x$, then $x = y$.

As usual, we write '$x < y$' to mean '$x \leqslant y$ and $y \nleqslant x$'; we also write '$x \geqslant y$' instead of '$y \leqslant x$', and '$x > y$' instead of '$y < x$'. The ordering is *total* if, in addition, any two elements are *comparable*, i.e.

O.4. $x \leqslant y$ or $y \leqslant x$ for all $x,y \in A$.

At the other extreme, an abstract set may be regarded as a *totally unordered* set, in which $x \leqslant y$ holds only when $x = y$; thus no two distinct elements are comparable.

If B is a subset of A, then the ordering of A, restricted to B, is an ordering of B. In this sense any subset of an ordered set is understood as an ordered set. If the ordering so defined in B is total, B is said to be a *chain* in A.

Let B be any subset of an ordered (or, more generally, preordered) set A. An element $a \in A$ with the property

$$x \leqslant a \qquad \text{for all } x \in B$$

is called an *upper bound* for B in A. When such elements exist, we say that B is *bounded above in* A. The lower bounds of B are defined similarly. An ordered set in which every finite subset has an upper bound is said to be *directed* (upwards). By induction on the number of elements it is enough to require that any pair of elements in the set have an upper bound. A set directed downwards is defined similarly; when nothing to the contrary is said, 'directed' will always mean 'directed upwards'.

If the ordered set A itself has an upper bound a, then a is clearly the only upper bound; it is called the *greatest element* of A. If $a \in A$ is such that none of the upper bounds in A of the singleton $\{a\}$ exceed a, then a is said to be *maximal* in A. Thus a is maximal in A whenever

$$a \nless x \text{ for all } x \in A.$$

Of course A may have more than one maximal element, or none at all. If it has a greatest element, then this is also the unique maximal element. The converse is not true in general, but it does hold in a chain, or more generally in a directed set: a maximal element in a directed set is also the greatest element. The *minimal* elements of A and the *least* element of A are defined correspondingly. If a subset B has a least upper bound, this is called the *supremum* of B and is written 'sup B'. Similarly the greatest lower bound, when it exists, is called the *infimum* of B and is written 'inf B'. All these definitions still apply when the set A is only preordered.

We can now state our final axiom:

A.11 (Zorn's lemma)

A nonempty ordered set in which every chain has an upper bound, has a maximal element.

Instead of A.11 one frequently postulates the

Axiom of choice

Given any set A, there is a function f from $\mathscr{B}(A)$ to A such that whenever $B \subseteq A$, $B \neq \emptyset$, then $Bf \in B$.

Usually, A.11 is proved by means of the axiom of choice (cf. e.g. Halmos [61] or Kelley [55]); but since the axiom of choice can in turn be derived from A.11 (cf. I.5), it is immaterial whether A.11 or the axiom of choice is assumed.

An ordered set A is said to satisfy the *minimum condition* if every nonempty subset has a minimal element. If this condition is satisfied and, moreover, the number of minimal elements of any subset is finite, A is said to be *partly well-ordered*; and if every nonempty subset has a unique minimal element, A is said to be *well-ordered*. It is easily seen that a partly well-ordered set is well-ordered if and only if it is totally ordered. For sets with minimum condition we have the

Generalized principle of induction (Noetherian induction)

Let A be an ordered set with minimum condition and B a subset of A which contains any element $a \in A$ whenever it contains all the elements $x \in A$ such that $x < a$. Then $B = A$.

For, the complement of B in A has no minimal element, and so must be empty. ∎

The special case of this principle when A is well-ordered is known as the *principle of transfinite induction*. This is itself a generalization of the principle of induction for natural numbers (cf. I.1). Thus the natural numbers N form a set which is well-ordered with respect to the relation $\subseteq$. In I.5 we shall see that every set can be well-ordered.

Corresponding to the above principle of proof by induction there is a principle of definition by induction which also applies to arbitrary ordered sets with minimum condition (cf. Kuroš [63], or, in the case of the natural numbers, also VII.1 below).

An ordered set in which any pair of elements a, b has a supremum $a \vee b$ (read: a cup b) and an infimum $a \wedge b$ (read: a cap b) is called a *lattice*. By induction it follows that in a lattice every nonempty finite subset has a supremum and an infimum; if this is true of *every* subset, the lattice is said to be *complete*. In particular, a complete lattice L always has a greatest element ($= \inf \emptyset = \sup L$) and a least element ($= \sup \emptyset = \inf L$).

Clearly, every lattice is directed; further, every totally ordered set is a lattice (though not necessarily complete). On the other hand, a finite lattice is always complete. An example of a complete lattice is $\mathscr{B}(A)$, where A is any set. More generally, let A be an ordered set; a subset X of A is said to be a *left segment* of A if for any $x \in X$, $y \leqslant x$ implies $y \in X$. Similarly, if $y \in X$ whenever $y \geqslant x$ for some $x \in X$, then X is called a *right segment* of A. The set $\mathscr{S}(A)$ of all left segments of A, ordered by inclusion, is a complete lattice, as is easily verified. We note that any abstract set A may be regarded as a totally unordered set; in this sense $\mathscr{S}(A)$ reduces to $\mathscr{B}(A)$. The following criterion is useful in verifying that a given ordered set is a complete lattice.

Proposition 4.1

If A is an ordered set such that every subset has an infimum, then A is a complete lattice.

For, given $X \subseteq A$, let Y be the set of all upper bounds of X in A and set $y = \inf Y$. Then any element of X is a lower bound of Y, hence $x \leqslant y$ for every $x \in X$; if also $x \leqslant z$ for every $x \in X$, then $z \in Y$, and hence $y \leqslant z$. Therefore $y = \sup X$. In particular, A has a least element, obtained as $\inf A$, while the greatest element is $\inf \emptyset$. ∎

Some care must be exercised in applying this proposition, for if e.g. A is a subset of a complete lattice L, and every subset of A has an infimum in A, then A is again a complete lattice, but the supremum of a subset of A will in general be different according to whether it is taken in A or in L; more precisely, for $X \subseteq A$ we have

$$\sup_L X \leqslant \sup_A X,$$

and equality need not hold.

By a *sublattice* of a lattice L we understand a subset A of L which contains with any pair a, b of elements of A also $a \vee b$ and $a \wedge b$. Thus a sublattice contains with any finite (nonempty) subset X also its supremum and infimum, taken in L. In a complete lattice a sublattice is required to contain the supremum and infimum of any of its subsets.

Let A and B be any ordered sets. A mapping $f: A \to B$ is said to be an *order-homomorphism* if

$$x \leqslant y \text{ implies } xf \leqslant yf \qquad \text{for all } x,y \in A.$$

If $f: A \to B$ is an order-homomorphism and its inverse f^{-1} is also a mapping and moreover an order-homomorphism (from B to A), then f is called an *order-isomorphism* between A and B, and we say that A is *order-isomorphic* to B. It is easily seen that if A is directed or totally ordered, then so is its image under an order-homomorphism. However, if f is an order-homomorphism between lattices, it need not preserve the supremum or infimum, i.e., it is not true in general that

$$(x \vee y)f = xf \vee yf, \quad (x \wedge y)f = xf \wedge yf \qquad \text{for all } x,y \in A.$$

When this condition is satisfied, f is called a *lattice-homomorphism.* Correspondingly, a *lattice-isomorphism* is a bijection f such that both f and f^{-1} are lattice-homomorphisms. Every order-isomorphism between lattices is in fact a lattice-isomorphism, even though the corresponding statement for homomorphisms is not true.

To compare ordered sets we shall make use of the following property of complete lattices, which can actually be used to characterize them (cf. Davis [55], Tarski [55]).

Proposition 4.2

Any order-homomorphism of a complete lattice into itself has a fixed point.

Proof:

Let L be the lattice, $f: L \to L$ the order-homomorphism, and put $I = \{x \in L \mid x \leqslant xf\}$, $a = \sup I$. For any $x \in I$, $x \leqslant xf$ and $x \leqslant a$; therefore $xf \leqslant af$, whence $x \leqslant af$ for all $x \in I$. Thus af is an upper bound for I; by definition of a, this means that

$$(1) \qquad a \leqslant af,$$

from which it follows that $a \in I$. By (1), $af \leqslant af^2$, i.e. $af \in I$. Therefore $af \leqslant a$ and together with (1) this shows that $af = a$. ∎

Theorem 4.3

Let A and B be ordered sets such that A is order-isomorphic to a left segment of B and B is order-isomorphic to a right segment of A. Then there *exists a bijection* $f: A \to B$ *which respects the ordering in the following weak sense:*

(2) $$x < y \text{ implies } xf \ngtr yf \text{ for all } x, y \in A.$$

Proof:

Let $g: A \to B_0$ and $h: B \to A_0$ be the given order-isomorphisms of A with a left segment B_0 of B and of B with a right segment A_0 of A. For any subset X of A, denote by X' its complement in A, and likewise for subsets of B. Clearly the complement of a left (right) segment is a right (left) segment, and in fact, $X \to X'$ (for $X \in \mathscr{S}(A)$) is an order-reversing mapping of the set of left segments of A onto the set of right segments of A. We define a mapping θ of $\mathscr{S}(A)$ into itself by the rule

$$X\theta = ((Xg)'h)'.$$

Since g, h preserve the ordering, while taking complements reverses it, θ is order-preserving, i.e. an order-homomorphism, and since $\mathscr{S}(A)$ is complete, θ has a fixed point (by Proposition 4.2), i.e. there exists a left segment A_1 of A such that $((A_1 g)'h)' = A_1$. Thus, if $B_1 = A_1 g$, then $B_1' h = A_1'$. Now define $f: A \to B$ by the rule

$$xf = \begin{cases} xg & \text{if } x \in A_1, \\ xh^{-1} & \text{if } x \in A_1'. \end{cases}$$

Since both g and h are injective, so is f, and moreover, if $x < y$, then $xf \neq yf$. So to verify (2) we need only show that $x < y$ implies $xf \ngtr yf$. If this is false, then for some $x, y \in A$ we have

(3) $$x < y \quad \text{and} \quad xf > yf.$$

If $y \in A_1$, then $x \in A_1$ and $xf = xg < yg = yf$, which contradicts (3). If $x \in A_1'$, then $y \in A_1'$ and $x = xfh > yfh = y$, which again contradicts (3). The only remaining possibility is that $x \in A_1$ and $y \in A_1'$; this means that $xf \in B_1$ and $yf \in B_1'$; but B_1' is a right segment of B, hence by (3) $xf \in B_1'$, which again is a contradiction. So (3) cannot hold. ∎

If we apply this theorem to totally unordered (i.e. abstract) sets, we obtain

Corollary 4.4 (Schröder-Bernstein theorem)

If A and B are any sets and $g: A \to B$, $h: B \to A$ are any injections, then there is a bijection between A and B. ∎

For totally ordered sets (2) implies that f is an order-isomorphism, and we obtain in this case

Corollary 4.5

Let A, B be ordered sets such that A is order-isomorphic to a left segment of B and B is order-isomorphic to a right segment of A. If at least one of A, B is totally ordered, then A is order-isomorphic to B.

For by symmetry we may take B to be totally ordered; applying the theorem, we obtain a bijection satisfying (2), i.e. an order-homomorphism. Since in fact A is also totally ordered (being isomorphic to a segment of B), the bijection is actually an order-isomorphism. ∎

Corollary 4.5 does not remain true for arbitrary ordered sets, as may be shown by examples (cf. Exercise 7).

With every element a of an ordered set A we can associate the left segment $S_a = \{x \in A \mid x \leqslant a\}$, and it is not hard to verify that the mapping $a \to S_a$ is an order-homomorphism of A into $\mathscr{S}(A)$. Since the mapping is clearly injective, this shows that A is always order-isomorphic to a subset of $\mathscr{S}(A)$. The following result, due to Dilworth and Gleason [62], shows that A can never be order-isomorphic to $\mathscr{S}(A)$.

Theorem 4.6

Let A be an ordered set and $f: A_0 \to \mathscr{S}(A)$ an order-homomorphism from a subset A_0 of A to $\mathscr{S}(A)$. Then f is not surjective.

For, assume that f is surjective; put $B = \{x \in A_0 \mid x \notin xf\}$, and let $\bar{B}$ be the left segment generated by B, i.e. the set of all $y \in A$ such that $y \leqslant x$ for some $x \in B$. By hypothesis, $\bar{B} = bf$ for some $b \in A_0$. If $b \notin \bar{B}$, then $b \in B \subseteq \bar{B}$, by the definition of B, which is a contradiction. Hence $b \in \bar{B}$, i.e. $b \leqslant x$ for some $x \in B$, and so $\bar{B} = bf \subseteq xf$. But we have $x \in B \subseteq \bar{B} \subseteq xf$, and this means, by the definition of B, that $x \notin B$, which is again a contradiction. ∎

The proof generalizes the well-known Cantor diagonal argument which is used to show that A is not equipotent to $\mathscr{B}(A)$. In order to obtain this result, we regard A as a totally unordered set; then any mapping from A (or any subset of A) to $\mathscr{S}(A)$ will be an order-homomorphism and we obtain

Corollary 4.7

If A is any set, there exists no bijection from A to $\mathscr{B}(A)$. ∎

For ordered sets in general we obtain

Corollary 4.8

If A is any ordered set, then A is not order-isomorphic to $\mathscr{S}(A)$. ∎

For certain types of ordered sets a graphical representation is useful, which will now be explained. By a *graph* one understands a figure consisting of points, the *vertices* of the graph, together with a number of *edges* joining certain pairs of vertices. We shall only be concerned with *oriented graphs*, in which each edge has a given orientation, corresponding to an ordering of the end-points of the edge, and which may be represented by an arrow drawn along the edge. Thus the commutative diagrams of mappings in I.3 are examples of oriented graphs. If Γ is any oriented graph, the set $V(\Gamma)$ of its vertices may be preordered by writing, for any $a,b \in V(\Gamma)$, '$a \leqslant b$' if and only if we can pass from a to b by going along edges of the graph (taking the orientation into account). Conversely, to represent a given preordered set A, we represent its elements by vertices, with an edge from a to b whenever $a \leqslant b$. The resulting oriented graph is denoted by $\Gamma(A)$ and is called the *graph of A*. Of course for many ordered sets this is not very useful for purposes of illustration (e.g. the set of all functions from N to itself, ordered by inclusion), and most actual diagrams will refer to finite graphs, where for clarity only those edges are drawn which connect elements a, b such that b *covers* a, i.e. b is a minimal element such that $b > a$.

A graph is said to be *connected* if any two vertices a, b can be connected, i.e., there exists a finite sequence of vertices $a_0 = a, a_1, \cdots, a_n = b$ such that a_{i-1} and a_i are joined by an edge (disregarding orientation). In general, the graph of an ordered set need not be connected; for example, a totally unordered set has a 'totally disconnected' graph, in which no two distinct vertices are connected. In any graph Γ, the relation: *a is connected to b* is obviously an equivalence on $V(\Gamma)$; its equivalence classes are called the *connected components* of the graph Γ.

The following reduction theorem (Newman [42]) will be useful later on (in III.9) when considering the normal form of elements in a given presentation.

Theorem 4.9 (Diamond lemma)

Let A be a preordered set and assume that

(i) *for each $a \in A$ there exists a positive integer $k = k(a)$ such that every descending chain through a,*

$$a = a_0 \geqslant a_1 \geqslant \cdots$$

where $a_{i-1} \neq a_i$, has at most k terms, and

(ii) *if a covers* b_1 *and* b_2, *then the set* $\{b_1,b_2\}$ *is bounded below in A.*

Then there is a bijection between the connected components of the graph of A and the minimal elements of A:

$$\Gamma_i \leftrightarrow a_i \qquad (i \in I).$$

If A_i *is the set of vertices of* Γ_i, *then* $a_i \in A_i$ *and* a_i *is in fact the least element of* A_i.

Proof:

Let Γ_i $(i \in I)$ be the connected components of the graph of A and A_i the corresponding subsets of A. From (i) it follows in the first place that A is actually ordered (not merely preordered) and that each A_i contains a minimal element; further, every minimal element of A_i is clearly minimal in A. If we can show that A_i contains only *one* minimal element, then, since by (i) every element is $\geqslant$ a minimal element, it follows that the unique minimal element of A_i must also be the least element of A_i.

Let a, b then be minimal elements of A_i. By definition of A_i, there exists a sequence $a_0 = a,a_1,\cdots,a_n = b$ such that for each $i = 1,\cdots,n$ either $a_i \leqslant a_{i-1}$ or $a_i \geqslant a_{i-1}$. If $a \neq b$, then by omitting repetitions we may assume that $a_i \neq a_{i-1}$; and inserting extra terms if necessary, we may assume (by (i)) that one of a_i, a_{i-1} covers the other. Now for any $x \in A$, denote by $h(x)$ the maximum length of any descending chain through x; this length is finite by (i), and clearly if $x < y$, then $h(x) < h(y)$. We now use double induction, (a) on max $h(a_i)$ and (b) on the number of a_i for which this maximum is attained. If max $h(a_i) = 0$, then all the a_i are minimal and therefore all are equal, whence $a = b$. Now let max $h(a_i) > 0$ and suppose that the maximum is attained for $i = j$. Then a_j covers a_{j-1} and a_{j+1}; hence by (ii), there exists $c \in A$ such that $c \leqslant a_{j-1}$, $c \leqslant a_{j+1}$, and using c in place of a_j together with insertions as before, we obtain a sequence (a'_k) from a to b, possibly longer than the last, but in which either max $h(a'_k)$ has a smaller value than max $h(a_i)$, or the two maxima are the same but are attained by fewer elements a'_k than elements a_i. This contradicts the induction hypothesis, and so $a = b$. ∎

We remark that condition (i) is considerably stronger than the minimum condition for A. Actually the result still holds when (i) is replaced by the minimum condition (cf. Newman [42]), but the above version is adequate for our purpose.

EXERCISES

1. Show that any intersection of orderings on a set is an ordering.

2. (Hausdorff.) Show that the following assumption is equivalent to Zorn's lemma: Every ordered set contains a maximal chain (i.e. every ordered set A contains a chain which is maximal in the set of all chains of A, ordered by inclusion).

3. If a directed set has a maximal element, this is its greatest element. Deduce that a lattice has at most one maximal element and at most one minimal element.

4. Give an example of an order-homomorphism between lattices which is not a lattice-homomorphism.

5. Show that an ordered set is partly well-ordered if and only if it has no infinite descending chains and no infinite totally unordered subsets (cf. III.2).

6. Let A be an ordered set in which any chain has at most m elements and any subset of pairwise incomparable elements has at most n elements. Show that A has at most mn elements.

7. Let A be the set of couples (m,n) of natural numbers such that $m \leqslant n$, ordered by the rule: $(m,n) \leqslant (m',n')$ if and only if $m \leqslant m'$ and $n = n'$; and let B be the complement of $\{(0,0)\}$ in A. Verify that A is order-isomorphic to a left segment of B and B is order-isomorphic to a right segment of A, but that A is not order-isomorphic to B.

8. If L is any complete lattice and f an order-homomorphism of L into itself, show that the fixed points of f form again a complete lattice with respect to the ordering induced by L. (Hint: Let F be the set of fixed points; if $X \subseteq F$, show that sup I is an infimum of X in F, where

$$I = \{y \in L \mid y \leqslant yf, \text{ and } y \leqslant x \text{ for all } x \in X\},$$

and use Proposition 4.1.) Is this lattice necessarily a sublattice of L?

9. For any ordered set A, verify that the set $\mathscr{S}(A)$ of left segments of A, ordered by inclusion, is a complete lattice.

10. Let A be an ordered set and for each $a \in A$ write

$$S_a = \{x \in A \mid x \leqslant a\}.$$

Show that $a \to S_a$ defines a monomorphism $A \to \mathscr{S}(A)$ which preserves the ordering. Deduce that any ordered set can be embedded in a complete lattice in such a way as to preserve any suprema or infima which exist in A.

11. Show that the preordering defined on the set of vertices of a graph Γ is an ordering if and only if Γ contains no closed paths with more than one vertex.

5. CARDINALS AND ORDINALS

In this section we briefly review the facts which we need about cardinals and ordinals, referring the reader to Hausdorff [14], Kelley [55], or Sierpiński [58] for further details. Unless otherwise stated, all sets are taken to lie in a fixed but arbitrary universe U.

With every ordered set A is associated an object called its *order type* and denoted by $o(A)$ such that

$$o(A) = o(B) \text{ if and only if } A \text{ is order-isomorphic to } B.$$

In detail this means that we partition the class of all ordered sets (in the universe U) into classes of pairwise order-isomorphic ones, and with each class associate a member of the given universe U.[9]

Two cases are of particular importance:

(i) A is an abstract set, regarded as totally unordered. In this case we write '$|A|$' instead of '$o(A)$' and call it the *power* or the *cardinal number* of A.

(ii) A is a well-ordered set. Then $o(A)$ is called the *ordinal number* of A.

Generally, by a *cardinal number*, or an *ordinal number*, one understands the order type of a totally unordered set, or of a well-ordered set, respectively. We now define a relation among order types by putting

$$o(A) \leqslant o(B) \text{ if and only if } A \text{ is order-isomorphic to a left segment of } B.$$

It is easily verified that this relation is a preordering (on the class of order types occurring in U). In general, it is not an ordering, as may be shown by using the sets in Exercise 4.7. We do however obtain an ordering, in fact a total ordering, if we limit ourselves to cardinals, or to ordinals. To begin with we note

Lemma 5.1

If A is an ordered set with minimum condition and $f: A \to A$ an injective order-homomorphism, then $xf \nless x$ for all $x \in A$.

[9] The class itself will in general not be a member of U and so cannot be used.

For if there exists $x \in A$ such that $xf < x$, let $a \in A$ be a minimal element with this property. Since $af < a$, we have $af^2 < af$, which shows that af also has the property; but $af < a$, which contradicts the definition of a. ∎

In the case where A is well-ordered, the conclusion of the lemma states that $xf \geqslant x$ for all $x \in A$.

Theorem 5.2

If α and β are two cardinal numbers, or two ordinal numbers, such that $\alpha \leqslant \beta$ and $\beta \leqslant \alpha$, then $\alpha = \beta$.

For cardinals, this follows from Corollary 4.4. If α, β are ordinals, say $\alpha = o(A)$, $\beta = o(B)$, then A is order-isomorphic to a left segment of B and B is order-isomorphic to a left segment of A. Combining these mappings, we obtain an order-isomorphism $f: A \to A_0$ of A with a left segment of itself, and the conclusion will follow if we show that $A_0 = A$. Suppose that $A_0 \neq A$ and let $a \in A$, $a \notin A_0$; then $af < a$, because $af \geqslant a$ would imply that $a \in A_0$ (since $af \in A_0$). But this contradicts Lemma 5.1; hence $A_0 = A$, and it follows that B is order-isomorphic to A, whence $\beta = \alpha$. ∎

We now show that the ordering of ordinal numbers is total.

Theorem 5.3

The ordinal numbers are totally ordered with respect to the relation $\leqslant$.

Proof:

Let A and B be well-ordered sets; to prove the theorem we need only show that one of A, B is order-isomorphic to a left segment of the other. Let R be the set of all functions defining an order-isomorphism between a left segment of A and a left segment of B. Then R is ordered by inclusion, and since any union of left segments of A is again a left segment of A, and likewise for B, it follows that any chain in R has an upper bound (in fact the union of the chain will again belong to R). By Zorn's lemma, R has a maximal element f, which is a function defining an order-isomorphism between a left segment A_0 of A and a left segment B_0 of B. Of course it is not excluded at this stage that $A_0 = B_0 = f = \emptyset$. If both $A_0 \neq A$ and $B_0 \neq B$, let a be the least element of A which is not in A_0 and b be the least element of B not in B_0; then we could replace f by $f \cup \{(a,b)\}$ and so obtain a proper extension of f, contradicting the maximality. Hence either $A_0 = A$ or $B_0 = B$ (or both) and accordingly $o(A) \leqslant o(B)$ or $o(B) \leqslant o(A)$. ∎

To prove the corresponding property for cardinal numbers, we need

Theorem 5.4

Every set can be well-ordered.

Proof:

Let A be any set and W the collection of all well-ordered subsets of A; thus each element of W is a subset X of A together with an ordering of X which makes X into a well-ordered set. For X, $Y \in W$ we put $X \leqslant Y$ if X is a subset of Y and the inclusion mapping $X \to Y$ is an order-isomorphism of X with a left segment of Y. This means in particular that if $X \leqslant Y$, then the ordering on X is that induced by the ordering on Y. It is easily verified that the relation $\leqslant$ so defined is an ordering of W. Now if $(C_i)_{i \in I}$ is any chain in W, then the set $D = \bigcup C_i$ has an ordering which is uniquely determined by the fact that it induces the ordering given on each C_i. Moreover, it is a well-ordering, for if $X \subseteq D$, $X \neq \emptyset$, let $x \in X$. Then $x \in C_i$ for some $i \in I$, hence $X \cap C_i \neq \emptyset$. Let a be the least element of $X \cap C_i$ (in the ordering of C_i); then a is also the least element of $X \cap C_i$ in the ordering of D and hence it is the least element of X in D, because every element $y \in X$ such that $y \leqslant a$ must lie in C_i. Thus $D \in W$ and is an upper bound for the chain (C_i). By Zorn's lemma there exists a maximal element B in W. If $B \neq A$, let $c \in A$, $c \notin B$, and consider the set $B^* = B \cup \{c\}$ with the ordering which extends the ordering on B and is such that $x < c$ for all $x \in B$. With this ordering B^* becomes a well-ordered set such that $B < B^*$, which contradicts the maximality of B. Hence $B = A$, i.e., A can be well-ordered. ∎

Corollary 5.5

The cardinal numbers are totally ordered with respect to the relation $\leqslant$.

For if A and B are any sets, then by Theorem 5.4 they can be well-ordered, and applying Theorem 5.3 we find that $o(A) \leqslant o(B)$ or $o(B) \leqslant o(A)$; accordingly we have $|A| \leqslant |B|$ or $|B| \leqslant |A|$. ∎

From Theorem 5.3 and Corollary 5.5 it is not hard to deduce that the ordinal numbers (and likewise the cardinal numbers) are well-ordered with respect to $\leqslant$. This observation leads to the following alternative definition of ordinals. We recall that ordinal numbers were defined as order types of well-ordered sets. A more explicit way of defining them would be to take a certain well-ordered set for each type and regard this as representing a class of order-isomorphic sets. This is essentially how the natural numbers were defined in I.1 and the same method may be used here.

Thus one obtains all the different isomorphism types of well-ordered sets by starting with $\emptyset$, adjoining one element at a time, and taking unions of ascending chains. In particular, the finite ordinal numbers are just the natural numbers, and the first infinite ordinal number is the set

$$\{0,1,2,\cdots\}$$

which is usually denoted by ω. We shall not enter into the details of this construction (cf. e.g. Kelley [55]) but merely note that if α is any given ordinal number, then the ordinal numbers less than α correspond by definition to the proper left segments of a well-ordered set of type α, and they are themselves well-ordered of type α. With each ordinal number we can associate a cardinal number $|\alpha|$, namely the cardinal number of a well-ordered set of type α. Often one identifies this cardinal number with the least ordinal number to which it belongs; however, the cardinal number of N is usually denoted by $\aleph_0$ (read: aleph-zero). A set of cardinal $\aleph_0$ is also said to be *countable*. The existence of uncountable sets follows from

Theorem 5.6

For any ordered set A, $o(A) \neq o(\mathscr{S}(A))$. Moreover, if A is well-ordered, then

$$o(A) < o(\mathscr{S}(A)),$$

and if A is totally unordered, then

$$|A| < |\mathscr{B}(A)|.$$

Proof:

The first part follows from Theorem 4.6, and in view of Corollaries 4.7 and 4.8 we need only show that $o(A) \leqslant o(\mathscr{S}(A))$ and $|A| \leqslant |\mathscr{B}(A)|$ to complete the proof. Thus we have to define an order-isomorphism $f: A \to S_0$, where S_0 is a left segment of $\mathscr{S}(A)$. It is easily verified that for well-ordered A, the mapping f given by $af = \{x \in A \mid x < a\}$ is such an order-isomorphism. If A is totally unordered, we put $af = \{a\}$ and remark that this defines an injection of A into $\mathscr{B}(A)$, and hence an order-isomorphism of A with a left segment of $\mathscr{B}'(A)$, where $\mathscr{B}'(A)$ is the set of all nonempty subsets of A. Thus $o(A) \leqslant o(\mathscr{B}'(A))$, and therefore $|A| \leqslant |\mathscr{B}'(A)| \leqslant |\mathscr{B}(A)|$. ∎

Corollary 5.7

Among the ordinal numbers of a given universe (and likewise among the cardinal numbers) there is no greatest one. ∎

Another way of expressing this corollary would be to say that the class of all ordinal numbers (or the class of all cardinal numbers) is not a member of the given universe.

If A and B are any two sets of cardinals α and β respectively, then $|A \times B|$ depends only on α and β and not on A, B themselves, as is easily seen. We write $|A \times B| = \alpha\beta$ and call $\alpha\beta$ the *product* of α and β. In a similar way the *sum* $\alpha + \beta$ may be defined as $|A \cup B|$, where A, B are disjoint sets of cardinals α, β respectively. From the properties of unions and Cartesian products it follows easily that the commutative and associative laws hold for both sum and product. On the basis of these definitions it is possible to develop cardinal arithmetic (cf. e.g. Sierpiński [58]); we shall not do so, but merely note the equations

$$\alpha + \beta = \alpha\beta = \max(\alpha, \beta), \qquad (\alpha, \beta \neq 0),$$

valid for any two cardinal numbers of which at least one is infinite (cf. Exercises 7 and 8). For the present we shall prove the second equation in the special case when $\alpha = \aleph_0$ and β is infinite.

Proposition 5.8

For every infinite cardinal number α,

$$\aleph_0\alpha = \alpha. \tag{1}$$

Proof:

In the case where $\alpha = \aleph_0$, (1) states that $N \times N$ is equipotent with N; this follows e.g. by enumerating the pairs (m,n) according the value of $m + n$ and pairs with the same value for $m + n$ according to m. Secondly, if α is of the form $\aleph_0\gamma$, then by what has been proved,

$$\aleph_0\alpha = \aleph_0(\aleph_0\gamma) = \aleph_0^2\gamma = \aleph_0\gamma = \alpha,$$

which proves (1) for this case. Now we complete the proof by showing that every infinite cardinal number is of the form $\aleph_0\gamma$. This amounts to showing that every infinite set A is equipotent to a product $N \times C$, where C is a suitably chosen set. Let A be an infinite set; then by Theorem 5.4, A can be well-ordered. In well-ordered form A consists of a totally ordered (in fact well-ordered) set of countable sequences, followed by a finite (possibly empty) sequence. Since A is infinite, at least one infinite sequence occurs, and we may rearrange A by taking the finite set from the end and putting it in front of the first sequence. The set A now consists entirely of countable sequences, i.e. well-ordered sets isomorphic to N; if they are indexed by a set C, it follows that A is equipotent with $N \times C$. ∎

We conclude this section with a remark on Zorn's lemma which will be needed later. The hypothesis of Zorn's lemma considers an ordered set A in which each chain has an upper bound. This is true in particular if each chain in A has a supremum in A. It is possible to express this condition in a weaker form, as well as a stronger form.

Proposition 5.9

Let A be an ordered set; then the following three conditions on A are equivalent:

(i) *Every nonempty directed subset of A has a supremum.*
(ii) *Every nonempty chain of A has a supremum.*
(iii) *Every nonempty well-ordered chain of A has a supremum.*

In (iii) the chain is understood to be well-ordered in the ordering induced by A. We remark that (iii) is used only to facilitate the proof of the equivalence of (i) and (ii) and will not be used again.

Proof:

Any well-ordered chain is a chain and any chain is directed, hence (i) $\Rightarrow$ (ii) $\Rightarrow$ (iii); to complete the proof we show that (iii) $\Rightarrow$ (i). Thus when (iii) holds and we have a directed subset $D \neq \emptyset$ of A, we have to show that sup D exists. The idea of the proof is to try and reach sup D by means of well-ordered chains in D. For this purpose one has to enlarge D; we state this step separately as a

Lemma. *Let A be an ordered set satisfying condition* (iii) *of Proposition* 5.9. *Further, let D be a nonempty directed set in A; then there is a directed set E in A with the properties:*

(a) $E \supseteq D$.
(b) *Any upper bound of D is an upper bound of E.*
(c) *Every well-ordered chain in E has a supremum which again belongs to E.*

To prove the lemma, consider the set of directed subsets E of A which satisfy (a) and (b). There are such subsets, e.g. D itself. If we have any chain of directed subsets satisfying (a) and (b), their union clearly is again of this form; hence, by Zorn's lemma, there is a maximal directed subset

E in A which satisfies (a) and (b). We assert that E also satisfies (c); if not, let λ be the least ordinal number such that a well-ordered chain in E of length λ does not have a supremum belonging to E. Let E' be the set of all elements of the form $\sup(a_\mu)$, where $(a_\mu)_{\mu<\lambda}$ is any well-ordered chain of length λ in E. Then E' is a directed subset of A satisfying (a) and (b).

To establish this fact, take any $a, b \in E'$, say $a = \sup(a_\mu)$, $b = \sup(b_\mu)$ $(a_\mu, b_\mu \in E)$; we define a family $(c_\mu)_{\mu<\lambda}$ inductively such that

$$(2) \qquad c_{\mu'} \leqslant c_\mu \ (\mu' < \mu), \qquad a_\mu \leqslant c_\mu, \qquad b_\mu \leqslant c_\mu, \qquad c_\mu \in E \ (\mu < \lambda).$$

For c_0 take any element of E such that $a_0 \leqslant c_0$, $b_0 \leqslant c_0$. If α satisfies $0 < \alpha < \lambda$ and c_μ is defined for $\mu < \alpha$ so as to satisfy (2), then by hypothesis $c'_\alpha = \sup(c_\mu) \in E$, and since E is directed, there exists $c_\alpha \in E$ such that

$$a_\alpha \leqslant c_\alpha, \qquad b_\alpha \leqslant c_\alpha, \qquad c'_\alpha \leqslant c_\alpha.$$

This means that (2) holds for $\mu = \alpha$. By transfinite induction we get a well-ordered chain $(c_\mu)_{\mu<\lambda}$ satisfying (2) for every $\mu < \lambda$. If $c = \sup(c_\mu)$, then $c \in E'$ and $c \geqslant a_\mu$, $c \geqslant b_\mu$ for all $\mu < \lambda$, hence $c \geqslant a$, $c \geqslant b$, which proves that E' is directed. Now any $a \in E$ can be written as $\sup(a_\mu)$ with $a_\mu = a$ for all $\mu < \lambda$; hence

$$(3) \qquad E \subseteq E'.$$

Further, since every element of E' is a supremum of a family of elements in E, every upper bound of E is an upper bound of E', and so by (b), every upper bound of D is an upper bound of E'. Thus E' satisfies (a) and (b); by maximality we conclude that $E' = E$. This means that every well-ordered chain of length λ has a supremum in E, which is a contradiction. Therefore E must satisfy (c) also and the lemma is established.

If U, V are well-ordered chains, write $U \leqslant V$ if U is a left segment of V. By Zorn's lemma there exists a maximal well-ordered chain (a_λ) in E; putting $a = \sup(a_\lambda)$, we have $a \in E$. We assert that $\sup D = a$; this will complete the proof. In the first place, we note that a is maximal in E, for if it were not, then we could extend the well-ordered chain (a_λ), contradicting its maximality. Since E is directed, a is actually the greatest element of E. Thus a is an upper bound for E, and hence for D (because $D \subseteq E$). If b is any upper bound for D, then by construction, b is an upper bound for E too, and so $b \geqslant a$. Therefore a is the least upper bound for D, i.e. $\sup D = a$. ∎

EXERCISES

1. Show that two finite, totally ordered sets are order-isomorphic if and only if they have the same cardinal number.

2. Show that a totally ordered set is finite if and only if both the given ordering and its opposite are well-orderings.

3. Let Ω be the set of all ordinal numbers in a given universe, with the total ordering defined as in the text. Show that for any $\alpha \in \Omega$, precisely one of the following is true: (i) $\alpha = 0$, (ii) the set of ordinals β such that $\beta < \alpha$ has a maximal element (this element is called the *immediate predecessor* of α), (iii) $0 \neq \alpha = \sup \{\beta \in \Omega \mid \beta < \alpha\}$ (in this case α is called a *limit ordinal*).

4. If A and B are two disjoint ordered sets, define their ordered sum $A + B$ as the set $A \cup B$, ordered so that $x \leqslant y$ holds for every couple (x,y) in $A \times B$, for no couple in $B \times A$, and for couples in A^2 (or in B^2) if and only if $x \leqslant y$ in A^2 (or in B^2). Now, for any order types α, β define $\alpha + \beta$ as the order type of $A + B$, where A, B are any disjoint ordered sets of types α, β respectively. Verify that $\alpha + \beta$ depends only on α, β and not on A, B, and that in the case of cardinal numbers it reduces to the definition given in the text. Do the commutative and associative laws hold for this operation?

5. Show that every ordinal number can be uniquely expressed in the form $\lambda + n$, where λ is a limit ordinal (Exercise 3) or 0, and n is a natural number.

6. If A, B are any ordered sets, define their lexicographic product as $A \times B$, ordered by the rule: $(a,b) \leqslant (a',b')$ if and only if $a < a'$ or $a = a'$ and $b \leqslant b'$. If A, B are well-ordered, show that their lexicographic product is again well-ordered. For any order types α, β define $\alpha\beta$ as the order type of the lexicographic product of ordered sets A and B, of types α and β respectively. Verify that $\alpha\beta$ depends only on α, β and not on A, B, and that it reduces to the definition given in the text for cardinal numbers. Do the commutative and associative laws hold for this operation?

7. Show that addition of cardinal numbers, and multiplication of cardinal numbers, both satisfy the commutative and associative laws. Show also that if $\alpha \leqslant \alpha'$, $\beta \leqslant \beta'$, then $\alpha + \beta \leqslant \alpha' + \beta'$ and $\alpha\beta \leqslant \alpha'\beta'$.

8. Show that if α, β are cardinal numbers of which at least one is infinite, then $\alpha + \beta = \max(\alpha,\beta)$. Assuming the equation $\gamma^2 = \gamma$ for infinite cardinal numbers (cf. VI.6), show that further if α, $\beta \neq 0$, then $\alpha\beta = \max(\alpha, \beta)$.

9. A set A is said to be *densely ordered* if it is totally ordered and for any $a,b \in A$ such that $a < b$, there exists $c \in A$ such that $a < c < b$. Show that there are precisely four types of countable densely ordered sets. (Hint: Show that any countable densely ordered set is order-isomorphic to a closed, open, or half-open interval of rational numbers.)

10. Prove Theorem 5.3 without using Zorn's lemma. (Hint: Use transfinite induction.)

6. CATEGORIES AND FUNCTORS

In algebra as well as topology we often have to consider sets with a certain structure, together with mappings between the sets which preserve the structure. There are a number of basic notions common to all these situations, and it is convenient to define these in a more general setting.

A *category* $\mathscr{K}$ is a class of *$\mathscr{K}$-objects*, together with a class of *$\mathscr{K}$-morphisms*, which are related in the following way:

C.1. *With each couple of objects a, b there is associated a set* Hom(a, b) *of $\mathscr{K}$-morphisms, such that each $\mathscr{K}$-morphism belongs to* Hom (a, b) *for just one couple of objects a, b.*

C.2. *If* $\alpha \in \text{Hom}(a,b)$ *and* $\beta \in \text{Hom}(b,c)$, *there is a unique element of* Hom(a,c) *called the composition or product of* α *and* β *and denoted by* $\alpha\beta$.

C.3. *Given* $\alpha \in \text{Hom}(a,b)$, $\beta \in \text{Hom}(b,c)$, $\gamma \in \text{Hom}(c,d)$, *so that* $(\alpha\beta)\gamma$ *and* $\alpha(\beta\gamma)$ *are defined, then*

$$(\alpha\beta)\gamma = \alpha(\beta\gamma). \tag{1}$$

C.4. *To each $\mathscr{K}$-object a there corresponds a $\mathscr{K}$-morphism* $\varepsilon_a \in \text{Hom}(a,a)$ *called the identity morphism such that for any* $\alpha \in \text{Hom}(b,a)$ *and* $\beta \in \text{Hom}(a,c)$,

$$\alpha\varepsilon_a = \alpha, \qquad \varepsilon_a\beta = \beta. \tag{2}$$

The classes of all $\mathscr{K}$-objects and $\mathscr{K}$-morphisms are also denoted by Ob $\mathscr{K}$ and Hom $\mathscr{K}$ respectively. Instead of '$\alpha \in \text{Hom}(a,b)$' we also write '$\alpha : a \to b$' and say '$\alpha$ goes from a to b'; further, we shall use commutative diagrams to illustrate the composition of morphisms in the same way as for mappings between sets. In this connexion we note that for any $\mathscr{K}$-morphisms $\alpha : a \to b$ and $\beta : c \to d$, the product $\alpha\beta$ is defined if and only if $b = c$.

It is easily verified that the identity morphism $\varepsilon_a : a \to a$ is uniquely determined by the properties (2). Thus the mapping

$$a \to \varepsilon_a$$

establishes a bijection between Ob $\mathscr{K}$ and the class of identity morphisms, so that $\mathscr{K}$ is completely determined by its morphisms. It is not difficult to express the definition of a category entirely in terms of morphisms (cf. Exercise 1); we shall not adopt this point of view, but we note the possibility of doing so, as it may be used to shorten some of the definitions.

The category $\mathscr{K}$ is said to have a *zero*, if

C.5. *There is a* $\mathscr{K}$*-object* 0, *called a zero object, such that for any* $a \in$ Ob $\mathscr{K}$, *the sets* Hom$(a,0)$ *and* Hom$(0,a)$ *each consist of a single element.*

In a category with zero, let us denote the unique morphism in Hom$(0,a)$ by ω_{0a}, that of Hom$(a,0)$ by ω_{a0}, and generally define the *zero morphism* from a to b by the equation

(3) $$\omega_{ab} = \omega_{a0}\omega_{0b};$$

then for any morphisms $\alpha: c \to a$, $\beta: b \to d$, we have $\alpha\omega_{a0} = \omega_{c0}$, $\omega_{0b}\beta = \omega_{0d}$, and hence

(4) $$\alpha\omega_{ab} = \omega_{cb}, \qquad \omega_{ab}\beta = \omega_{ad}.$$

While a category may have more than one zero object, the zero morphisms ω_{ab} are uniquely determined by (4), for if $\omega'_{ab}: a \to b$ is another zero morphism, defined in terms of another zero object $0'$, then by (4), $\omega_{cb} = \omega_{ca}\omega'_{ab} = \omega'_{cb}$.

Examples of Categories

(i) All nonempty sets in a given universe and all mappings between them form a category, denoted by St*.

(ii) All sets in a given universe, and all mappings between them, including, for each set B, the mapping $\emptyset \to B$ defined by the empty function $\emptyset$. This is a category denoted by St.

(iii) The category of all groups and homomorphisms.

(iv) The category of all topological spaces and continuous mappings.

(v) The category of all ordered sets and all order-homomorphisms.

(vi) The category of all lattices and all lattice-homomorphisms.

An *isomorphism* between categories $\mathscr{K}$ and $\mathscr{K}'$ is a bijection $\alpha \to \alpha'$ between Hom $\mathscr{K}$ and Hom $\mathscr{K}'$ such that $\alpha\beta$ is defined if and only if $\alpha'\beta'$ is defined, and then

(5) $$(\alpha\beta)' = \alpha'\beta'.$$

It follows from this that identity morphisms correspond, and hence the $\mathscr{K}$-objects correspond to $\mathscr{K}'$-objects. More precisely, if $a \in$ Ob $\mathscr{K}$, then

ε'_a is an identity morphism, for by (2) and (5), $\varepsilon_a^2 = \varepsilon_a$, hence $\varepsilon_a'^2 = \varepsilon'_a$, so that if $\varepsilon'_a : u \to v$, then $v = u$, and if η_u is the identity morphism of u, let $\lambda \in \operatorname{Hom} \mathscr{K}$ be such that $\lambda' = \eta_u$; then $\varepsilon_a \lambda = \lambda$, hence

$$\eta_u = \varepsilon'_a \eta_u = \varepsilon'_a.$$

Writing $u = a'$, we obtain a bijection $a \to a'$ between $\operatorname{Ob} \mathscr{K}$ and $\operatorname{Ob} \mathscr{K}'$ such that $\alpha : a \to b$ if and only if $\alpha' : a' \to b'$.

An *anti-isomorphism* from $\mathscr{K}$ to $\mathscr{K}'$ is a bijection between $\operatorname{Hom} \mathscr{K}$ and $\operatorname{Hom} \mathscr{K}' : \alpha \to \alpha'$ such that $\alpha\beta$ is defined if and only if $\beta'\alpha'$ is defined and

(6) $$(\alpha\beta)' = \beta'\alpha'.$$

Again this establishes a bijection between $\operatorname{Ob} \mathscr{K}$ and $\operatorname{Ob} \mathscr{K}'$, say $a \to a'$, but this is such that $\alpha : a \to b$ if and only if $\alpha' : b' \to a'$.

With every category $\mathscr{K}$ we associate a category $\mathscr{K}^\circ$, the *opposite* of $\mathscr{K}$, whose morphisms are the $\mathscr{K}$-morphisms, but with multiplication

$$\alpha * \beta = \beta \alpha,$$

whenever the right-hand side is defined. It follows that $\mathscr{K}^\circ$ is anti-isomorphic with $\mathscr{K}$. More generally, a category $\mathscr{L}$ is anti-isomorphic with $\mathscr{K}$ if and only if it is isomorphic with $\mathscr{K}^\circ$.

The notion of isomorphism between categories is a special case of the notion of functor. By a *functor* from a category $\mathscr{K}$ to a category $\mathscr{L}$ one understands a couple of mappings, from $\operatorname{Ob} \mathscr{K}$ to $\operatorname{Ob} \mathscr{L}$ and $\operatorname{Hom} \mathscr{K}$ to $\operatorname{Hom} \mathscr{L}$, both denoted by the same letter, F say, for simplicity, such that:

(i) $(\varepsilon_a)F = \varepsilon_{aF}$.

(ii) If $\alpha\beta$ is defined in $\mathscr{K}$, then $(\alpha F)(\beta F)$ is defined in $\mathscr{L}$ and

$$(\alpha\beta)F = (\alpha F)(\beta F).$$

More precisely, a functor as just defined is said to be *covariant*; by a *contravariant* functor from $\mathscr{K}$ to $\mathscr{L}$ one understands a covariant functor from $\mathscr{K}$ to $\mathscr{L}^\circ$, or equivalently, from $\mathscr{K}^\circ$ to $\mathscr{L}$.

A *subcategory* $\mathscr{L}$ of $\mathscr{K}$ consists of a subclass of $\operatorname{Ob} \mathscr{K}$ and a subclass of $\operatorname{Hom} \mathscr{K}$, denoted by $\operatorname{Ob} \mathscr{L}$ and $\operatorname{Hom} \mathscr{L}$ respectively, such that

(i) If $a \in \operatorname{Ob} \mathscr{L}$, then $\varepsilon_a \in \operatorname{Hom} \mathscr{L}$.

(ii) If $\alpha, \beta \in \operatorname{Hom} \mathscr{L}$ and $\alpha\beta$ is defined in $\mathscr{K}$, then $\alpha\beta \in \operatorname{Hom} \mathscr{L}$.

(iii) If $\alpha \in \operatorname{Hom} \mathscr{L}$ and $\alpha : a \to b$, then $a, b \in \operatorname{Ob} \mathscr{L}$.

If $a,b \in \text{Ob } \mathscr{L}$, then one often writes '$\text{Hom}_{\mathscr{L}}(a,b)$' in place of '$\text{Hom}_{\mathscr{K}}(a,b) \cap \text{Hom } \mathscr{L}$'. The subcategory $\mathscr{L}$ of $\mathscr{K}$ is said to be *full* if $\text{Hom}_{\mathscr{L}}(a,b) = \text{Hom}_{\mathscr{K}}(a,b)$, for any $a,b \in \text{Ob } \mathscr{L}$. Thus a full subcategory of $\mathscr{K}$ is completely determined once the class of its objects is given; for example, abelian groups and homomorphisms form a full subcategory of the category of groups and homomorphisms. Lattices and lattice-homomorphisms form a subcategory of the category of ordered sets and order-homomorphisms, which however is not full, as we saw in I.4.

Let $\mathscr{K}$ be any category; then a $\mathscr{K}$-morphism $\alpha : a \to b$ is said to be *invertible* or an *equivalence*, if there is a $\mathscr{K}$-morphism $\beta : b \to a$ such that $\alpha\beta = \varepsilon_a$, $\beta\alpha = \varepsilon_b$. Such a morphism β, when it exists, is uniquely determined by α, as is easily seen. It is denoted by α^{-1} and called the *inverse* of α. Two $\mathscr{K}$-objects a and b are said to be *equivalent*: $a \sim b$, if there exists an equivalence $\alpha : a \to b$. Thus e.g., any two zero objects of a category with zero are equivalent.

Given two categories $\mathscr{K}$ and $\mathscr{L}$, let F, G be two functors from $\mathscr{K}$ to $\mathscr{L}$. Then a *natural transformation* of functors $\tau : F \to G$ is a function which associates with each $a \in \text{Ob } \mathscr{K}$ an $\mathscr{L}$-morphism $\tau(a) : aF \to aG$ such that for any $\alpha : a \to b$, the following diagram commutes:

$$\begin{array}{ccc} aF & \xrightarrow{\alpha F} & bF \\ {\scriptstyle \tau(a)}\big\downarrow & & \big\downarrow{\scriptstyle \tau(b)} \\ aG & \xrightarrow[\alpha G]{} & bG \end{array}$$

If $\tau(a)$ is an $\mathscr{L}$-equivalence for each $a \in \text{Ob } \mathscr{K}$, then τ is called a *natural equivalence*. In particular, when $\mathscr{L} = \mathscr{K}$ and F is the identity functor, a natural transformation is a $\mathscr{K}$-morphism $\tau(a)$ for which the above diagram commutes. This is the case which occurs most frequently in the sequel. As a simple illustration, the usual construction of the field of fractions of an integral domain is a functor from the category of integral domains and monomorphisms to the category of fields and homomorphisms, and it may be verified that the embedding of an integral domain in its field of fractions is a natural transformation. This no longer holds if we take as our categories the integral domains and all homomorphisms, and fields and their homomorphisms, respectively; to obtain a functor one then has to extend the latter category to the category of all fields and places (cf. Lang [58]).

EXERCISES

1. Show that for any category $\mathscr{K}$, Hom $\mathscr{K}$ satisfies the following conditions:

(i) For certain couples $\alpha, \beta \in \text{Hom } \mathscr{K}$, an element $\alpha\beta \in \text{Hom } \mathscr{K}$ is defined.

(ii) If $\alpha\beta$ and $\beta\gamma$ are defined, then $(\alpha\beta)\gamma$ and $\alpha(\beta\gamma)$ are defined and are equal. Moreover, $\beta\gamma$ is defined whenever $(\alpha\beta)\gamma$ is defined for some α or $\beta(\gamma\delta)$ is defined for some δ.

(iii) Every element of Hom $\mathscr{K}$ has a left unit and a right unit. (An element $\varepsilon \in \text{Hom } \mathscr{K}$ is called a *left* (*right*) *unit* for α if $\varepsilon\alpha(\alpha\varepsilon)$ is defined and $\varepsilon\beta = \beta$, $\gamma\varepsilon = \gamma$, whenever the left-hand side is defined.)

Conversely, show that any class satisfying these conditions forms the class of morphisms of some category.

2. Let $\mathscr{S}_U$ be the category of all sets and mappings in a given universe U. Show that the subcategories of $\mathscr{S}_U$ with functors as morphisms form a category.

3. A category is said to be *self-dual* if it is isomorphic to its opposite. Show that the category of all sets and correspondences between them (in a given universe) is self-dual.

4. (a) In the category of sets and mappings, show that a mapping α is injective if and only if it has a right inverse (i.e., if $\alpha : a \to b$, there exists $\beta : b \to a$ such that $\alpha\beta = \varepsilon_a$) and is surjective if and only if it has a left inverse (i.e. there exists $\gamma : b \to a$ such that $\gamma\alpha = \varepsilon_b$).

(b) For any mapping α show that the following are equivalent: (i) α is invertible, (ii) α has a unique right inverse, (iii) α has a unique left inverse, (iv) α has a right inverse and a left inverse.

5. In the semigroup S_P of mappings from an infinite set P to itself, show that

(a) If a mapping has more than one right inverse, it has infinitely many,

(b) There exists a mapping with exactly two left inverses.

Deduce that S_P, qua category, is not self-dual. Is S_P self-dual when P is finite?

6. Let $\mathscr{K}$ be any category. A $\mathscr{K}$-morphism μ is said to be *right regular* if $\alpha\mu = \beta\mu$ implies that $\alpha = \beta$. If $\mu : b \to a$ is a right regular $\mathscr{K}$-morphism, then b is called a *subobject* of a with the $\mathscr{K}$-morphism μ. Show that the subobjects of a are preordered by the rule: $c \leqslant b$ if $\nu : c \to a$ and $\mu : b \to a$ are the given right regular morphisms and $\nu = \alpha\mu$ for some α. Show also that $c \leqslant b$ and $b \leqslant c$ hold if and only if $b \sim c$. Define *quotient objects* of a by duality, in terms of left regular morphisms, and prove corresponding statements for them.

7. Let $\mathscr{K}$ be any category and for $a, b \in \text{Ob } \mathscr{K}$, write '$a \leqslant b$' if and only if $\text{Hom}(a,b) \neq \emptyset$. Show that the relation so defined is a preordering on Ob $\mathscr{K}$.

Chapter II

Algebraic Structures

An algebraic structure on a set A is essentially a collection of finitary operations on A; the set A with this structure is also called an *algebra*. Most of the notions introduced for sets, such as subset, mapping, equivalence, have analogues for algebras, namely subalgebra, homomorphism, congruence. The mapping theorems of I.3 then correspond to the isomorphism theorems, which are probably best known in the case of groups. The analogy is less complete for the Jordan-Hölder and Krull-Schmidt theorems, which are therefore first considered in their abstract setting in lattice theory, and then for algebras.

In addition the set of subalgebras of a given algebra plays an important role, and to a lesser extent the set of all congruences; they form complete lattices with certain characteristic properties on which the applicability of Zorn's lemma depends. We begin therefore by studying these properties in the abstract.

1. CLOSURE SYSTEMS

Let A be any set and $\mathscr{B}(A)$ its Boolean, i.e. the set of all its subsets. We wish to consider certain subsets of $\mathscr{B}(A)$, or as we shall say, *systems* of subsets of A. A system $\mathscr{C}$ of subsets of A is said to be a *closure system* if $\mathscr{C}$ is closed under intersections, i.e.

$$\text{for any subsystem } \mathscr{D} \subseteq \mathscr{C}, \text{ we have } \bigcap \mathscr{D} \in \mathscr{C}.$$

In particular, taking $\mathscr{D} = \emptyset$, we see that A always belongs to $\mathscr{C}$. Since a closure system admits arbitrary intersections, it follows by Proposition I.4.1 that it is a complete lattice (with respect to the ordering by inclusion). However, it need not be a sublattice of $\mathscr{B}(A)$, since the cup operation in $\mathscr{C}$ is in general different from that of $\mathscr{B}(A)$ (cf. the examples below).

The most important examples of closure systems are the following:

(i) In a group G, the system of all subgroups of G is a closure system. This case will be generalized later.

(ii) Let X be a topological space and $\mathscr{T}$ the system of closed subsets. Then $\mathscr{T}$ is a closure system which has the further property:

(1) $$\text{For any } A,B \in \mathscr{T}, \qquad A \cup B \in \mathscr{T}.$$

Thus $\mathscr{T}$ is a sublattice of $\mathscr{B}(X)$, though not in general complete. Any closure system satisfying (1) is said to be *topological.* If we are given a topological closure system $\mathscr{T}$ on a set X which in addition includes $\emptyset$, then we can define a topology on X by declaring the members of $\mathscr{T}$ to be the closed sets. Of course the resulting topology will not in general be separated.

A second notion which we require (and which will turn out to be equivalent to that of a closure system) is that of a closure operator on a set. A *closure operator* or *join operator* on a set A is a mapping J of $\mathscr{B}(A)$ into itself with the properties:

J.1. If $X \subseteq Y$, then $J(X) \subseteq J(Y)$,

J.2. $X \subseteq J(X)$,

J.3. $JJ(X) = J(X)$,

for all X, $Y \in \mathscr{B}(A)$. For every closure system $\mathscr{C}$ we can define a closure operator J by the equation

(2) $$J(X) = \bigcap \{Y \in \mathscr{C} \mid Y \supseteq X\}.$$

This operator satisfies J.1–2 by definition. Further, we have

(3) $$J(X) = X \text{ if and only if } X \in \mathscr{C},$$

because $\mathscr{C}$ is a closure system; since $J(X) \in \mathscr{C}$, this proves J.3.

Conversely, given a closure operator J (satisfying J.1–3), we put

(4) $$\mathscr{C} = \{X \subseteq A \mid J(X) = X\}.$$

If $(X)_{i \in I}$ is any family in $\mathscr{C}$ and $\bigcap X_i = X$, then $X \subseteq X_i$; hence by J.1, $J(X) \subseteq J(X_i) = X_i$ for all i, and so

$$J(X) \subseteq \bigcap X_i = X.$$

Together with J.2 this shows that $J(X) = X$, i.e. $X \in \mathscr{C}$. Thus we have obtained a closure system $\mathscr{C}$ from J, and incidentally we have done so without using J.3. We now use J.3 to show that the correspondence $\mathscr{C} \to J$ is bijective.

First, let $\mathscr{C}$ be any closure system, J the operator defined by (2), and $\mathscr{C}'$ the closure system defined in terms of J by (4). Then $\mathscr{C}' = \mathscr{C}$ by (3). Next take a closure operator J, and let $\mathscr{C}$ be the closure system defined in terms of J by (4) and J' the operator defined by (2) in terms of $\mathscr{C}$. By what has just been shown, $\mathscr{C}$ is then also defined by J', hence

(5) $$J(X) = X \text{ if and only if } J'(X) = X.$$

By J.3, $JJ(X) = J(X)$; hence (5) gives $J'J(X) = J(X)$. But $X \subseteq J(X)$, and applying J' we obtain $J'(X) \subseteq J'J(X) = J(X)$. This shows that $J'(X) \subseteq J(X)$, and the reverse inclusion follows by symmetry. Thus we have proved

Theorem 1.1

Every closure system $\mathscr{C}$ on a set A defines a closure operator J on A by the rule

$$J(X) = \bigcap \{Y \in \mathscr{C} \mid Y \supseteq X\}.$$

Conversely, every closure operator J on A defines a closure system by

$$\mathscr{C} = \{X \subseteq A \mid J(X) = X\},$$

and the correspondence $\mathscr{C} \leftrightarrow J$ between closure systems and closure operators thus defined is bijective. ∎

We note that closure systems and operators may be defined on any complete lattice L and the relations between them expressed in Theorem 1.1 still subsist; in fact Theorem 1.1 is just the special case $L = \mathscr{B}(A)$.

The members of $\mathscr{C}$ are called the *$\mathscr{C}$-sets* or *closed sets* of A, and $J(X)$ is called the *closure* of X in A (it is indeed closed, by J.3). As we have noted, $\mathscr{C}$ is a complete lattice with respect to $\subseteq$. Explicitly, given any family $(X_i)_{i \in I}$ in $\mathscr{C}$, the set $\bigcap X_i$ is the greatest closed set contained in all the X_i, and $\bigcap \{Y \in \mathscr{C} \mid Y \supseteq X_i \text{ for all } i \in I\}$ is the least closed set containing all the X_i.

We now give a third example of closure systems, which will be of importance in what follows.

Let A and B be any sets and Φ a correspondence from A to B, i.e. a subset of $A \times B$. For any subset X of A we define a subset X^* of B by the equation

$$X^* = \{y \in B \mid (x,y) \in \Phi \text{ for all } x \in X\},$$

and similarly, for any subset Y of B, we define a subset Y^* of A by

$$Y^* = \{x \in A \mid (x,y) \in \Phi \text{ for all } y \in Y\}.$$

Thus we have the mappings

$$X \to X^*, \qquad Y \to Y^* \tag{6}$$

of $\mathscr{B}(A)$, $\mathscr{B}(B)$ into each other, with the properties

$$\begin{cases} \text{If } X_1 \subseteq X_2 & \text{then } X_1^* \supseteq X_2^*, \\ \text{If } Y_1 \subseteq Y_2 & \text{then } Y_1^* \supseteq Y_2^*, \end{cases} \tag{7}$$

$$X \subseteq X^{**}, \qquad Y \subseteq Y^{**}, \tag{8}$$

$$X^{***} = X^*, \qquad Y^{***} = Y^*. \tag{9}$$

Conditions (7) and (8) follow immediately from the definitions; if (7) is applied to (8), we get $X^* \supseteq X^{***}$, while (8) applied to X^* gives the reverse inequality. Thus any mappings (6) which satisfy (7) and (8) also satisfy (9).

A pair of mappings (6) between $\mathscr{B}(A)$ and $\mathscr{B}(B)$, or more generally, between any ordered sets, is called a *Galois connexion* if it satisfies (7), (8) (and hence (9)). Most Galois connexions encountered in practice arise from a correspondence between sets in the way described above (cf. also the study by Ore [44]).

Examples of Galois Connexions

(i) Let F be a (commutative) field and G the group of all automorphisms of F. Then the pairs $(x, \alpha) \in F \times G$ such that $x^\alpha = x$ form a correspondence which establishes a Galois connexion between certain subfields of F and certain subgroups of G. If G is not the group of all automorphisms of F, but merely a finite group of automorphisms of F, and E is the subfield of F of elements left fixed by G, then the correspondence is between all subgroups of G and all fields between F and E. (This is the subject of Galois theory from which Galois connexions take their name.)

(ii) Let A be a simple (finite-dimensional linear associative) algebra and consider the correspondence of A with itself defined by the relation $xy = yx$. This establishes a Galois connexion of the set of subalgebras of A with itself (cf. Artin, Nesbitt, & Thrall [44]).

(iii) Let R be a commutative ring with a unit element, and define a correspondence in R by the rule $x \neq y$. This establishes in particular a Galois connexion between the prime ideals of R and certain multiplicatively closed subsets of R (cf. Zariski & Samuel [58]).

To establish the link with closure systems we observe that in any Galois connexion the mapping $X \to X^{**}$ is a closure operator in A and $Y \to Y^{**}$ is a closure operator in B (by (7)–(9)). Moreover, the mappings (6) give a bijection between these two closure systems.

A closure operator J on a set A is said to be *algebraic*, if for any $X \subseteq A$ and $a \in A$,

$$\text{if } a \in J(X), \text{ then } a \in J(X_f) \text{ for some finite subset } X_f \text{ of } X.$$

Now a closure system is said to be *algebraic* whenever the corresponding closure operator is algebraic. In order to have a more direct description of algebraic closure systems we need another definition: A nonempty system $\mathscr{C}$ of subsets of A is called *inductive* if every chain in $\mathscr{C}$ has a supremum in $\mathscr{C}$. By Proposition I.5.9 (applied to $\mathscr{C}$) we can replace here the word 'chain' by 'directed set'. Now we have the following characterization of algebraic closure systems, due to Schmidt [52]:

Theorem 1.2

A closure system is algebraic if and only if it is inductive.

Proof:

Let $\mathscr{C}$ be an algebraic closure system on a set A, $\mathscr{K}$ a chain in $\mathscr{C}$, and $K = \sup \mathscr{K}$ in $\mathscr{B}(A)$. We shall show that $K \in \mathscr{C}$, then sup K in $\mathscr{C}$ exists and equals K. To show that $K \in \mathscr{C}$ we need only show that $J(K_f) \subseteq K$ for each finite subset K_f of K. Let $K_f = \{x_1, \ldots, x_n\}$; then each x_i belongs to some term of $\mathscr{K}$, and because $\mathscr{K}$ is a chain, we can find $L \in \mathscr{K}$ to contain them all. Then $K_f \subseteq L \subseteq K$ and $L \in \mathscr{C}$, hence $J(K_f) \subseteq J(L) = L \subseteq K$, i.e. $J(K_f) \subseteq K$, as we wished to show. Conversely, let $\mathscr{C}$ be an inductive closure system on A and J the corresponding closure operator. We have to show, for any $X \subseteq A$,

$$J(X) = \sup\{J(X_f) \mid X_f \subseteq X, X_f \text{ finite}\}, \text{ taken in } \mathscr{C}.$$

Fix $X \subseteq A$ and put $\mathscr{K} = \{J(X_f) \mid X_f \subseteq X, X_f \text{ finite}\}$. Given $Y, Z \subseteq A$, we have $J(Y) \cup J(Z) \subseteq J(Y \cup Z)$ and if Y, Z are finite subsets of X, then so is $Y \cup Z$. Hence $\mathscr{K}$ is directed and so has a supremum K in $\mathscr{C}$. Now $J(X_f) \subseteq J(X)$ for all X_f, hence $K \subseteq J(X)$, but $X \subseteq K$, so $J(X) \subseteq K$, and hence $K = J(X)$ as asserted. ∎

Corollary 1.3

If $\mathscr{C}$ is an algebraic closure system on A, and $\mathscr{K}$ is a directed subsystem of $\mathscr{C}$, then $\sup \mathscr{K} \in \mathscr{C}$. ∎

We note that Zorn's lemma implies that every nonempty inductive system of subsets of a set A contains a maximal subset. This leads to the following corollary of Theorem 1.2 which includes most of the important applications of Zorn's lemma to algebra.

Theorem 1.4

Let $\mathscr{C}$ be an algebraic closure system in A and let A_0, A_1, B be subsets of A such that $B \in \mathscr{C}$ and $B \cap A_1 = A_0$. Then $\mathscr{C}$ contains an element C which is maximal in $\mathscr{C}$ with respect to the properties $C \supseteq B$, $C \cap A_1 = A_0$.

To prove this assertion we take $\mathscr{C}'$ to be the system of all sets $X \in \mathscr{C}$ such that $X \supseteq B$ and $X \cap A_1 = A_0$, and show that $\mathscr{C}'$ has a maximal element. In the first place $\mathscr{C}' \neq \emptyset$, because $B \in \mathscr{C}'$. Now let (X_i) be any chain in $\mathscr{C}'$ and put $X = \sup X_i$. Then $X \in \mathscr{C}$, because $\mathscr{C}$ is inductive. Further, $X \supseteq B$ and $X \cap A_1 = A_0$; therefore $X \in \mathscr{C}'$. Thus $\mathscr{C}'$ is inductive, and by Zorn's lemma, $\mathscr{C}'$ has a maximal element. ∎

EXERCISES

1. Let $X \to H(X)$ be any mapping of $\mathscr{B}(A)$ into itself. Show that $J(X) = H(X) \cup X$ defines a closure operator if and only if $X \subseteq J(Y)$ implies $J(X) \subseteq J(Y)$.

2. Let J_0 be any mapping of $\mathscr{B}(A)$ into itself which satisfies J.1, and define $J_1(X) = J_0(X) \bigcup X$. Verify that J_1 satisfies J.1 and J.2; if $\mathscr{C}$ is the closure system associated with J_1, describe the closure operator J associated with $\mathscr{C}$.

3. Show that the collection of all algebraic closure systems on a given set A is a closure system on $\mathscr{B}(A)$. Is this closure system ever algebraic?

4. Let $\mathscr{C}$ be an algebraic closure system on A. Then every nonempty subsystem of $\mathscr{C}$ has a maximal element if and only if each $X \in \mathscr{C}$ is the closure of a finite subset.

5. (G. Higman.) Let Γ be the set of all closure systems on A and define an operation $\mathscr{C} \to \mathscr{C}^*$ on Γ by the equation

$$J_{\mathscr{C}^*}(X) = \bigcup \{J_{\mathscr{C}}(X_f) \mid X_f \subseteq X,\ X_f \text{ finite}\}.$$

Show that this defines a closure operator on Γ and that $\mathscr{C}^*$ is in fact the least algebraic closure system containing $\mathscr{C}$.

6. (P. J. Higgins.) Show that an inductive system $\mathscr{C}$ of subsets of A which includes all finite subsets must coincide with $\mathscr{B}(A)$. (Hint: For any $X \subseteq A$ construct a maximal subset Y of X such that $Y \cup A_f \in \mathscr{C}$ for all finite sets A_f and show that $Y = X$.) Hence deduce Theorem 1.2 without using Proposition I.5.9. (Hint: If $\mathscr{C}$ is inductive, show that the system of all $X \subseteq A$ such that $J_{\mathscr{C}}(X) = J_{\mathscr{C}^*}(X)$ [in the notation of Exercise 5] is inductive and includes all finite subsets.)

7. In any Galois connexion $X \to X^*$, $Y \to Y^*$, establish the identities

$$(\bigcup X_i)^* = \bigcap X_i^*,$$

$$(\bigcap X_i^{**})^* = (\bigcup X_i^*)^{**}$$

for an arbitrary family of subsets $(X_i)_{i \in I}$.

8. Show that the relation $x \leqslant y$ in an ordered set A establishes a Galois connexion between the left segments and the right segments of A, and that the mapping $x \to \{x\}^{**}$ of A into $\mathscr{S}(A)$ provides an embedding of A in a complete lattice (note that when A is taken to be the set of rational numbers, this is the construction of real numbers by Dedekind cuts).

9. A system $\mathscr{C}$ of subsets of A is said to be of *finite character* (of *character n*) if there is a system $\mathscr{F}$ of finite subsets (subsets with at most n elements) of A such that for each $X \subseteq A$, whether or not $X \in \mathscr{C}$ is uniquely determined by its intersections $F \cap X$, for all $F \in \mathscr{F}$. Show that a closure system is algebraic if and only if it is of finite character.

Give an example of a system of finite character which is not of character n, for any n.

10. Show that the set of all preorderings on a set A is an algebraic closure system on A^2. Is the same true for the set of all orderings?

2. Ω-ALGEBRAS

As already indicated, an algebra is to be thought of as a set with certain operations defined on it. If we wish to compare different algebras we first have to establish a correspondence between their operations. The most convenient way of doing this is to index the operations in each algebra by a given indexing set, which is kept constant in any problem under discussion. This has the further advantage that no notational complications arise when the correspondence between the operations of the two algebras is many-many, as it may well be. The only restriction imposed is that n-ary operations correspond to n-ary operations. Thus we have

Definition (1)

An *operator domain* is a set Ω with a mapping $a: \Omega \to N$; the elements of Ω are called *operators*, and if $\omega \in \Omega$, then $a(\omega)$ is called the *arity* of ω. If $a(\omega) = n$ we also say that ω *is n-ary*, and we write

$$\Omega(n) = \{\omega \in \Omega \mid a(\omega) = n\}.$$

Definition (2)

Let A be a set and Ω an operator domain; then an Ω*-algebra structure* on A is a family of mappings

$$\Omega(n) \to A^{A^n} \qquad (n \in N).$$

Thus with each $\omega \in \Omega(n)$ an n-ary operation on A is associated. The set A with this structure is also called an Ω*-algebra* and is sometimes written A_Ω to emphasize its dependence on Ω.

The underlying set A is also called the *carrier* of A_Ω.

Given an Ω-algebra A and $\omega \in \Omega(n)$, then ω applied to an n-tuple $(a_1, \cdots, a_n)$ from A gives an element of A which we write as $a_1 a_2 \cdots a_n \omega$. In the case $n = 0$ this merely states that ω is an element of A; thus a 0-ary operator picks out a certain distinguished element in the algebra. For this reason a 0-ary operator is sometimes called a *constant operator*. For instance, in defining groups we may use a 0-ary operator whose value is the unit element (see below). This operator may be denoted by 1, so that we are justified in denoting the unit element in all groups by the same symbol.

We now consider Ω-algebras for a fixed domain Ω and introduce some standard notions.

Given Ω-algebras A_Ω and B_Ω, we say that B is a *subalgebra* of A if the carrier of B is a subset of the carrier of A; and if $\omega \in \Omega$ defines operations ω_A, ω_B in A and B respectively, then B admits ω_A and

$$\omega_A \mid B = \omega_B \qquad \text{for each } \omega \in \Omega.$$

Thus any subset of the carrier of A which admits each $\omega \in \Omega$ can be defined in just one way as an Ω-subalgebra of A. The set of all subalgebras of A is denoted by $\mathscr{B}_\Omega(A)$. This set always contains A, whereas $\emptyset$ is a member if and only if Ω has no constant operators. A subalgebra of A is said to be *proper* if it is distinct from A itself.

Given Ω-algebras A and B, a mapping $f: A \to B$, and $\omega \in \Omega(n)$, we say that f is *compatible* with ω, if for all $a_1, \cdots, a_n \in A$,

$$(1) \qquad (a_1 f)\cdots(a_n f)\omega = (a_1 \cdots a_n \omega)f.$$

If f is compatible with each $\omega \in \Omega$, then f is said to be a *homomorphism* or *homomorphic mapping* from A to B. A homomorphism $f: A \to B$ with an inverse $f^{-1}: B \to A$ which is also a homomorphism is called an *isomorphism* between A and B. If there is an isomorphism from A to B we say that A and B are *isomorphic* and write $A \cong B$. Other special cases of homomorphisms are named as follows: an injective homomorphism is called a *monomorphism*; a surjective homomorphism is called an *epimorphism*; a homomorphism in which source and target are the same algebra is called an *endomorphism*; and an endomorphism which is also an isomorphism is called an *automorphism*. Given Ω-algebras A and B, if there is a monomorphism from A to B, we say that A *can be embedded* or is *embeddable* in B; if there is an epimorphism from A to B, then B is said to be a *homomorphic image* of A.

With any family $(A_\lambda)_{\lambda \in \Lambda}$ of Ω-algebras a direct product is associated, which is defined as follows. Let P be the Cartesian product of the A_λ regarded as sets, with projections $\varepsilon_\lambda: P \to A_\lambda$. Then any element $a \in P$ is completely determined by its components $a\varepsilon_\lambda$, and any choice of elements $a(\lambda) \in A_\lambda$ defines a unique element a of P by $a\varepsilon_\lambda = a(\lambda)$ $(\lambda \in \Lambda)$. Therefore if $a_1, \cdots, a_n \in P$ and $\omega \in \Omega(n)$, we can define $a_1 \cdots a_n \omega$ by the equation

$$(2) \qquad (a_1 \cdots a_n \omega)\varepsilon_\lambda = (a_1 \varepsilon_\lambda)\cdots(a_n \varepsilon_\lambda)\omega.$$

In this way an Ω-algebra structure is defined on P, and it is clear from the form of equation (2) that the projections are homomorphisms. The algebra so defined is called the *direct product* of the A_λ and is denoted by $\prod A_\lambda$. We remark that the algebras A_λ need not be distinct. For example, if $A_\lambda = A$ for all $\lambda \in \Lambda$, we obtain the *direct power* of A, whose carrier is A^Λ and which is itself denoted by A^Λ. If the elements of A^Λ are regarded as functions from Λ to A, then the operations in A^Λ are carried out componentwise; for example, if there is an addition defined in A, then in A^Λ we have

$$(f+g)(\lambda) = f(\lambda) + g(\lambda) \qquad (\lambda \in \Lambda).$$

On a given set A one can in general define different Ω-algebra structures, leading to Ω-algebras which may or may not be isomorphic, but if A has only one element, there is only one way of defining an Ω-algebra structure, because for any integer n, there is only one mapping from A^n to A. An

Ω-algebra with only one element is called *trivial*; from what has been said it follows that all trivial Ω-algebras are isomorphic.

Let U be any nonempty universe; then the Ω-algebras with carrier in U, together with all homomorphisms between them, form a category, which we shall denote by $(\Omega)_U$, or simply by (Ω), since U will usually be fixed throughout the discussion. Here it is understood that two Ω-algebras are equal if and only if they have the same carrier and the identity mapping is an isomorphism between them; from this definition it follows that on a given set there will in general be more than one Ω-algebra structure, so that (Ω) is not a subcategory of St, the category of all sets and mappings (in U).

We conclude this section with some examples of algebraic structures.

(i) *Groupoids*. A *groupoid* is a set with a single binary operation. Here Ω consists of a single element μ of arity two. A groupoid A satisfying the associative law:

(3) $$xy\mu z\mu = xyz\mu\mu \qquad \text{for} \quad \text{all } x,y,z \in A,$$

is called a *semigroup*. Semigroups and their homomorphisms form a full subcategory of the category of groupoids. By a *neutral element* in a groupoid one understands an element e such that

(4) $$xe\mu = ex\mu = x \qquad \text{for all } x \in A.$$

It is easily verified that a groupoid can have at most one neutral element.

(ii) *Groupoids with* 1. A *groupoid with* 1 is a set A with two operations, one binary μ, and one 0-ary 1, such that

(5) $$x1\mu = 1x\mu = x \qquad \text{for all } x \in A.$$

While the difference between a 'groupoid with neutral element' and a 'groupoid with 1' is only a formal one, there is a real difference when we come to consider subgroupoids and homomorphisms. For if A is a groupoid with a neutral element e, then a subgroupoid of A need not contain e and a homomorphism between such groupoids need not map the neutral element of one to that of the other. But if we regard A as 'groupoid with 1', with e as the value of the constant operator 1, then only subgroupoids containing e and only homomorphisms preserving the neutral elements can be admitted. In particular, groupoids with 1 and their homomorphisms form a subcategory of the category of all groupoids (and homomorphisms), which however is not full.

(iii) *Groups*. A *group* is a nonempty semigroup in which the equations $ax\mu = b$, $ya\mu = b$ have solutions x, y for any choice of a and b. As is well known, these equations then have uniquely determined solutions (cf. e.g. Kuroš [63], ch. 2). An alternative definition, in terms of a binary operator μ, a unary operator θ, and a 0-ary operator 1, is as follows: A group is a $(\mu, \theta, 1)$-algebra satisfying (3) and (5) above and in addition

$$xx\theta\mu = x\theta x\mu = 1, \tag{6}$$

for all $x \in A$. The proof of the equivalence of these definitions is again well known and may be left to the reader (cf. Kuroš [63], ch. 2). Of course it is more customary to write 'xy' for '$xy\mu$' and 'x^{-1}' for '$x\theta$' and to call 1 the *unit element*; in particular examples we shall use this notation also, so that (3), (5), and (6) may be rewritten as

$$(xy)z = x(yz), \tag{3'}$$

$$x1 = 1x = x, \tag{5'}$$

$$xx^{-1} = x^{-1}x = 1. \tag{6'}$$

Note that whereas (3) is unambiguous as it stands, in (3′) we had to put parentheses to distinguish the two sides, and this would still have been necessary if we had denoted the operation by some symbol, such as a dot, placed *between* the elements on which it operates. In III.2 it will be proved more generally that parentheses are unnecessary when all operators are written *on the right* of the arguments (or on the left). The category of all groups and homomorphisms will be denoted by Gp.

Frequently one uses the additive notation for groups, which consists in writing '$x + y$' for '$xy\mu$', '$-x$' for '$x\theta$', and '0' for '1'. The laws (3), (5), (6) then read:

$$(x + y) + z = x + (y + z), \tag{3''}$$

$$x + 0 = 0 + x = x, \tag{5''}$$

$$x + (-x) = (-x) + x = 0. \tag{6''}$$

The additive notation is used especially, but not exclusively, for *abelian* groups, i.e., groups satisfying the commutative law:

$$x + y = y + x. \tag{7}$$

(iv) *Groups with operators*. Let G be a group with a set Ω of unary operators such that

$$(xy)\omega = (x\omega)(y\omega) \qquad (x,y \in G,\ \omega \in \Omega). \tag{8}$$

Then G is called a *group with operators.* More generally, a *group with multiple operators* (Higgins [56]) is a group with a set of operators ω not necessarily unary, such that

$$11\cdots 1\omega = 1 \qquad (\omega \in \Omega).$$

Thus any Ω-algebra which is also a group, with 1 as Ω-subalgebra, may be regarded as a group with multiple operators. It turns out that many of the general structure theorems for groups go over to groups with multiple operators with little change (cf. Higgins [56]).

(v) *Quasigroups. A quasigroup* is a groupoid in which the equations $xa = b$, $ay = b$ each have a unique solution, for every couple of elements a,b. A quasigroup with neutral element is called a *loop.*

(vi) *Rings.* A *ring* is an abelian group (written additively) which is also a semigroup relative to a binary operator, called *multiplication* and usually denoted by juxtaposition, where these operators are related by the distributive laws:

$$x(y + z) = xy + xz,$$
$$(x + y)z = xz + yz.$$

If, further, there is a 0-ary operator 1 which acts as neutral element with respect to multiplication, $1x = x1 = x$, we speak of a ring with unit element 1. If Rg denotes the category of rings and their homomorphisms, and Rg* denotes the category of rings with 1 and their homomorphisms, then Rg* is a subcategory of Rg which is not full: an Rg-morphism between Rg*-objects R and S is an Rg*-morphism if and only if it maps the 1 of R to the 1 of S.

(vii) *Modules over rings.* A *module over a ring R*, or *R-module*, is an abelian group M with a unary operator ω_a for each $a \in R$, such that the operation defined by ω_a is an endomorphism of M and

$$(9) \qquad \omega_{ab} = \omega_a\omega_b, \quad \omega_{a+b} = \omega_a + \omega_b \qquad (a,b \in R).$$

Note that different operators may define the same operation. In case R is a ring with 1, one usually requires the module to be *unital*; this means that the endomorphism ω_1 defined by 1 is the identity on M. If instead of the first equation (9) we have $\omega_{ab} = \omega_b\omega_a$, we speak of a *left R*-module. Examples of an R-module and a left R-module are obtained by taking M to be the additive group of R with right or left multiplication, respectively:

$$\rho_a : x \to xa, \qquad \lambda_a : x \to ax.$$

When R is a ring with 1, the resulting modules are unital.

Sometimes one defines a module over a group G, or G-module, as a set P with a unary operator ω_x for each $x \in G$ such that

$$\omega_{xy} = \omega_x \omega_y, \quad \omega_1 = 1 \qquad (x,y \in G).$$

(viii) *Linear algebras.* Let K be a commutative ring with 1, then a *K-linear algebra A* is a unital K-module which is also a ring, such that

$$(ab)\alpha = (a\alpha)b = a(b\alpha) \qquad (a,b \in A,\ \alpha \in K).$$

If A also has a neutral element, denoted by e, say, then the mapping $\alpha \to e\alpha$ ($\alpha \in K$) is a homomorphism $K \to A$ which determines the K-module structure of A completely.

(ix) *Sets.* A set may be regarded as a $\emptyset$-algebra. In this sense everything that is said about Ω-algebras applies in the special case of sets.

(x) *The natural numbers.* Let Ω consist of a 0-ary operator denoted by 0 and a unary operator denoted by $'$. The natural numbers form a special case of such an algebra (cf. I.1 and VII.1).

(xi) *Lattices.* A *lattice* may be regarded as an algebra with two binary operators (cf. II.4), but a complete lattice is *not* an algebra as defined here, since it involves infinitary operations in general.

(xii) *Ordered sets.* An ordered set is not an algebra as it stands: it is defined by a relation ($x \leqslant y$) rather than by operations. But we can, at the risk of some artificiality, describe an ordered set by operations. E.g., for each couple (a,b) such that $a \leqslant b$ we may introduce a unary operator $\lambda = \lambda(a,b)$ defined by

$$x\lambda = \begin{cases} b & \text{if } x = a, \\ x & \text{otherwise.} \end{cases}$$

We note that here the operators (and not merely the operations defined by them) depend on the carrier. A more natural way of taking account of relations in an algebra will be described in Chapter V.

(xiii) *Fields.* A *field* is not an algebra, since the operations usually employed are not everywhere defined: x^{-1} is defined only for $x \neq 0$. We can overcome this difficulty by setting

$$x\theta = \begin{cases} x^{-1} & \text{if } x \neq 0, \\ 0 & \text{if } x = 0, \end{cases}$$

and so make formally an algebra out of our field, but this definition has other disadvantages, e.g., the equation $x \cdot x\theta = 1$ holds only when $x \neq 0$, whereas $x \cdot x^{-1} = 1$ holds whenever both sides are defined. We shall see later, in Chapter IV, that there is no way of defining fields as algebras satisfying certain laws, analogous to the definition of groups and rings given above.

If Ω is an operator domain, the structure obtained by associating with every $\omega \in \Omega(n)$ a mapping from a subset of A^n into A is called a *partial* Ω-*algebra*. As in the case of fields, a partial algebra may always be regarded as an algebra by taking a fixed element $c \in A$, and for any $a \in A^n$ such that $a\omega$ is undefined, putting $a\omega = c$. In this way, any partial algebra with nonempty carrier can be regarded as an algebra in the sense defined here.

(xiv) *Extensions.* Let C be a fixed Ω-algebra; then an Ω-*algebra over* C is an Ω-algebra A with a homomorphism

$$\lambda : C \to A.$$

If λ is injective, C may be identified with a subalgebra of A, and A is then called an *extension* of C; this is often written A/C. Equivalently, an Ω-algebra over C may be defined as an algebra with operators Ω and a constant operator $\gamma(a)$ for each $a \in C$, such that

$$\gamma(a_1)\cdots\gamma(a_n)\omega = \gamma(a_1\cdots a_n\omega) \qquad (\omega \in \Omega(n)).$$

(xv) *Projective planes.* A *projective plane* is a set whose elements are of two kinds ('lines' and 'points') with a binary operation (subject to certain rules; cf. e.g. M. Hall [59]) which is defined only on pairs of like elements, with a value unlike the arguments; thus the product of two (distinct) points is a line, the product of two (distinct) lines is a point, and other products are not defined. This is again a partial algebra.

(xvi) *Topological spaces.* A topological space may be defined as a set with certain infinitary operations: for each subset A and each point x of $\bar{A}$, the closure of A, we introduce an operator $\tau(A,x)$ which associates x with A. Thus a topological space fails to be an algebra because $\tau(A,x)$ is infinitary whenever A is infinite. In fact, this definition allows us to regard a topological space as an algebra whenever the system of all closed subsets is an algebraic closure system. This is a special case of Theorem 5.2 below.

These examples of algebras and nonalgebras serve to illustrate the definition and the limitations imposed by it. Thus we have limited ourselves to single-valued relations which are everywhere defined, i.e. to *operations*; as we shall see in Chapters V and VI, relations require an essentially different treatment. The limitation to finitary operations is necessary if we want the subalgebras to form an algebraic closure system (cf. II.5), and in fact, the study of infinitary operations has received little attention until recently (cf. Słomiński [59]). Finally, it may be noticed that with few exceptions the operators occurring in the examples are at most binary. This is no accident, for in a certain sense all finitary operators may be built up from binary ones (III.7). However, there may be no particularly natural way of doing this in any given instance, and besides, the gain in simplicity would not be very great. We shall therefore allow n-ary operators, for arbitrary finite n.

EXERCISES

1. Show that a category is a partial semigroup.

2. Verify that the permutations of a set P form a group $\sum(P)$, the *symmetric group* on P, and that a G-module structure on P is completely specified by a homomorphism $G \to \sum(P)$.

3. Let P be a G-module, where G is a group, and for any $p \in P$ define the *stabilizer* of p as $G_p = \{x \in G \mid px = p\}$. Show that the stabilizer of any $p \in P$ is a subgroup of G. What are the stabilizers of the elements of G when G is regarded as a G-module by right multiplication?

4. Define an S-module, where S is a semigroup with 1, and show that S itself may be regarded as an S-module by right multiplication. Show that the endomorphisms of S, qua S-module, are just the left multiplications $\lambda_a : x \to ax$. (Hint: An endomorphism of S qua S-module is a mapping $\theta: S \to S$ such that $(ab)\theta = (a\theta)b$ for all $a,b \in S$.)

5. Show that any group is the automorphism group of some algebraic structure. (Hint: Use Exercise 4; cf. also Birkhoff [35].)

6. Show that an algebra with a finite carrier having n elements has at most $n!$ automorphisms and at most n^n endomorphisms.

7. For any group G define a left G-module as a set P with operators ω_a satisfying $\omega_{ab} = \omega_b\omega_a$, $\omega_1 = 1$. Show that every left G-module may be regarded as a G-module.

8. (G. Birkhoff.) Show that the set $\mathscr{C}(A)$ of all equivalences on a set A is a complete lattice with respect to the ordering by inclusion. If A is finite show that the group of automorphisms of $\mathscr{C}(A)$ is isomorphic to the symmetric group on A. (Hint: Note that the *singular* equivalences, i.e., those whose classes are a 1-element set and its complement, are permuted among themselves by any automorphism of $\mathscr{C}(A)$, and show that every equivalence on A can be expressed in terms of singular ones.)

9. The *centre* of a ring R is defined as

$$Z(R) = \{a \in R \mid ax = xa \quad \text{for all } x \in R\}.$$

Show that the centre of a K-linear algebra A is a subalgebra of A. Moreover, if A has a unit element e, then the mapping $\alpha \to e\alpha$ is a homomorphism of K into $Z(A)$; and conversely, given a ring R, any homomorphism $K \to Z(R)$ defines a K-linear algebra structure on R.

10. If (A_λ) is any family of Ω-algebras, show that the Ω-algebra structure on the direct product $\prod A_\lambda$ is uniquely determined by requiring the projections to be homomorphisms.

11. Let A be a groupoid whose operation is written xy, and possibly with other operations. Then A is said to be the *inner direct product* of B and C if B and C are subalgebras of A such that the mapping $(b,c) \to bc$ is an isomorphism between $B \times C$ and A. Write down explicit conditions for A to be the inner direct product of B and C. (Cf. Jónsson & Tarski [47].)

12. Let a and b be two lines in the plane, A_i and B_i ($i = 1,2,3$) any triples of points on a and b, respectively, and C_i the intersection of the cross-joins A_jB_k and A_kB_j (for any permutation ijk of 123). By Pappus's theorem the points C_i are collinear, in a line c say; the configuration consisting of the nine points A_i, B_i, C_i and the nine lines a,b,c and cross-joins A_jB_k, A_kB_j is called a *Pappus configuration.* Show that if P and Q are any two distinct points of a Pappus configuration, then there is exactly one point R of the configuration (distinct from P and Q) such that either (i) PQR is a line of the configuration or (ii) PRS is not a line of the configuration for any point $S \neq Q$.

Let Γ be the set consisting of the nine points of a Pappus configuration and a tenth point E, and for any $P,Q \in \Gamma$ define

$$P \cdot Q = \begin{cases} P & \text{if } Q = E, \\ Q & \text{if } P = E, \\ E & \text{if } P = Q, \\ R & \text{otherwise,} \end{cases}$$

where R is the point determined above. Show that the algebraic structure so defined on Γ is a commutative loop which is not a group. Show also that the subloop generated by any two elements is a group.

3. THE ISOMORPHISM THEOREMS

There is another way of defining homomorphisms which is useful in some applications. We recall that a mapping from A to B is defined by a function from A to B, i.e. a certain type of subset of $A \times B$. Now we can say that a homomorphism $f: A \to B$ is a mapping whose function is also a subalgebra of $A \times B$. For, to say that f admits $\omega \in \Omega(n)$ means that given $(a_i, b_i) \in f$, $i = 1, \cdots, n$, we have

$$(a_1 \cdots a_n \omega, b_1 \cdots b_n \omega) \in f,$$

which states that

$$(a_1 \cdots a_n \omega) f = b_1 \cdots b_n \omega = (a_1 f) \cdots (a_n f) \omega.$$

But this is just the compatibility condition for ω. In order to use the new definition we need a lemma.

Lemma 3.1

Let A, B, C be Ω-algebras, A' a subalgebra of A, and Φ, Ψ subalgebras of $A \times B$ and $B \times C$, respectively. Then $A'\Phi$, Φ^{-1}, and $\Phi \circ \Psi$ again admit Ω (i.e. they are subalgebras of B, $B \times A$, and $A \times C$, respectively).

To prove the first assertion, let $b_i \in A'\Phi$ $(i = 1, \cdots, n)$; then there exists $a_i \in A'$ such that $(a_i, b_i) \in \Phi$ $(i = 1, \cdots, n)$, and since A' and Φ both admit Ω, we have for any $\omega \in \Omega(n)$

$$a_1 \cdots a_n \omega \in A' \quad \text{and} \quad (a_1 \cdots a_n \omega, b_1 \cdots b_n \omega) \in \Phi,$$

whence $b_1 \cdots b_n \omega \in A'\Phi$. The remaining assertions are proved similarly. ∎

This lemma has the following consequences:

Proposition 3.2

The product of homomorphisms is again a homomorphism, and similarly for isomorphisms, endomorphisms or automorphisms. ∎

Proposition 3.3

The image of a homomorphism is a subalgebra of the target. ∎

Definition

An equivalence on an Ω-algebra A which is at the same time a subalgebra of A^2 is called a *congruence* on A. The set of all congruences on A is denoted by $\mathscr{C}_\Omega(A)$.

With this definition we have

Proposition 3.4

The kernel of a homomorphism is a congruence.

For the kernel of f is $f \circ f^{-1}$. This is a subalgebra by Lemma 3.1 and an equivalence by Theorem I.3.1, hence it is a congruence. ∎

We note briefly the interpretation of congruences in the special case of groups. Let G be a group with unit element e, and $\mathfrak{q}$ a congruence on G in the sense defined above. Then $N = e^{\mathfrak{q}}$ is a normal subgroup of G and the $\mathfrak{q}$-classes are the cosets of N in G. In fact $a^{\mathfrak{q}} = Na$, for any $a \in G$, so that $\mathfrak{q}$ is completely determined by N. Now if f is a homomorphism of G (into some group) then the kernel of f, in the sense of group theory, is a normal subgroup of G, namely the class containing e of the congruence $f \circ f^{-1}$. This justifies calling this class (and not $f \circ f^{-1}$) the kernel of f in this case.

Rings may be regarded as a special case of groups; the 0-class of a congruence on a ring R is a subgroup $\mathfrak{a}$ of the additive group of R such that

(1) $\qquad xa \in \mathfrak{a} \qquad$ for all $x \in R$ and $a \in \mathfrak{a}$

and

(2) $\qquad ax \in \mathfrak{a} \qquad$ for all $x \in R$ and $a \in \mathfrak{a}$.

An additive subgroup $\mathfrak{a}$ satisfying (1) and (2) is called an *ideal* of R. If only (1) holds (or only (2)) we obtain a left (or right) ideal of R. This arises as the 0-class of a congruence on R, regarded as a left (right) R-module.

To every normal subgroup N of a group G there corresponds a homomorphism with N as kernel; we need only take the natural homomorphism onto the factor group G/N. In the same way we can, for every Ω-algebra A and every congruence $\mathfrak{q}$ on A, define a quotient algebra $A/\mathfrak{q}$ and a natural homomorphism $A \to A/\mathfrak{q}$ with kernel $\mathfrak{q}$. This is the content of

Theorem 3.5

Let A be an Ω-algebra and $\mathfrak{q}$ a congruence on A. Then there exists a unique Ω-algebra structure on the quotient set $A/\mathfrak{q}$ such that the natural mapping $A \to A/\mathfrak{q}$ is a homomorphism.

We denote the resulting algebra again by $A/\mathfrak{q}$ and call it the *quotient algebra* of A by $\mathfrak{q}$, with the *natural homomorphism* $A \to A/\mathfrak{q}$. A quotient of a subalgebra of A is also called a *factor* of A.

Proof:

Let $\theta = \text{nat}\, \mathfrak{q}$ be the natural mapping $A \to A/\mathfrak{q}$. This induces for each $n = 1, 2 \cdots$ a mapping θ_n: $A^n \to (A/\mathfrak{q})^n$ in an obvious fashion, and to prove the theorem we have only to show that for $\omega \in \Omega(n)$, there is just one way of completing

$$\begin{array}{ccc} A^n & \xrightarrow{\;\;\omega\;\;} & A \\ {\scriptstyle\theta_n}\big\downarrow & & \big\downarrow{\scriptstyle\theta} \\ (A/\mathfrak{q})^n & & A/\mathfrak{q} \end{array}$$

to a commutative diagram. If there is a mapping $\bar{\omega}:(A/\mathfrak{q})^n \to A/\mathfrak{q}$ to do this, it must satisfy

(3) $$a_1^{\mathfrak{q}} \cdots a_n^{\mathfrak{q}} \bar{\omega} = (a_1 \cdots a_n \omega)^{\mathfrak{q}},$$

and this equation evidently defines $\bar{\omega}$ uniquely provided we can show that the right-hand side is independent of the choice of a_i in its $\mathfrak{q}$-class. Thus let $(a_i, a_i') \in \mathfrak{q}$; then since $\mathfrak{q}$ is a subalgebra,

$$(a_1 \cdots a_n \omega,\, a_1' \cdots a_n' \omega) \in \mathfrak{q},$$

i.e.

$$(a_1 \cdots a_n \omega)^{\mathfrak{q}} = (a_1' \cdots a_n' \omega)^{\mathfrak{q}}$$

as asserted. ∎

As an example, we note that every Ω-algebra A has the congruences Δ and A^2. In the first case, we get an isomorphism:

$$A/\Delta \cong A,$$

while A/A^2 is the trivial algebra. Any congruence on A other than A^2 is said to be *proper*, and a congruence different from Δ is called *nontrivial.*

Let $\mathfrak{q}$ be a congruence on an algebra A. If S is a subset of A which meets each $\mathfrak{q}$-class in at most one element, i.e.

(4) $$\mathfrak{q} \cap S^2 = \Delta_S,$$

then we say that $\mathfrak{q}$ *separates* S; if S meets each $\mathfrak{q}$-class in exactly one element, i.e., (4) and the union of all $\mathfrak{q}$-classes meeting S is A:

(5) $$S^{\mathfrak{q}} = A,$$

then we say that S is a *transversal* for $A/\mathfrak{q}$ in A.

We can now state the celebrated three isomorphism theorems. These theorems, first stated explicitly by E. Noether (cf. v.d. Waerden [37]), are derived from Theorems I.3.1, I.3.3, and I.3.4 on sets and mappings, and apply under very general conditions to sets with a structure and mappings between the sets preserving the structure. Thus the natural place of these theorems is in category theory where they figure as axioms (cf. the axioms for abelian categories, MacLane [63]). We shall not pursue this more general line, but restrict ourselves to algebraic structures. Here it is possible to say slightly more than in the general case; for we have the following obvious lemma, which has no analogue for more general structures such as topological groups (cf. Bourbaki [51]):

Lemma 3.6

A bijective homomorphism is necessarily an isomorphism.

Proof:

Let $f: A \to B$ be a bijective homomorphism; then $f^{-1}: B \to A$ is also a mapping, and by Lemma 3.1 it is again a homomorphism, hence f is an isomorphism. ∎

Theorem 3.7 (first isomorphism theorem)

Let $f: A \to B$ be any homomorphism of Ω-algebras with kernel $\mathfrak{q}$. Then there is a decomposition

$$f = \varepsilon f' \mu$$

where $\varepsilon = \text{nat}\ \mathfrak{q}$ is the natural homomorphism $A \to A/\mathfrak{q}$, μ is the inclusion mapping $Af \to B$, and

$$f': A/\mathfrak{q} \to Af$$

is an isomorphism.

This follows easily from Theorem I.3.1, we need only check that the mapping f' there defined is a homomorphism, and then apply Lemma 3.6. ∎

The following corollary, derived from the factor theorem (Theorem I.3.3) is often useful.

Corollary 3.8

Let $f: A \to B$ be any homomorphism of Ω-algebras and $\mathfrak{q}$ a congruence on A which is contained in the kernel of f. Then there is a unique homomorphism $\bar{f}: A/\mathfrak{q} \to B$ such that the diagram

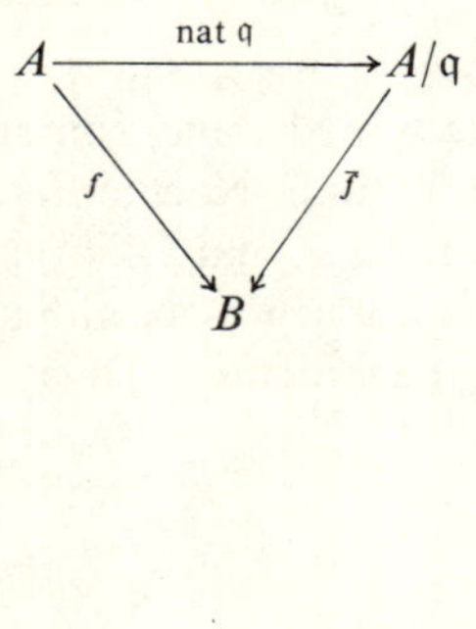

is commutative.

This is an immediate consequence of Theorem I.3.3. ∎

From Theorem 3.5 we see that every Ω-algebra A has itself and the trivial algebra as homomorphic images. If it has no others and is non-trivial it is said to be *simple*. By Theorem 3.7 an Ω-algebra A is simple if and only if it has precisely two congruences, namely A^2 and Δ.

Theorem 3.9 (second isomorphism theorem)

Let A be an Ω-algebra, A_1 a subalgebra of A and $\mathfrak{q}$ a congruence on A. Then $A_1^{\mathfrak{q}}$ is a subalgebra of A, $\mathfrak{q}_1 = \mathfrak{q} \cap A_1^2$ is a congruence on A_1, and

$$A_1/\mathfrak{q}_1 \cong A_1^{\mathfrak{q}}/\mathfrak{q}. \tag{6}$$

Proof:

Let $\theta = \operatorname{nat} \mathfrak{q}$ be the natural mapping and $\theta_1 = \theta | A_1$ its restriction to A_1. Then θ_1 is a homomorphism of A_1 into $A/\mathfrak{q}$, the image $A_1\theta_1$ is the set of $\mathfrak{q}$-classes meeting A_1 (i.e., $A_1^{\mathfrak{q}}/\mathfrak{q}$), and the kernel is $\mathfrak{q} \cap A_1^2 = \mathfrak{q}_1$. Hence by Theorem 3.7 we obtain (6), and the rest is clear by direct verification. ∎

The special case $\mathfrak{q}_1 = \Delta_{A_1}$ is worth stating separately:

Corollary 3.10

If A is an Ω-algebra, A_1 a subalgebra, and $\mathfrak{q}$ a congruence separating A_1, then the inclusion mapping $A_1 \to A$ induces an isomorphism

$$A_1 \cong A_1^{\mathfrak{q}}/\mathfrak{q}. \quad ∎$$

Theorem 3.9 states in effect that any subalgebra of a quotient algebra of A is isomorphic to a factor of A. Of course the converse is not true, as the existence of nonabelian simple groups shows.

By translating Theorem I.3.4 we obtain

Theorem 3.11 (third isomorphism theorem)

Let A be an Ω-algebra, and $\mathfrak{q}$, $\mathfrak{r}$ any congruences on A such that $\mathfrak{q} \subseteq \mathfrak{r}$. Then there is a unique homomorphism $\theta: A/\mathfrak{q} \to A/\mathfrak{r}$ such that $(\operatorname{nat} \mathfrak{q})\theta = \operatorname{nat} \mathfrak{r}$. *If* $\ker \theta = \mathfrak{r}/\mathfrak{q}$, *then $\mathfrak{r}/\mathfrak{q}$ is a congruence on $A/\mathfrak{q}$ and θ induces an isomorphism*

$$\theta' : (A/\mathfrak{q})/(\mathfrak{r}/\mathfrak{q}) \to A/\mathfrak{r},$$

such that $\theta = (\operatorname{nat} \mathfrak{r}/\mathfrak{q})\theta'$. ∎

Keeping $\mathfrak{q}$ fixed and varying $\mathfrak{r}$, we obtain

Corollary 3.12

Let A be an Ω-algebra and $\mathfrak{q}$ a congruence on A. If we interpret $\operatorname{nat} \mathfrak{q}$ *as a mapping $A^2 \to (A/\mathfrak{q})^2$, the pullback along* $\operatorname{nat} \mathfrak{q}$ *induces a mapping*

$$\mathfrak{r} \to \mathfrak{r}/\mathfrak{q} \tag{7}$$

which establishes a bijection between the set of congruences on A which contain $\mathfrak{q}$ *and the set of congruences on* $A/\mathfrak{q}$, *and moreover,*

$$A/\mathfrak{r} \cong (A/\mathfrak{q})/(\mathfrak{r}/\mathfrak{q}). \quad \blacksquare$$

The remark following Corollary 3.8 may now be restated as

Corollary 3.13

Let A be any Ω*-algebra and* $\mathfrak{q}$ *any congruence on A; then* $\mathfrak{q}$ *is a maximal proper congruence if and only if* $A/\mathfrak{q}$ *is simple.* ∎

In many problems the algebras given by the data are only determined up to isomorphism and the answer is only required up to isomorphism. In these circumstances the following result is often useful.

Proposition 3.14

Let A be an Ω*-algebra and B a subalgebra of A. If* B' *is an* Ω*-algebra isomorphic to B, then there exists an* Ω*-algebra* A' *isomorphic to A, with* B' *as subalgebra. More precisely, if* $\theta: B' \to B$ *is an isomorphism, then there exists an* Ω*-algebra* A' *containing* B' *as subalgebra and an isomorphism* $\bar{\theta}: A' \to A$ *such that* $\bar{\theta}|B' = \theta$.

Proof:

We may assume that $A \cap B' = \emptyset$, replacing A by an isomorphic copy of itself, if necessary (e.g. $A \times \{u\}$, for a suitable choice of the element u, is disjoint from B' and isomorphic to A). We now define $A' = B' \cup (A \backslash B)$ and extend θ to a mapping $\bar{\theta}: A' \to A$ by defining it as the identity on $A \backslash B$. Now there is a unique algebra structure on A' for which $\bar{\theta}$ is an isomorphism, and it is easily seen that with this definition B' is a subalgebra of A'. ∎

EXERCISES

1. State and prove analogues of the isomorphism theorems for the category of Ω-algebras and all correspondences admitting Ω (a correspondence from A to B is said to admit Ω if it is a subalgebra of $A \times B$). (Cf. Exercise I.3.7).

2. State and prove analogues of the isomorphism theorems for the category of topological spaces and continuous mappings.

3. Give a description of congruences on a group with multiple operators.

4. Give a description of congruences on a semigroup.

5. If G is any group and P is a G-module, show that the relation on P defined by

$$p \sim q \text{ if and only if } p = qx \text{ for some } x \in G$$

is an equivalence on P and that the equivalence class containing p is the set $pG = \{px \in P \mid x \in G\}$. (This set is called the *orbit* of p under the action of G.)

6. If G is a group and H is a subgroup, then G may be regarded as a (left) H-module under left multiplication by H. Denoting this module by ${}_HG$, show that the set of orbits of ${}_HG$ is a G-module under right multiplication by G. If this G-module is denoted by G/H, show that G/H consists of a single orbit, and that every orbit of a G-module is isomorphic to a module of the form G/H, for some subgroup H of G. More precisely, show that the orbit of p is isomorphic to G/G_p, where G_p is the stabilizer of p; deduce the relation

$$|pG| \cdot |G_p| = |G|.$$

7. Classify all congruences on **N** (cf. Ex. VII. 1.1, p. 251 and X. 3, p. 339f.)

4. LATTICES

In the study of general algebras one often encounters lattices, and it may be worthwhile to establish their properties here for later reference, particularly as lattices may themselves be regarded as an algebraic structure. In I.4 a lattice was defined as an ordered set in which any pair of elements has a supremum and an infimum. Since the sup and inf are uniquely determined, they are in effect two binary operators, and the lattice may be defined in terms of them as an algebra:

Proposition 4.1

Let L be any lattice; then for all $a,b,c \in L$,

(1) $a \vee (b \vee c) = (a \vee b) \vee c, \quad a \wedge (b \wedge c) = (a \wedge b) \wedge c$ (*associative law*),

(2) $a \vee b = b \vee a, \quad a \wedge b = b \wedge a$ (*commutative law*),

(3) $a \wedge (a \vee b) = a, \quad a \vee (a \wedge b) = a$ (*absorptive law*).

Conversely, if L is an algebra with two binary operators $\vee$ and $\wedge$, satisfying (1)–(3), *then an ordering may be defined on L by the rule*

(4) $\qquad a \leqslant b$ *if and only if* $a \vee b = b$,

and relative to this ordering L is a lattice such that $\sup(a,b) = a \vee b$, $\inf(a,b) = a \wedge b$.

Proof:

By the definition of $a \vee b$ as supremum we see that the unique supremum of a and b can be written either as $a \vee b$ or as $b \vee a$, and sup(a,b,c) can be written as $a \vee (b \vee c)$ or $(a \vee b) \vee c$; a similar remark applies to $a \wedge b$, and this shows that (1) and (2) hold. Further, (3) holds because $a \leqslant a \vee b$.

Now let L be an algebra with two operators $\vee$, $\wedge$ satisfying (1)–(3). Then

$$a \vee b = b \text{ if and only if } a \wedge b = a. \tag{5}$$

For if $a \vee b = b$, then $a = a \wedge (a \vee b) = a \wedge b$ by (3), and the converse follows by symmetry. If we define a relation $\leqslant$ by (4), then this must be an ordering: given $a \leqslant b$, $b \leqslant c$, we have $a \vee b = b$, $b \vee c = c$, and hence by (1), $a \vee c = a \vee (b \vee c) = (a \vee b) \vee c = b \vee c = c$, i.e. $a \leqslant c$. Next we have $a \wedge (a \vee a) = a$ by (3); therefore, $a \vee a = a \vee (a \wedge (a \vee a)) = a$, and so $a \leqslant a$. Thirdly, if $a \leqslant b$, $b \leqslant a$, then $b = a \vee b = b \vee a = a$ by (2), and this shows $\leqslant$ to be an ordering. By the definition of $a \vee b$ and (2), $a \vee b$ is an upper bound for $\{a,b\}$; if c is also an upper bound for $\{a,b\}$, then $a \leqslant c$, $b \leqslant c$, whence $c = a \vee c = b \vee c$, and so $c = a \vee (b \vee c) = (a \vee b) \vee c$; this states that $a \vee b \leqslant c$, i.e. $a \vee b$ is the least upper bound of a and b. Since the definition of $\leqslant$ is symmetrical in $\vee$ and $\wedge$, by (5), it also follows that inf(a,b) $= a \wedge b$. ∎

By this result lattices may be regarded as algebraic structures, and the definitions of sublattice and lattice-homomorphism in I.4 are seen to be special cases of the general definitions in II.3.

The symmetry referred to in the proof of Proposition 4.1 is based on the observation that the axioms for a lattice are permuted among themselves if $\vee$ and $\wedge$ are interchanged as well as $\leqslant$ and $\geqslant$. Therefore, when these changes are made in a theorem, we obtain again a theorem, called the *dual* of the first. This principle, which almost halves the work of proving theorems in lattice theory, is called the *principle of duality*, familiar from projective geometry (which is essentially about the lattice of subspaces of a vector space). It does not quite cut the work by half because for some propositions it may not yield anything new (namely those that are self-dual).

In any lattice L, let $a,b \in L$ be given such that $a \leqslant b$; the subset

$$[a,b] = \{x \in L \mid a \leqslant x \leqslant b\}$$

is called the *interval* defined by a and b. Such an interval need not be a chain, but it is always a sublattice of L, with greatest element b and least element a. With any interval $I = [a,b]$ we associate two ways of mapping L into I, the *projection operators* λ and ρ, defined by

$$\lambda : x \to (x \wedge b) \vee a, \qquad \rho : x \to (x \vee a) \wedge b.$$

It is easily verified that the assertions (i)$x \in I$, (ii) $x\lambda = x$, and (iii) $x\rho = x$ are all equivalent; therefore,

$$\lambda\rho = \lambda^2 = \lambda, \qquad \rho\lambda = \rho^2 = \rho.$$

Moreover, since $x \wedge b \leqslant x\rho$ and $a \leqslant x\rho$, it follows that $x\lambda \leqslant x\rho$, i.e.

(6) $$(x \wedge b) \vee a \leqslant (x \vee a) \wedge b \qquad \text{for all } x \in L.$$

If equality holds in (6) for all $x \in L$, then the interval I is said to be *modular*. Of particular importance are the lattices in which all intervals are modular; they are the *modular* lattices or *Dedekind* lattices (Dedekind [00]). Thus L is modular if and only if

(7) $$(c \vee a) \wedge b \leqslant (c \wedge b) \vee a \text{ for all } a,b,c \in L \text{ such that } a \leqslant b.$$

In view of (6) this is equivalent to the condition

(8) $$(c \vee a) \wedge b = (c \wedge b) \vee a \text{ for all } a,b,c \in L \text{ such that } a \leqslant b.$$

Condition (8) is sometimes called the *modular law*; strictly speaking, it is not a law in the usual sense (cf. IV.1 below), but it is equivalent to a law (cf. Exercise 1).

Probably the most important example of a modular lattice is the set of normal subgroups of a group. The set $\mathcal{N}(G)$ of all normal subgroups of a group G is ordered by inclusion and is easily seen to be a lattice, with $A \wedge B = A \cap B$:

$$A \vee B = AB = \{ab \in G \mid a \in A, b \in B\},$$

To say that $\mathcal{N}(G)$ is modular is to assert that

(9) $$CA \cap B \subseteq (C \cap B)A \qquad \text{whenever } A \subseteq B.$$

To verify that this holds, let $b \in CA \cap B$, say $b = ca(c \in C, a \in A)$; then since $a^{-1} \in A \subseteq B$, it follows that $c = ba^{-1} \in B$, and so $c \in C \cap B$, whence $b \in (C \cap B)A$. This still holds for groups with (unary) operators; in particular, the lattice of all submodules of an R-module (where R is any ring) is modular.

We next derive a criterion for a given lattice to be modular. First some general definitions: In any lattice L with least element 0 and greatest element 1, two elements x and y are said to be *complementary*, whenever

$$x \wedge y = 0, \qquad x \vee y = 1.$$

An element complementary to x is also called a *complement* of x in L. Two elements which have a common complement x in L are said to be *x-related*, or simply *related*, in L. This relation is symmetric, but in general it is neither transitive nor even reflexive.

Proposition 4.2

A lattice L is modular if and only if, for each interval I of L, any two elements of I which are comparable and are related in I are equal.

Proof:

As we have seen, L is nonmodular precisely if the inequality

$$(c \wedge b) \vee a \leqslant (c \vee a) \wedge b \tag{10}$$

is strict for at least one triple a,b,c such that $a \leqslant b$. When $a = b$, the two sides of (10) are equal by the absorptive law, so we may assume that $a < b$.

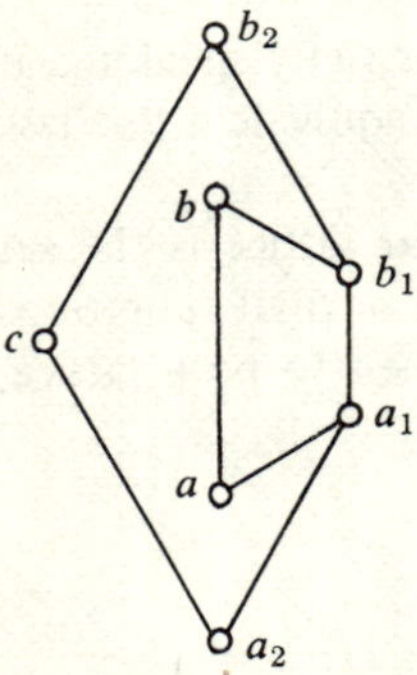

Figure 1

Suppose first that strict inequality holds in (10): put $a_1 = (c \wedge b) \vee a$, $b_1 = (c \vee a) \wedge b$; then $a \leqslant a_1 < b_1 \leqslant b$; hence $c \wedge b_1 \leqslant (c \wedge b) \vee a = a_1$, $c \vee a_1 \geqslant (c \vee a) \wedge b = b_1$; therefore $c \wedge b_1 = c \wedge a_1 = a_2$, $c \vee a_1 = c \vee b_1 = b_2$ say, and so a_1, b_1 are comparable and are related in $[a_2,b_2]$, although not equal. Conversely, if a', b' are distinct elements which are comparable and related in $[a,b]$, say $a' \wedge c = b' \wedge c = a$, $a' \vee c = b' \vee c = b$, and $a \leqslant a' < b' \leqslant b$, then $(c \wedge b') \vee a' = a' < b' = (c \vee a') \wedge b'$; therefore L is not modular. ∎

The property stated in this proposition involves only five elements, namely, the end-points of the interval, the given element, and its two complements. By considering the sublattice formed from these elements, we obtain

Corollary 4.3

A lattice is modular if and only if it does not contain a sublattice isomorphic to the five-element lattice of Fig. 2. ∎

A second important class of lattices is formed by the distributive lattices. A lattice L is said to be *distributive*, if it satisfies the laws

(11) $(a \vee b) \wedge c = (a \wedge c) \vee (b \wedge c)$ for all $a,b,c, \in L$

(12) $(a \wedge b) \vee c = (a \vee c) \wedge (b \vee c)$ for all $a,b,c \in L$.

Actually it is enough to assume that one of (11), (12) holds; this implies that the other one holds too, by the next proposition, which also shows that every distributive lattice is modular.

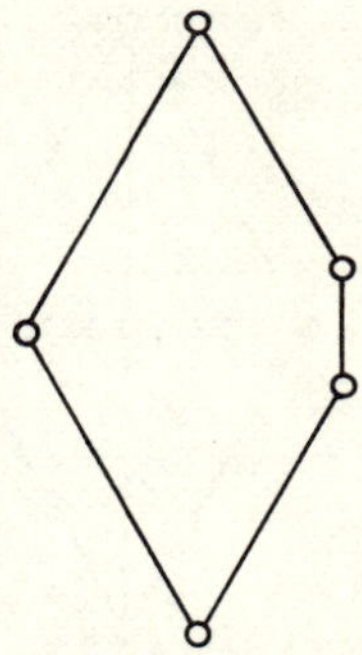

Figure 2

Proposition 4.4

In any lattice L, the following three conditions are equivalent: (i) *condition* (11), (ii) *condition* (12), *and*

(iii) $(c \vee a) \wedge b \leqslant (c \wedge b) \vee a$ for all $a,b,c \in L$.

In particular, every distributive lattice satisfies (iii) *and is therefore modular.*

Proof:

If (i) holds, then

$$(c \vee a) \wedge b = (c \wedge b) \vee (a \wedge b) \leqslant (c \wedge b) \vee a.$$

Conversely, assume that (iii) holds; then

$$(a \vee b) \wedge c \leqslant (b \wedge c) \vee a;$$

applying $\wedge c$ to both sides, we obtain

(13) $$(a \vee b) \wedge c \leqslant [(b \wedge c) \vee a] \wedge c \leqslant (a \wedge c) \vee (b \wedge c).$$

Since $a \vee b \geqslant (a \wedge c) \vee (b \wedge c)$ and $c \geqslant (a \wedge c) \vee (b \wedge c)$, we have $(a \vee b) \wedge c \geqslant (a \wedge c) \vee (b \wedge c)$, which together with (13) establishes (i). Thus (i) $\Leftrightarrow$ (iii), and by duality, (ii) $\Leftrightarrow$ (iii); it follows that (i), (ii), and (iii) are equivalent. Moreover, any lattice satisfying (iii) also satisfies (7), and so is modular. ∎

There is a criterion analogous to Proposition 4.2 for a lattice to be distributive.

Proposition 4.5

A lattice L is distributive if and only if for each interval I of L, any two elements of I which are related in I are equal.

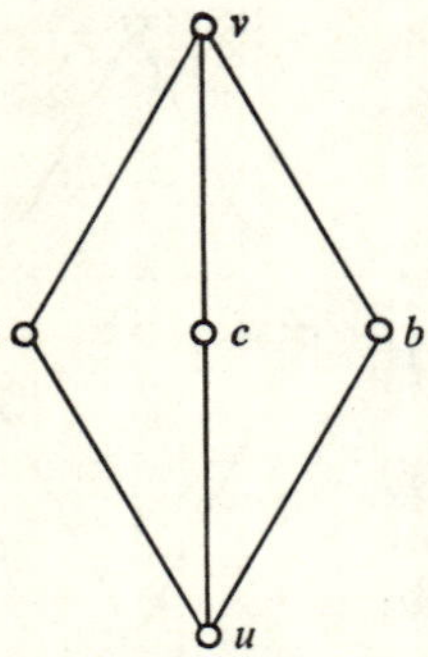

Figure 3

Proof:

Let L be distributive and assume that a,b are c-related in $[u,v]$. Then

$$a = a \wedge (b \vee c) = (a \wedge b) \vee u = a \wedge b;$$

hence $a \leqslant b$, and since L is modular (Proposition 4.4), it follows that $a = b$ by Proposition 4.2.

Conversely, assume that L satisfies the given conditions; then L is modular, by Proposition 4.2. Take any three elements $c_1,c_2,c_3 \in L$; write $a_1 = c_2 \wedge c_3$, $b_1 = c_2 \vee c_3$, and define a_2,a_3,b_2,b_3 by permuting 1,2,3 cyclically. Since $a_i \leqslant b_i$, we have by the modular law

$$(c_i \vee a_i) \wedge b_i = (c_i \wedge b_i) \vee a_i = d_i \qquad \text{say;}$$

we also put $u = a_1 \vee a_2 \vee a_3$ and $v = b_1 \wedge b_2 \wedge b_3$. Then, since $c_2 \wedge b_2 \leqslant b_1$ and $c_1 \leqslant b_2$, we have (using the modular law twice)

$$\begin{aligned}(c_1 \wedge b_1) \vee (c_2 \wedge b_2) &= [c_1 \vee (c_2 \wedge b_2)] \wedge b_1\\ &= [(c_2 \vee c_1) \wedge b_2] \wedge b_1\\ &= b_3 \wedge b_2 \wedge b_1\\ &= v.\end{aligned}$$

Therefore,

$$\begin{aligned}d_1 \vee d_2 &= (c_1 \wedge b_1) \vee a_1 \vee (c_2 \wedge b_2) \vee a_2\\ &= v \vee a_1 \vee a_2 = v.\end{aligned}$$

Permuting 1, 2, and 3 cyclically, we find that

$$d_1 \vee d_2 = d_2 \vee d_3 = d_3 \vee d_1 = v, \tag{14}$$

and by duality,

$$d_1 \wedge d_2 = d_2 \wedge d_3 = d_3 \wedge d_1 = u. \tag{15}$$

This shows that d_i and d_j are related in $[u,v]$ for $i,j = 1,2,3$, $i \neq j$, and hence, $d_1 = d_2 = d_3$. By (14) and (15), $u = d_1 = v$, and so

$$c_1 \vee a_1 \geqslant u = v \geqslant c_2 \wedge b_2.$$

Writing this out we find that

$$c_1 \vee (c_2 \wedge c_3) \geqslant c_2 \wedge (c_1 \vee c_3).$$

Since the c_i are arbitrary elements of L, this proves distributivity, by Proposition 4.4. ∎

Again the condition of this proposition may be expressed in terms of a certain five-element lattice, and we have

Corollary 4.6

A lattice is distributive if and only if it is modular and does not contain a sublattice isomorphic to the five-element lattice of Fig. 4. ∎

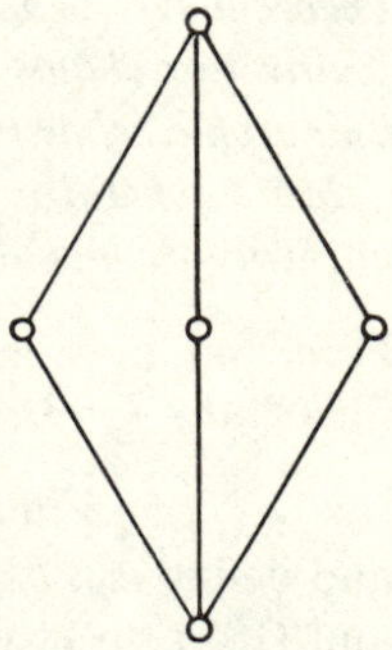

Figure 4

Let L be a modular lattice and a,b any two elements in L; then there is an isomorphism between the intervals $I = [a \wedge b, a]$ and $J = [b, a \vee b]$, as lattices, which may be established as follows. Given $x \in I$, we have $x \vee b \in J$, and since $a \wedge b \leqslant x \leqslant a$, it follows that $(x \vee b) \wedge a = x \vee (b \wedge a) = x$. Dually, if $y \in J$, then $y \wedge a \in I$ and $(y \wedge a) \vee b = y$. Thus the mappings

$$x \to x \vee b \quad (x \in I), \qquad y \to y \wedge a \quad (y \in J) \tag{16}$$

are inverse to each other. Since each is order-preserving, they establish a lattice-isomorphism between I and J. Any two intervals related in this way are said to be in *perspective*, under the perspectivity (16). More generally, two intervals I and J are said to be *projective* if there is a chain of perspectivities from I to J, i.e. a chain $I_0 = I, I_1, \cdots, I_n = J$ of intervals such that I_{i-1} and I_i are perspective. E.g. if a and b are any related elements in a modular lattice with 0 and 1, then the intervals $[0,a]$ and $[0,b]$ are projective.

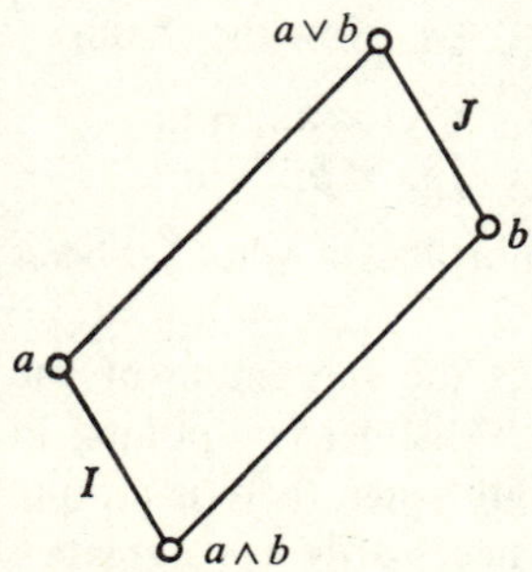

Figure 5

Since any two intervals in perspective are isomorphic, it follows that any two projective intervals are isomorphic. These facts may be regarded as an analogue of the second isomorphism theorem, and as in group theory, they may be used to establish a refinement theorem for chains. Two chains in a lattice L,

$$e = a_0 \leqslant a_1 \leqslant \cdots \leqslant a_m = a, \tag{17}$$

$$e = b_0 \leqslant b_1 \leqslant \cdots \leqslant b_n = a \tag{18}$$

between the same elements e and a are said to be *isomorphic*, if $m = n$ and there is a permutation π of $1, \cdots, n$ such that the interval $[a_{i-1}, a_i]$ is isomorphic to $[b_{i_\pi-1}, b_{i_\pi}]$. Any chain obtained from (17) by inserting further terms is called a *refinement* of (17). We can now state

Theorem 4.7 (refinement theorem for chains)

Any two chains between the same two points in a modular lattice have isomorphic refinements.

More generally, the result holds in any lattice for two chains all of whose intervals are modular.

Proof:

For any $i = 1, \cdots, m$, $j = 1, \cdots, n$ we write

$$a_{ij} = (a_i \wedge b_j) \vee b_{j-1}, \qquad b_{ji} = (b_j \wedge a_i) \vee a_{i-1},$$

and define a_{0j}, b_{0i} likewise, where a_i, b_j are the terms of the chains (17) and (18) respectively. Then

$$a_{i-1j} \wedge a_i \wedge b_j = (a_{i-1} \vee b_{j-1}) \wedge b_j \wedge a_i$$

and

$$\begin{aligned} a_{i-1j} \vee (a_i \wedge b_j) &= [(a_{i-1} \wedge b_j) \vee b_{j-1}] \vee (a_i \wedge b_j) \\ &= (a_i \wedge b_j) \vee b_{j-1} = a_{ij}. \end{aligned}$$

Hence the interval $[(a_{i-1} \vee b_{j-1}) \wedge a_i \wedge b_j, a_i \wedge b_j]$ is in perspective with $[a_{i-1j}, a_{ij}]$, and by symmetry also with $[b_{j-1i}, b_{ji}]$. Thus $[b_{j-1i}, b_{ji}]$ is isomorphic to $[a_{i-1j}, a_{ij}]$ for $i = 1, \cdots, m$ and $j = 1, \cdots, n$. Now the chains

$$\begin{aligned} e &= a_{01} \leqslant a_{11} \leqslant \cdots \leqslant a_{m1} \leqslant a_{12} \leqslant \cdots \leqslant a_{m2} \leqslant a_{13} \leqslant \cdots \leqslant a_{mn} = a, \\ e &= b_{01} \leqslant b_{11} \leqslant \cdots \leqslant b_{n1} \leqslant b_{12} \leqslant \cdots \leqslant b_{n2} \leqslant b_{13} \leqslant \cdots \leqslant b_{nm} = a \end{aligned}$$

refine (17) and (18) respectively, and they are isomorphic, by what has been shown. ∎

For any lattice L we define the *length* of L as the supremum of the number of nontrivial intervals (i.e. intervals with distinct end-points) in any chain. In particular, a lattice is of finite length when there is a finite bound on the lengths of its chains; such a lattice necessarily has a greatest and least element. From Theorem 4.7 we now obtain the following analogue of the Jordan-Hölder theorem for groups:

Corollary 4.8

In a modular lattice of finite length, any chain can be refined to a maximal chain and any two maximal chains between two given end-points have the same length.

For when a maximal chain is refined, its length remains unchanged, and isomorphic maximal chains clearly have the same length. ∎

If L is a modular lattice of finite length, any sublattice of L is again of finite length; in particular, for each $a \in L$, the length of $[0,a]$ is denoted by $l(a)$ and is also called the *length* of a. Thus the length of L itself is $l(1)$, and clearly for any $a \in L$,

$$l(a) \leqslant l(1),$$

with equality if and only if $a = 1$. Moreover, for any $a,b \in L$, we obtain by comparing the lengths of the isomorphic intervals $[a \wedge b, a]$ and $[b, a \vee b]$,

$$l(a) + l(b) = l(a \wedge b) + l(a \vee b). \tag{19}$$

An interval is said to be *prime* if it is nontrivial and contains no elements apart from its end-points. Clearly any two prime intervals are isomorphic; therefore, to assert that two maximal chains between the same end-points (in a modular lattice) are isomorphic is merely to say that they have the same length. This fact makes Corollary 4.8 appear very much weaker than the Jordan-Hölder theorem for groups. The corollary could in fact be strengthened by using the fact (established in the course of proving Theorem 4.7) that corresponding intervals are projective and not merely isomorphic. However, it is not worth doing this because even then we should not be able to obtain the Jordan-Hölder theorem for groups as a special case, for this theorem is concerned with chains of subgroups each normal in its successor, but not necessarily in the whole group; so we may not be dealing with a modular lattice. For instance, the lattice of all subgroups of a finite p-group need not be modular, although every subgroup occurs in some chain of subgroups each normal in its successor (cf. Suzuki [56]).

In contrast with the Jordan-Hölder theorem, which concerns maximal chains, the Krull-Schmidt theorem, which concerns maximal direct decompositions, can be stated entirely within the lattice of normal subgroups of the whole group, and it is therefore easier to give a purely lattice-theoretical formulation of the Krull-Schmidt theorem than of the Jordan-Hölder theorem (although the latter is easier to prove). We shall therefore formulate it as a lattice result and give the application to general algebras later (II.6).

Let L be a modular lattice with least element 0. Then the finite subset $\{a_1, a_2, \ldots, a_n\}$ of L is said to be *independent* if $a_i \neq 0$ $(i = 1, \ldots, n)$ and

$$a_i \wedge (a_1 \vee \cdots \vee a_{i-1} \vee a_{i+1} \vee \cdots \vee a_n) = 0 \qquad (i = 1, \cdots, n). \tag{20}$$

When an element a is represented as the join of an independent set,

$$a = a_1 \vee \cdots \vee a_n \qquad (a_1, \cdots, a_n \text{ independent}),$$

a is called the *direct join* of the elements $a_1,\cdots,a_n$, and we write

$$a = a_1 \times \cdots \times a_n.$$

E.g., if L_i is a lattice with least element 0_i and greatest element 1_i $(i = 1,2)$, then in the direct product of L_1 and L_2 we have

$$(1_1,1_2) = (1_1,0_2) \times (0_1,1_2).$$

More generally, in any lattice L with 0 and 1, any pair of complementary elements a and b define a representation of 1 as a direct join:

$$1 = a \times b.$$

Of course this does not mean that L is necessarily a direct product (cf. e.g. Fig. 4).

We note that for a set $\{a_1,\cdots,a_n\}$ to be independent, it is sufficient that

(21) $$(a_1 \vee \cdots \vee a_{i-1}) \wedge a_i = 0 \qquad (i = 2,\cdots,n).$$

This may be proved by induction on n (in any modular lattice). We shall not give a separate proof of this result (the reader may easily supply one), because we need the result only for modular lattices of finite length; in this case it is an immediate consequence of

Proposition 4.9

In a modular lattice of finite length, any subset $\{a_1,\cdots,a_n\}$ *satisfies the inequality*

(22) $$l(a_1 \vee \cdots \vee a_n) \leqslant \sum l(a_i),$$

with equality holding if and only if the a's are independent.

Proof:

For $n = 1$ the assertion is vacuous; we therefore use induction on n. Thus let $n > 1$; then, by (19)

(23) $$l(a_1 \vee \cdots \vee a_n) + l((a_1 \vee \cdots \vee a_{n-1}) \wedge a_n) = l(a_1 \vee \cdots \vee a_{n-1}) + l(a_n).$$

Observing that $l((a_1 \vee \cdots \vee a_{n-1}) \wedge a_n) \geqslant 0$ and applying the induction hypothesis on the right, we see that (22) holds in any case. If we have equality in (22), it follows in particular that $l((a_1 \vee \cdots \vee a_{n-1}) \wedge a_n) = 0$, whence

(24) $$(a_1 \vee \cdots \vee a_{n-1}) \wedge a_n = 0;$$

by the symmetry of the hypothesis in $a_1,\cdots,a_n$, (20) holds; therefore the

a's are independent. Conversely, when the a's are independent, then $a_1,\cdots,a_{n-1}$ are independent; therefore

$$l(a_1 \vee \cdots \vee a_{n-1}) = l(a_1) + \cdots + l(a_{n-1}),$$

by the induction hypothesis; we also have (24), and inserting this into (23) we see that equality holds in (22). ∎

One often wants to use the observation that if a occurs as a factor in a direct decomposition of b, then any element of the interval $[a,b]$ can be written as a direct join with a as factor. We state this as

Lemma 4.10

In a modular lattice with 0 *and* 1, *if*

$$1 = a \times a' \qquad \text{and } a \leqslant b,$$

then $b = a \times (b \wedge a')$.

For $a \wedge (b \wedge a') = a \wedge a' = 0$, and since $a \leqslant b$, we have $a \vee (b \wedge a') = b \wedge (a \vee a') = b$. ∎

An element a is said to be *indecomposable* if $a \neq 0$ and it admits no direct decomposition $a = b \times c$ in which $b \neq a$ and $c \neq a$. If a is written as a direct join of indecomposable elements, we speak of a *complete decomposition* for a.

Theorem 4.11 (Krull-Schmidt theorem for lattices)

In a modular lattice L of finite length, if

$$1 = a_1 \times \cdots \times a_m \tag{25}$$

and

$$1 = b_1 \times \cdots \times b_n \tag{26}$$

are two complete decompositions of 1, *then each* a_i *is related to some* b_j.

Proof:

Write $a_i' = a_1 \times \cdots \times a_{i-1} \times a_{i+1} \times \cdots \times a_m$, and define b_j' similarly in terms of $b_1,\cdots,b_n$. We shall prove that each a_i is a_i'-related to some b_j, using induction on the length of L. To begin with we note that for any element $c \in L$,

$$c \leqslant \bigvee_i ((c \vee a_i') \wedge a_i). \tag{27}$$

For if we write $c_i = c \vee a_i'$ for brevity, then $c_i \geqslant a_i' \geqslant a_k$ for $i \neq k$, hence

$$
\begin{aligned}
\bigvee_i (c_i \wedge a_i) &= (c_1 \wedge a_1) \vee (c_2 \wedge a_2) \vee \cdots \vee (c_m \wedge a_m) \\
&= c_1 \wedge [a_1 \vee (c_2 \wedge a_2) \vee \cdots \vee (c_m \wedge a_m)] \\
&= c_1 \wedge c_2 \wedge [a_1 \vee a_2 \vee (c_3 \wedge a_3) \vee \cdots \vee (c_m \wedge a_m)] \\
&\quad \cdots \quad \cdots \\
&= c_1 \wedge c_2 \wedge \cdots \wedge c_m \wedge (a_1 \vee a_2 \vee \cdots \vee a_m) \\
&\geqslant c,
\end{aligned}
$$

which proves (27).

We now distinguish two cases.

(i) $a_1 \vee b'_j < 1$ *for some* j.

Write $c_j = (a_1 \vee b'_j) \wedge b_j$, $c = c_1 \vee \cdots \vee c_n$; then since $c_j \leqslant b_j$, the c_j are again independent; therefore,

(28) $$c = c_1 \times \cdots \times c_n.$$

If we had $c_j = b_j$ for all j, then $a_1 \vee b'_j \geqslant b_j$ for all j, and hence $a_1 \vee b'_j \geqslant b_j \vee b'_j = 1$, against our assumption; thus, $c_j < b_j$ for some j, and it follows that $l(c) < l(1)$, so the induction hypothesis may be used for $[0,c]$. Now by (27), $c \geqslant a_1$; hence (Lemma 4.10),

(29) $$c = a_1 \times (c \wedge a'_1).$$

If we decompose each c_j in (28) so as to obtain a complete decomposition of c, and compare this with (29), we see that the indecomposable factor a_1 is $(c \wedge a'_1)$-related to an indecomposable factor of some c_j in (28); say for $j = 1$ we have $c_1 = d \times e$, where d is indecomposable, and also

(30) $$c = d \times (c \wedge a'_1).$$

Now

$$
\begin{aligned}
d \vee a'_1 &= d \vee (c \wedge a'_1) \vee a'_1 \\
&= a_1 \vee (c \wedge a'_1) \vee a'_1 \\
&= 1.
\end{aligned}
$$

Since a_1 and d are related, they are of the same length, and since $d \vee a'_1 = a_1 \vee a'_1 = 1$, it follows by (19) that $d \wedge a'_1 = a_1 \wedge a'_1 = 0$; thus

(31) $$1 = d \times a'_1.$$

Now $d \leqslant c_1 \leqslant b_1$, hence $b_1 = d \times (b_1 \wedge a'_1)$ (by Lemma 4.10), and since b_1 is indecomposable, it follows that $b_1 = d$; thus a_1 is a'_1-related to b_1. We remark that since $b_1 = d = c_1 = (a_1 \vee b'_1) \wedge b_1$, we also have

(32) $$a_1 \vee b'_1 = 1.$$

(ii) $a_1 \vee b'_j = 1$ *for all* j.

If $b_j \vee a'_1 < 1$ for all j, then by applying (i) with the a's and b's interchanged, we see that each b_j is related to some $a_{j'}$. Thus we may in (26) replace b_1 by $a_{1'}$, then b_2 by $a_{2'}$ and so on until we reach j such that $j' = 1$. Such j must occur, for if not, then after exhausting the b's we should have a complete decomposition of 1 into some of the a's, with a_1 not occurring, which would contradict (25). But when $j' = 1$, we have by (32) $b_j \vee a'_1 = 1$, which contradicts our assumption. Hence we must have $b_j \vee a'_1 = 1$ for some j, say for $j = 1$. Using this and the fact that $a_1 \vee a'_1 = 1$ we obtain, by (19),

$$\begin{aligned} l(b_1) + l(a'_1) &= l(1) + l(b_1 \wedge a'_1) \\ &= l(a_1) + l(a'_1) + l(b_1 \wedge a'_1), \end{aligned}$$

i.e.

$$l(b_1) - l(a_1) = l(b_1 \wedge a'_1) \geqslant 0. \tag{33}$$

But by assumption we also have $a_1 \vee b'_1 = 1$; hence, interchanging the roles of a_1 and b_1, we obtain

$$l(a_1) - l(b_1) = l(a_1 \wedge b'_1) \geqslant 0. \tag{34}$$

A comparison of (33) and (34) shows that $l(a_1) = l(b_1)$, and inserting this into (33), we find that $l(b_1 \wedge a'_1) = 0$; hence, $b_1 \wedge a'_1 = 0$, and so $1 = b_1 \times a'_1$, i.e. a_1 is a'_1-related to b_1. ∎

If in this theorem we take the decomposition (25) and replace a_1 by some b_j related to it, say by b_1, and then replace a_2 by some b_j (by another application of the theorem), then this b_j is clearly different from b_1, and by renumbering, may be taken as b_2. By induction we obtain the decomposition

$$1 = b_1 \times b_2 \times \cdots \times b_{i-1} \times a_i \times \cdots \times a_m$$

for $i = 1,2,\cdots,m$. Taking $i = m$ we see in particular that $n \geqslant m$, and by symmetry, $m \geqslant n$. Thus we have

Corollary 4.12

Given two complete decompositions (25) *and* (26) *of* 1 *in a modular lattice of finite length, we have* $m = n$ *and for some permutation* π *of* $1,\cdots,n$, *the interval* $[0,a_i]$ *is projective with* $[0,b_{i\pi}]$. ∎

In a distributive lattice, related elements are equal, and so we get

Corollary 4.13

In a distributive lattice of finite length, the factors in a complete decomposition of 1 *are unique except for their order.* ∎

In some contexts one is interested in decompositions of the form

(35) $$1 = a_1 \vee \cdots \vee a_m,$$

where the a's are not necessarily independent, but instead, each a_i is *join-irreducible*, i.e., a_i cannot be written as the join of two elements both different from a_i. It is natural to limit oneself to *irredundant* decompositions, i.e. decompositions (35) in which no a_i can be omitted. Such a decomposition does not determine each a_i so precisely as a direct decomposition would, so we cannot expect as strong a uniqueness assertion as in the Krull-Schmidt theorem; however, we have again the exchange property expressed in

Theorem 4.14 (Kuroš-Ore Theorem)

In a modular lattice L, let

(36) $$1 = p_1 \vee \cdots \vee p_r$$

and

(37) $$1 = q_1 \vee \cdots \vee q_s$$

be two representations of 1 *as an irredundant join of join-irreducible elements. Then* $r = s$ *and the* p_i *may be exchanged against the* q_j, *i.e., after suitable renumbering of the q's we have*

(38) $$1 = q_1 \vee \cdots \vee q_i \vee p_{i+1} \vee \cdots \vee p_r \qquad (i = 1, \cdots, r).$$

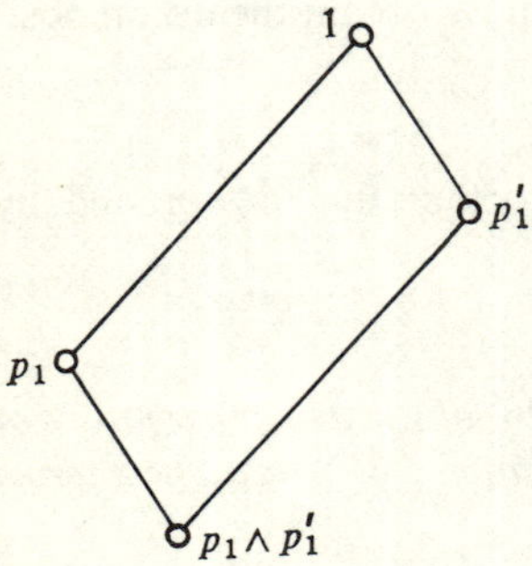

Figure 6

Proof:

Let $p'_1 = p_2 \vee \cdots \vee p_r$; then, by the irredundancy, $p'_1 < 1$ and $p'_1 \vee p_1 = 1$. Write $a_j = p'_1 \vee q_j$; then $1 \geqslant a_j \geqslant p'_1$; hence $p_1 \geqslant p_1 \wedge a_j \geqslant p_1 \wedge p'_1$. Now p_1 is join-irreducible in $[p_1 \wedge p'_1, p_1]$, hence 1 is join-irreducible in $[p'_1, 1]$, and since $a_1 \vee \cdots \vee a_s = 1$, we have $a_j = 1$ for some j, say $j = 1$. But this means that

$$1 = q_1 \vee p_2 \vee \cdots \vee p_r.$$

This representation is again irredundant, for q_1 cannot be omitted, and if some p_i could be omitted, then by exchanging q_1 against a p_j in (36) we should get a shorter representation of 1 in terms of p's alone. Therefore when we repeat the process, exchanging p_2 against some q_j, we must have $j \neq 1$, and by renumbering the q's we may take $j = 2$. Continuing in this way we obtain (38); for $i = r$, this shows that $r \geqslant s$, and so by symmetry we conclude that $r = s$. ∎

This theorem is perhaps more familiar in the dual form, in which it was first established for ideals in Noetherian rings, by E. Noether [21] (uniqueness of primary decomposition). In the abstract form it is due to Kuroš [35] and Ore [36]. In the latter paper Ore also establishes the lattice form of the Krull-Schmidt theorem. There is a large literature on both these theorems (cf. e.g. Birkhoff [48]).

For distributive lattices we have again the strong uniqueness:

Proposition 4.15

In a distributive lattice, any irredundant representation of 1 *as a join of join-irreducible elements*

$$1 = p_1 \vee \cdots \vee p_r$$

is unique, except for the order of the terms.

Proof:

Assume that $1 = q_1 \vee \cdots \vee q_s$ is a second such decomposition; then $p_1 \leqslant 1 = q_1 \vee \cdots \vee q_s$, hence

$$\begin{aligned} p_1 &= p_1 \wedge (q_1 \vee \cdots \vee q_s) \\ &= (p_1 \wedge q_1) \vee (p_1 \wedge q_2) \vee \cdots \vee (p_1 \wedge q_s). \end{aligned}$$

Since p_1 is join-irreducible, $p_1 = p_1 \wedge q_j$ for some j, i.e.

$$p_1 \leqslant q_j.$$

By the same argument, $q_j \leqslant p_i$ for some i, thus $p_1 \leqslant p_i$, which is possible only if $i = 1$. Hence $p_1 = q_j$, and now the result follows by induction. ∎

EXERCISES

1. Show that a lattice L is modular if and only if, for all $a,b,c \in L$,

$$(c \wedge (a \vee b)) \vee b = (c \vee b) \wedge (a \vee b).$$

2. Show that a lattice L is distributive if and only if, for all $a,b,c \in L$, $a \wedge b \leqslant c$ and $a \leqslant b \vee c$ imply $a \leqslant c$.

3. Show that the lattice of subgroups of the direct product of two cyclic groups of order two (the four-group) is modular but not distributive.

4. Show that the subgroup lattice of the alternating group on four symbols is not modular.

5. An element a in a lattice L with least element 0 is said to be an *atom* if $[0, a]$ is a prime interval. If L is a modular lattice in which 1 is the direct join of two atoms, a and b say, show that L is the direct product of $[0, a]$ and $[0,b]$ if and only if L is distributive.

6. Let L be a modular lattice with least element 0. Given any three elements $c_1,c_2,c_3 \in L$ such that $c_1 \wedge c_2 = 0$ and $c_1 \wedge (c_2 \vee c_3) \neq 0$, show that $(c_1 \vee c_2) \wedge c_3 \neq 0$. (This is known as the *exchange property*; cf. VII.2.)

7. A lattice is said to be *complemented* if it has a 0 and 1 and every element has a complement. Show that a modular lattice of finite length is complemented if and only if 1 is the join of atoms. [In the lattice of all submodules of a unital module, these two conditions are equivalent without assuming finite length (cf. e.g. Cartan & Eilenberg [56], ch. I), but in general neither implies the other (see the next exercise and Exercise V.2.5).]

8. The positive integers form a lattice with respect to the ordering by divisibility, with least element 1. If a greatest element U is adjoined, defined as the supremum of all the elements in the lattice, show that U can be written as (infinite) join of atoms, but that the resulting lattice is not complemented.

9. A commutative integral domain is called a *Bezout ring*, if the principal ideals form a sublattice of the lattice of all ideals. Show that in a Bezout ring the lattice of principal ideals is distributive. (Hint: Observe that the lattice is modular and apply Proposition 4.5. It is shown in Zariski & Samuel [58] ch. V that the distributivity of the lattice of ideals in a ring is equivalent to the Chinese remainder theorem.)

10. Let L be the set of principal ideals in a commutative integral domain, ordered by inclusion. If $a,b \in L$ have an infimum, show that they have a supremum, but that the converse need not hold. (The ordering here is opposite to that in Exercise 8. Hint: Consider the ring of polynomials in x with integer coefficients and even coefficient of x.) If every pair of elements has a supremum, show that L is a lattice.

11. Show that in any lattice L, the sublattice generated by two chains with modular intervals is distributive.

12. Give a direct proof of Corollary 4.13 (without using Theorem 4.11).

13. Show that a modular lattice L is distributive if and only if, for any a, $b_1, b_2 \in L$, $a \wedge b_1 = a \wedge b_2 = c$, say, implies $a \wedge (b_1 \vee b_2) = c$.

14. Define the *centralizer* of an equivalence $\mathfrak{q}$ on a set S as

$$Z_{\mathfrak{q}} = \{\mathfrak{r} \in \mathscr{C}(S) \mid \mathfrak{r} \circ \mathfrak{q} = \mathfrak{q} \circ \mathfrak{r}\}.$$

Show that $Z_{\mathfrak{q}}$ is a sublattice of $\mathscr{C}(S)$.

15. Let L be a modular lattice with 0 and 1. Given $a \leqslant x \leqslant b$, if x' is a complement of x in L, show that $(x' \vee a) \wedge b$ is a complement of x in $[a,b]$.

16. Show that a modular lattice has finite length if and only if it satisfies the maximum and the minimum condition.

17. Show that a distributive lattice of finite length is finite. (Remark that any atom a has as a complement any maximal element $b \ngeqslant a$. Now apply Proposition 4.5 and use induction on n.)

5. THE LATTICE OF SUBALGEBRAS

We now take a fixed Ω-algebra A and consider the set $\mathscr{B}_\Omega(A)$ of all its subalgebras. First we notice that $\mathscr{B}_\Omega(A)$ is a closure system on A, and hence a complete lattice. For if $(A_\lambda)_{\lambda \in \Lambda}$ is a family of subalgebras of A, then for each $\omega \in \Omega$, each A_λ admits ω, and so the same holds for $\bigcap A_\lambda$. We denote the corresponding closure operator by J_Ω; thus $J_\Omega(X)$ is the intersection of all subalgebras containing X. The following proposition gives an explicit construction of $J_\Omega(X)$, sometimes also denoted by $\langle X \rangle$.

Proposition 5.1

Let A be any Ω-algebra and X a subset of A. Define a subset X_k of A by induction on k:

$$X_0 = X,$$
$$X_{k+1} = \{x \in A \mid x \in X_k \text{ or } x = a\omega \text{ for some } a \in X_k^n \text{ and } \omega \in \Omega(n)\};$$

then

$$\bigcup_{k=0}^{\infty} X_k = J_\Omega(X).$$

Proof:

If we put $U = \bigcup X_k$, then by definition of X_k, we have $X_0 \subseteq J_\Omega(X)$, and if $X_k \subseteq J_\Omega(X)$, then $X_{k+1} \subseteq J_\Omega(X)$; by induction it follows that $X_k \subseteq J_\Omega(X)$ for all k, and so $U \subseteq J_\Omega(X)$. On the other hand, U is a subalgebra of A, for if $a \in U^n$, say $a = (a_1, \cdots, a_n)$ where $a_i \in X_{k_i}$, then if

$k = \max\{k_1,\cdots,k_n\}$, it follows that $a \in X_k^n$, whence $a\omega \in X_{k+1} \subseteq U$. This shows that U is a subalgebra; since it contains X, it must contain $J_\Omega(X)$, and this shows that $U = J_\Omega(X)$. ∎

We call $J_\Omega(X)$ the *join* of X, or the *subalgebra generated by* X, and X is a *generating set* of $J_\Omega(X)$. In particular, a generating set of A is a subset X of A such that $J_\Omega(X) = A$. If A has a finite generating set, we say that A is *finitely generated*. Every Ω-algebra has a least subalgebra, namely $J_\Omega(\emptyset)$, the subalgebra generated by the empty set. This is called the *minimal subalgebra*; clearly it is empty if and only if Ω has no constant operators.[1]

From Proposition 5.1 we can infer that J_Ω is algebraic. For if $a \in J_\Omega(X)$, then $a \in X_k$ for some k. Take the least k for which there is an element a in X_k which does not belong to $J_\Omega(X')$ for any finite subset X' of X; then $k > 0$, and such an element cannot belong to X_{k-1}. Hence by the definition of X_k, there exists $\omega \in \Omega$, and if $a(\omega) = n$, there are $b_1,\cdots,b_n \in X_{k-1}$ such that $a = b_1\cdots b_n\omega$. By definition of k we have $b_i \in J_\Omega(Y_i)$ where Y_i is a finite subset of X. Hence $a \in J_\Omega(Y)$, where $Y = \bigcup Y_i$ is again finite. This contradicts the definition of a, and we conclude that every element of $J_\Omega(X)$ belongs to $J_\Omega(X')$ for some finite subset X' of X, i.e. J_Ω is algebraic.

The converse also holds:

If $\mathscr{C}$ is any algebraic closure system on a set A, then for a suitable Ω there is an Ω-algebra structure on A such that $\mathscr{B}_\Omega(A) = \mathscr{C}$.

To prove this, let J be the closure operator for $\mathscr{C}$. Given any elements $a_1,\cdots,a_n \in A$ and $b \in J(\{a_1,\cdots,a_n\})$, we take a symbol

$$\omega = \omega(a_1,\cdots,a_n,b) \in \Omega \tag{1}$$

and define an n-ary operation on A by the rule

$$x_1\cdots x_n\omega = \begin{cases} b & \text{if } x_i = a_i (i = 1,\cdots,n) \text{ or if } n = 0, \\ x_1 & \text{otherwise.} \end{cases}$$

This defines an Ω-algebra structure on A, where for each integer n, the operators in $\Omega(n)$ are given by (1). Thus there are possibly infinitely many operators, but each of them is finitary. Let J_Ω be the closure operator corresponding to the system $\mathscr{B}_\Omega(A)$ of subalgebras; we assert that $J = J_\Omega$. Let $X \subseteq A$ and suppose first that X is finite. Then $J(X) = J_\Omega(X)$ by the definition of Ω. Now let X be an arbitrary subset of A; then since both J

[1] This shows that a trivial subalgebra need not be minimal; nor is the minimal subalgebra generally trivial.

and J_Ω are algebraic (the first by hypothesis and the second by what has been shown above), we have

$$J(X) = \bigcup J(X') = \bigcup J_\Omega(X') = J_\Omega(X),$$

where X' runs over the finite subsets of X. This proves the following result, due to P. Hall (unpublished) and J. Schmidt [52]:

Theorem 5.2

The system $\mathscr{B}_\Omega(A)$ of subalgebras of an Ω-algebra A is an algebraic closure system. Conversely, given an algebraic closure system $\mathscr{C}$ on a set A, then for a suitable operator domain Ω, an Ω-algebra structure may be defined on A such that $\mathscr{B}_\Omega(A) = \mathscr{C}$. ∎

Of course there are many ways of choosing the Ω-algebra structure in Theorem 5.2 to produce the given closure system. In particular, the choice of the operator domain can often be made in a more economical fashion than was done above.

Theorem 5.2 combined with Theorem 1.4 gives the following important consequence of Zorn's lemma (Neumann [37′]).

Theorem 5.3

Let A be an Ω-algebra, B a subalgebra and S a subset of A. Then there exists a maximal subalgebra C of A such that $C \supseteq B$ and $C \cap S = B \cap S$.

This is merely a translation of Theorem 1.4, bearing in mind that $\mathscr{B}_\Omega(A)$ is an algebraic closure system. ∎

If we try to apply Theorem 5.3 with $S = \emptyset$, we reach the obvious conclusion that A itself satisfies the conditions; clearly A is the (unique) greatest element of $\mathscr{B}_\Omega(A)$, and we do not need Theorem 5.3 to deduce this. On the other hand, the set of all *proper* subalgebras of A, i.e. all subalgebras distinct from A itself, need not have a greatest element, nor even a maximal element (cf. Exercise 3). However, there is a special case in which the existence of maximal proper subalgebras may be asserted (cf. Neumann [37′]).

Theorem 5.4

Let A be a finitely generated Ω-algebra and B a proper subalgebra of A; then there exists a maximal proper subalgebra which contains B.

Proof:

By hypothesis A has a finite generating set, $\{x_1,\cdots,x_r\}$ say. Let $\mathscr{S}$ be the set of all proper subalgebras of A which contain B. Then $\mathscr{S}$ is not empty, since $B \in \mathscr{S}$, and to prove the result we need only show that $\mathscr{S}$ is inductive. Let $\mathscr{C}$ be a chain in $\mathscr{S}$ and put $C = \bigcup \mathscr{C}$; if $C = A$, then for each $i = 1,\cdots,r$, $x_i \in C$, and hence $x_i \in C_i$ for some $C_i \in \mathscr{C}$. Since $\mathscr{C}$ is a chain, there is some suffix j such that $C_i \subseteq C_j$, and hence $x_i \in C_j$, for $i = 1,\cdots,r$; since $x_1,\cdots,x_r$ generate A, it follows that $C_j = A$, which contradicts the fact that C_j is proper. Thus, C must be proper, and it contains B, so $C \in \mathscr{S}$. This shows $\mathscr{S}$ to be inductive, and by Zorn's lemma it has a maximal element. ∎

A finitely generated algebra always has a minimal generating set: given any finite generating set X of A, we need only take a minimal set among the generating sets Y of A such that $Y \subseteq X$, and this is always possible. By contrast, if A cannot be finitely generated it need not have a minimal generating set (the additive group of rational numbers is a case in point); however, when it does have such a set, its cardinal is completely determined by the algebra:

Proposition 5.5

Let A be an Ω-algebra and X a minimal generating set of A. If X is infinite, of cardinal α, then any generating set of A has cardinal at least α. In particular, A cannot be finitely generated and any two minimal generating sets of A have the same cardinal.

Proof:

Let Y be any generating set of A and write $|Y| = \beta$. Every $y \in Y$ belongs to $A = J_\Omega(X)$, and hence there is a finite subset X_y of X such that

(2) $$y \in J_\Omega(X_y).$$

We assert that

(3) $$X = \bigcup X_y.$$

For clearly, $\bigcup X_y \subseteq X$, and by (2), $\langle \bigcup X_y \rangle \supseteq \langle Y \rangle = A$; thus $\bigcup X_y$ is a subset of X which also generates A, and so (3) follows by the minimality of X. Now if Y were finite, (3) expresses X as a union of a finite number of finite sets, which contradicts the fact that X is infinite. Hence Y is infinite, and from (3) we find that

$$\alpha = |X| \leqslant \sum |X_y| \leqslant \aleph_0 \beta = \beta.$$

This shows that Y has cardinal at least α. If Y is also minimal, then by interchanging the roles of X and Y we see that $\beta \leqslant \alpha$, and hence $\alpha = \beta$. ∎

For finitely generated algebras the result no longer holds; for example, the cyclic group of order six has a single generator a, say, and a minimal generating set consisting of two elements, namely $\{a^2,a^3\}$. Nevertheless there is an analogue in certain cases, namely, for free generating sets of certain free algebras (III.5).

In conclusion we prove a lattice property which shows the rather special role played by unary (and 0-ary) operators. We recall that $\mathscr{B}_\Omega(A)$, like any closure system, is a complete lattice, but of course it need not be a sublattice of $\mathscr{B}(A)$. To say that $\mathscr{B}_\Omega(A)$ is a sublattice of $\mathscr{B}(A)$ would mean that for any B, $C \in \mathscr{B}_\Omega(A)$, we have

$$B \cap C \in \mathscr{B}_\Omega(A), \qquad B \cup C \in \mathscr{B}_\Omega(A).$$

The first of these conditions always holds, but not the second; for example the union of two subgroups is not generally a subgroup. In fact the second condition just expresses that $\mathscr{B}_\Omega(A)$ is a topological closure system. This condition may also be expressed in terms of the operator domain, as follows.

Theorem 5.6

Let $\mathscr{C}$ be an algebraic closure system on a set A; then the following three conditions on $\mathscr{C}$ are equivalent:

(i) *$\mathscr{C}$ is a sublattice of $\mathscr{B}(A)$.*

(ii) *$\mathscr{C}$ is a topological closure system.*

(iii) *There is an operator domain Ω whose operators have arity at most one (i.e. they are all unary or 0-ary) with an Ω-algebra structure on A such that $\mathscr{B}_\Omega(A) = \mathscr{C}$.*

Proof:

The equivalence of (i) and (ii) follows from the definitions. Now assume that (iii) holds and let $B,C \in \mathscr{B}_\Omega(A)$; then any 0-ary operator gives an element of $B \cap C$, while a unary operator, applied to $b \in B$, gives an element of B, and applied to $c \in C$, gives an element of C. Thus $B \cup C$ admits Ω, i.e. $B \cup C \in \mathscr{B}_\Omega(A)$.

Conversely, if $\mathscr{C}$ is a topological closure system and J is the corresponding closure operator, we take for each $a \in J(\emptyset)$ a 0-ary operator $\lambda = \lambda(a)$ and put

$$\lambda = a.$$

Next, let $a \in A$; then for each $b \in J(\{a\})$ we take a unary operator $\mu = \mu(a,b)$ and put

$$x\mu = \begin{cases} b & \text{if } x = a, \\ x & \text{otherwise.} \end{cases}$$

Let Ω be the set of all λ, μ. We complete the proof by showing that $\mathscr{B}_\Omega(A) = \mathscr{C}$, or, what amounts to the same thing, that

$$J_\Omega(X) = J(X) \tag{4}$$

for all subsets X of A. If X is empty or consists of a single element, (4) follows by the definition of Ω. Now J and J_Ω are both topological, the first by hypothesis and the second by the first part of the proof; hence (4) holds for all finite subsets of A, by induction. Finally, J and J_Ω are both algebraic, and therefore (4) holds generally. ∎

Some care is needed in applying this theorem. Thus if A is an Ω-algebra for which $\mathscr{B}_\Omega(A)$ is topological, it does not follow that Ω has only operators of arity at most one; e.g. it might have a binary operator which on A happens to be independent of the second argument. Nor does it mean that $\mathscr{B}_\Omega(A)$ is definable in terms of the 0-ary and unary operators of Ω alone; it may happen that a new operator domain (consisting of 0-ary and unary operators only) will have to be constructed (cf. Exercise 4).

The following generalization of Theorem 5.6 is proved in the same way and will be used later. If Ω^* is an operator domain containing Ω, then we say that an Ω-algebra structure A is *enlarged* to an Ω^*-algebra structure, if an Ω^*-algebra structure is defined on A such that the action of any $\omega \in \Omega$ is the same as in A.

Theorem 5.7

Let A be an Ω-algebra and suppose that an Ω^*-algebra structure on A is given which enlarges the Ω-algebra structure on A, where $\Omega^* \supseteq \Omega$. Then $\mathscr{B}_{\Omega^*}(A) \subseteq \mathscr{B}_\Omega(A)$, and further, $\mathscr{B}_{\Omega^*}(A)$ is a sublattice of $\mathscr{B}_\Omega(A)$ provided that Ω can be enlarged to a domain Ω' by the adjunction of 0-ary and unary operators only and an Ω'-algebra structure can be defined on A enlarging the Ω-algebra structure, such that

$$\mathscr{B}_{\Omega'}(A) = \mathscr{B}_{\Omega^*}(A),$$

and for each $\omega \in \Omega(n)$, $\rho \in \Omega'\backslash\Omega$ there exist $\omega_1 \in \Omega(n)$ and $\rho_1, \ldots, \rho_n \in \Omega'\backslash\Omega$ such that $\omega\rho = \rho_1 \cdots \rho_n\omega_1$.

The proof is as for Theorem 5.6; on the other hand we obtain that theorem as a special case, by replacing Ω^* by Ω and Ω by $\emptyset$. ∎

EXERCISES

1 . If $f: A \to B$, $g: A \to B$ are homomorphisms of Ω-algebras, then the set $A_0 = \{x \in A \mid xf = xg\}$ is a subalgebra of A.

2. Let A be an Ω-algebra; an element $a \in A$ is said to be a *nongenerator* of A, if for any subset X of A, $J_\Omega(X \cup \{a\}) = A$ implies that $J_\Omega(X) = A$. Show that the set of all nongenerators of A is a subalgebra of A (called the *Frattini subalgebra*) which admits all automorphisms of A. Show also that the Frattini subalgebra is the intersection of all maximal proper subalgebras of A (if no maximal proper subalgebras of A exist, this intersection is equal to A, as the intersection of the empty family of subalgebras).

3. Show that the additive group of rational numbers has no maximal proper subgroups. Show also that it has no minimal generating set, but that it does have a minimal generating set, qua ring.

4. Let G be a group with more than one element and let ω be the binary operator defined on G by the rule

$$ab\omega = \begin{cases} ab^{-1} & \text{if one of } a, b \text{ is a power of the other,} \\ 1 & \text{otherwise.} \end{cases}$$

Taking $\Omega = \{\omega\}$, verify that the lattice of subalgebras of G, regarded as Ω-algebra, is topological, although Ω has no unary or 0-ary operators at all.

5. Let $\mathscr{C}$ be an algebraic closure system on a set A. Show that:

(a) $\mathscr{C}$ can be regarded as the lattice of all subalgebras with respect to an Ω-algebra structure on A, where Ω consists of unary operators only, if and only if $\mathscr{C}$ is topological and $\emptyset \in \mathscr{C}$;

(b) $\mathscr{C}$ can be regarded as the lattice of all subalgebras with respect to an Ω-algebra structure on A, where Ω consists of 0-ary operators only, if and only if $\mathscr{C}$ consists of all subsets of A containing a given fixed subset A_0 of A.

6. Show that in the lattice $\mathscr{B}_\Omega(A)$ of subalgebras of an Ω-algebra A, every interval whose end-points do not coincide contains a prime interval. (Hint: Use the fact that $\mathscr{B}_\Omega(A)$ is a complete lattice.)

7. Let A be an Ω-algebra and B a subalgebra of A; then B is finitely generated if and only if B is not of the form $\sup(B_\lambda)$, where $(B_\lambda)_{\lambda \in \Lambda}$ is a directed family of subalgebras $\neq B$ of A.

8. (Birkhoff & Frink.) Let L be a complete lattice; an element $a \in L$ is said to be *inaccessible* if it is not of the form $\sup(a_\lambda)$, where $(a_\lambda)_{\lambda \in \Lambda}$ is a directed family in L of elements $< a$. Show that a complete lattice L is isomorphic to the lattice $\mathscr{B}_\Omega(A)$ of all subalgebras of an Ω-algebra A (for some Ω) if and only if every element of L is expressible as a supremum of inaccessible elements and, further, L is *meet-continuous*, i.e., for any $a \in L$ and any directed family $(b_\lambda)_{\lambda \in \Lambda}$ in L,

$$a \wedge \Big(\bigvee_{\lambda \in \Lambda} b_\lambda\Big) = \bigvee_{\lambda \in \Lambda} (a \wedge b_\lambda).$$

(Hint: Take A to be the set of inaccessible elements of L and define the closure of $X \subseteq A$ as the least left segment of L which contains the sublattice generated by X.)

9. Show that an algebra A which is generated by $\emptyset$ has no proper subalgebras.

10. If A is any finitely generated algebra, show that any generating set of A contains a finite subset which generates A.

11. If X is any set, define a set Ω of finitary operators on X^2 such that the Ω-subalgebras are precisely the preorderings of X. (Cf. Exercise 1.10).

6. THE LATTICE OF CONGRUENCES

Let A be an Ω-algebra; together with the set $\mathscr{C}_\Omega(A)$ of all congruences we shall consider the set $\mathscr{C}(A)$ of equivalences on A (qua set). We remark that in the special case $\Omega = \emptyset$ we have $\mathscr{C}_\Omega(A) = \mathscr{C}(A)$, so that any result about $\mathscr{C}_\Omega$ also applies to $\mathscr{C}$. In general we have the inclusions

(1) $$\mathscr{C}_\Omega(A) \subseteq \mathscr{C}(A) \subseteq \mathscr{B}(A^2).$$

It is easily seen (and will also follow from Theorem 6.2) that $\mathscr{C}_\Omega(A)$ is always a complete lattice. In particular, $\mathscr{C}(A)$ is also a complete lattice, and our object is to obtain the relations between the lattices in (1). We shall find that although $\mathscr{C}(A)$ is not necessarily a sublattice of $\mathscr{B}(A^2)$, $\mathscr{C}_\Omega(A)$ is always a sublattice of $\mathscr{C}(A)$.

By definition, a congruence is an equivalence which admits the operators ω ($\omega \in \Omega$). Now each n-ary operator ω defines an n-ary operation on A:

(2) $$(x_1,\cdots,x_n) \to x_1 \cdots x_n\omega.$$

By giving fixed values in A to some of the arguments, we obtain r-ary operations for $r \leqslant n$; in particular, if we fix all the x_i except one, we obtain for any $n - 1$ elements $a_1,\cdots,a_{n-1} \in A$ and any $i = 1,\cdots,n$, a unary operation

(3) $$x \to a_1 \cdots a_{i-1} x a_i \cdots a_{n-1}\omega.$$

We shall say that the operation (3) is an *elementary translation* derived from (2) by *specialization* in A. Generally, a mapping $\theta: A \to A$ is said to be a *translation* if $\theta = 1$ or if θ can be expressed as a product of a finite number of elementary translations. As a special case of translations we have the operations on A defined by unary operators; but we note that whereas a subalgebra of A admits all the unary operators, it will not in general admit all translations; for instance, in a group G the only subgroup admitting all translations is G itself. For congruences the situation is rather different:

Proposition 6.1

An equivalence $\mathfrak{q}$ on an Ω-algebra A is a congruence if and only if it admits all translations; more precisely, a congruence admits all translations, while any equivalence admitting all elementary translations is a congruence.

Proof:

If $\mathfrak{q}$ is a congruence, then for any $\omega \in \Omega(n)$ and any $a_1,\cdots,a_n,b \in A$, if $a_i \equiv b \pmod{\mathfrak{q}}$, then

$$a_1 \cdots a_n\omega \equiv a_1 \cdots a_{i-1}ba_{i+1} \cdots a_n\omega \pmod{\mathfrak{q}};$$

hence $\mathfrak{q}$ admits all elementary translations and hence, by an easy induction, $\mathfrak{q}$ admits all translations. Conversely, assume that $\mathfrak{q}$ admits all elementary translations and let $a_i, a_i' \in A$ and $a_i \equiv a_i' \pmod{\mathfrak{q}}$ $(i = 1,\cdots,n)$; then we have (mod $\mathfrak{q}$),

$$\begin{aligned} a_1 \cdots a_n\omega &\equiv a_1'a_2 \cdots a_n\omega \\ &\equiv a_1'a_2'a_3 \cdots a_n\omega \\ &\cdots \\ &\equiv a_1'a_2' \cdots a_n'\omega. \end{aligned}$$

Thus $\mathfrak{q}$ admits ω, and since this holds for every $\omega \in \Omega(n)$, $(n = 0,1,\cdots)$, $\mathfrak{q}$ is in fact a congruence. ∎

To show that the congruences on A form a complete lattice, we shall prove the stronger assertion that $\mathscr{C}_\Omega(A)$ is an algebraic closure system on A^2. This is most easily done by describing an operator domain Γ and a Γ-algebra structure on A^2 such that $\mathscr{B}_\Gamma(A^2) = \mathscr{C}_\Omega(A)$. To be specific Γ is to consist of the following operators, acting on A^2 in the manner stated:

(i) For each $a \in A$, there is a 0-ary operator $\lambda = \lambda(a)$ such that

$$\lambda = (a,a).$$

(ii) There is one unary operator μ, such that

$$(x,y)\mu = (y,x).$$

(iii) There is one binary operator ν, such that

$$(x,y)(z,t)\nu = \begin{cases} (x,t) & \text{if } y = z, \\ (x,y) & \text{otherwise.} \end{cases}$$

(iv) For each $\omega \in \Omega(n+1)$, where $n = 0,1,\cdots$, each n-tuple $(a_1,\cdots,a_n)$ from A, and each integer $i = 1,2,\cdots,n+1$, there is a unary operator $\rho = \rho(\omega,a_1,\cdots,a_n,i)$ such that

$$(x,y)\rho = (a_1\cdots a_{i-1}xa_i\cdots a_n\omega, a_1\cdots a_{i-1}ya_i\cdots a_n\omega).$$

All these operators are finitary, and it is clear from the definition of an equivalence that an equivalence on A is just a subset of A^2 admitting all λ's, μ, and ν, while a congruence is a subset of A^2 admitting all λ's, μ, ν, and all ρ's. Thus we have established

Theorem 6.2

The set $\mathscr{C}_\Omega(A)$ of congruences on A is an algebraic closure system. ∎

It follows that $\mathscr{C}_\Omega(A)$ is a complete lattice.

Corollary 6.3

The set $\mathscr{C}(A)$ of equivalences on a set A is an algebraic closure system. ∎

If we apply Theorem 5.3 we obtain

Corollary 6.4

Let A be an Ω-algebra, Φ a correspondence in A and $\mathfrak{q}$ a congruence on A. Then there exists a maximal congruence $\bar{\mathfrak{q}}$ in A subject to the conditions $\bar{\mathfrak{q}} \supseteq \mathfrak{q}, \bar{\mathfrak{q}} \cap \Phi = \mathfrak{q} \cap \Phi$. ∎

This corollary is chiefly used in the case where $\mathfrak{q}$ is the diagonal on A and Φ consists of a couple of distinct elements of A. Then it states that for any two distinct elements a and b of A there is a maximal congruence separating a and b.

The congruences of A were obtained in the proof of Theorem 6.2 as equivalences admitting all the unary operators ρ. By Theorem 5.7 we therefore find

Corollary 6.5

The lattice of congruences $\mathscr{C}_\Omega(A)$ of an Ω-algebra A is a sublattice of $\mathscr{C}(A)$, the lattice of equivalences on A. ∎

This means that if $\mathfrak{q}$ and $\mathfrak{r}$ are two congruences on A, then in order to obtain the least congruence containing both $\mathfrak{q}$ and $\mathfrak{r}$, we need only form $\mathfrak{q} \vee \mathfrak{r}$ in $\mathscr{C}(A)$; in other words, the least equivalence containing $\mathfrak{q}$ and $\mathfrak{r}$ is already a congruence. We shall denote this equivalence by $\mathfrak{q} \vee \mathfrak{r}$ and also refer to it as the equivalence *generated* by $\mathfrak{q}$ and $\mathfrak{r}$. This is in accordance with the general terminology, regarding $\mathfrak{q} \vee \mathfrak{r}$ as the least subalgebra of A^2, qua Γ-algebra, containing $\mathfrak{q}$ and $\mathfrak{r}$.

In order to study the lattice operations in $\mathscr{C}_\Omega(A)$ more closely, we may, by Corollary 6.5, limit ourselves to the case $\Omega = \emptyset$; so, let A be any set. The lattice operations in $\mathscr{C}(A)$ may then be described as follows: $\mathfrak{q} \wedge \mathfrak{r} = \mathfrak{q} \cap \mathfrak{r}$, because $\mathscr{C}(A)$ admits intersections; in order to obtain $\mathfrak{q} \vee \mathfrak{r}$, we define $\mathfrak{s}_0 = \Delta_A$, and for $k \geqslant 0$, put $\mathfrak{s}_{k+1} = \mathfrak{s}_k \circ \mathfrak{r} \circ \mathfrak{q}$; then

$$\mathfrak{s}_0 \subseteq \mathfrak{s}_1 \subseteq \cdots \qquad \bigcup \mathfrak{s}_k = \mathfrak{q} \vee \mathfrak{r}. \tag{4}$$

It is easily verified that $\mathfrak{s} = \bigcup \mathfrak{s}_k$ is an equivalence containing $\mathfrak{q}$ and $\mathfrak{r}$, and in fact is the least such equivalence, from which (4) follows. The most important special case is that where the chain $\mathfrak{s}_k$ breaks off after one term:

Proposition 6.6

Let A be any set and $\mathfrak{q},\mathfrak{r} \in \mathscr{C}(A)$. Then $\mathfrak{q} \circ \mathfrak{r}$ is an equivalence if and only if $\mathfrak{q}$ and $\mathfrak{r}$ commute, and in that case

$$\mathfrak{q} \vee \mathfrak{r} = \mathfrak{q} \circ \mathfrak{r}.$$

Proof:

Clearly, $\mathfrak{q} \vee \mathfrak{r} \supseteq \mathfrak{q} \circ \mathfrak{r} \supseteq \mathfrak{q} \cup \mathfrak{r}$ for any equivalences $\mathfrak{q}$ and $\mathfrak{r}$, so if $\mathfrak{q} \circ \mathfrak{r}$ is an equivalence, it coincides with $\mathfrak{q} \vee \mathfrak{r}$, and moreover,

$$\mathfrak{q} \circ \mathfrak{r} = (\mathfrak{q} \circ \mathfrak{r})^{-1} = \mathfrak{r}^{-1} \circ \mathfrak{q}^{-1} = \mathfrak{r} \circ \mathfrak{q}.$$

Conversely, if $\mathfrak{q} \circ \mathfrak{r} = \mathfrak{r} \circ \mathfrak{q}$, then

$$(\mathfrak{q} \circ \mathfrak{r})^{-1} = \mathfrak{r}^{-1} \circ \mathfrak{q}^{-1} = \mathfrak{r} \circ \mathfrak{q} = \mathfrak{q} \circ \mathfrak{r},$$

hence $\mathfrak{q} \circ \mathfrak{r}$ is symmetric. It is clearly reflexive and

$$\mathfrak{q} \circ \mathfrak{r} \circ \mathfrak{q} \circ \mathfrak{r} = \mathfrak{q} \circ \mathfrak{q} \circ \mathfrak{r} \circ \mathfrak{r} = \mathfrak{q} \circ \mathfrak{r},$$

which shows $\mathfrak{q} \circ \mathfrak{r}$ to be also transitive, and therefore an equivalence. ∎

Corollary 6.7

Let A be an Ω-algebra and $\mathfrak{q},\mathfrak{r}$ any congruences on A; then $\mathfrak{q} \circ \mathfrak{r}$ is a congruence if and only if $\mathfrak{q}$ and $\mathfrak{r}$ commute, and in that case,

$$\mathfrak{q} \vee \mathfrak{r} = \mathfrak{q} \circ \mathfrak{r}. \quad ∎$$

If the set A has at least three elements, $\mathscr{C}(A)$ is not a sublattice of $\mathscr{B}(A^2)$, and we can even find a pair of noncommuting equivalences. However, when we are dealing with an Ω-algebra structure on A, the lattice $\mathscr{C}_\Omega(A)$

will in general be smaller than $\mathscr{C}(A)$, and if Ω is sufficiently complex it may happen that all congruences on A commute. A sufficient condition for this to be so is given by

Proposition 6.8

Let A be an Ω-algebra; then the congruences on A commute, provided that for every pair $a,b \in A$ there is a translation of A which interchanges a and b.

Proof:

Let A be an Ω-algebra satisfying the condition stated, and let $\mathfrak{q},\mathfrak{r} \in \mathscr{C}_\Omega(A)$. If $(a,b) \in \mathfrak{q} \circ \mathfrak{r}$, this means that for some $c \in A$,

$$a \equiv c \pmod{\mathfrak{q}} \qquad \text{and} \qquad c \equiv b \pmod{\mathfrak{r}}.$$

By hypothesis, there is a translation τ of A interchanging a and b, and since $\mathfrak{q}$ and $\mathfrak{r}$ admit τ (by Proposition 6.1), we have

$$b \equiv c^\tau \pmod{\mathfrak{q}} \qquad \text{and} \qquad c^\tau \equiv a \pmod{\mathfrak{r}};$$

hence $(b,a) \in \mathfrak{q} \circ \mathfrak{r}$, i.e., $(a,b) \in \mathfrak{r} \circ \mathfrak{q}$. This shows that $\mathfrak{q} \circ \mathfrak{r} \subseteq \mathfrak{r} \circ \mathfrak{q}$, and by symmetry, $\mathfrak{q} \circ \mathfrak{r} \supseteq \mathfrak{r} \circ \mathfrak{q}$, whence $\mathfrak{q} \circ \mathfrak{r} = \mathfrak{r} \circ \mathfrak{q}$. ∎

From this proposition it follows immediately that all congruences on a group commute, for if a, b are any elements of a group G, then $\tau : x \to ax^{-1}b$ is a translation which interchanges a and b. This shows more generally that congruences on any group with multiple operators commute.

In Chapter III, we shall meet other conditions for the congruences on an Ω-algebra to commute. The importance of such conditions is not only practical, in facilitating the calculation of equivalence-joins (by Proposition 6.6), but also theoretical, in establishing the modularity of $\mathscr{C}_\Omega(A)$:

Proposition 6.9

If all congruences on an Ω-algebra A commute, then the lattice $\mathscr{C}_\Omega(A)$ of congruences on A is modular.

The proof is as in the special case of groups: Given $\mathfrak{p},\mathfrak{q},\mathfrak{r} \in \mathscr{C}_\Omega(A)$, where $\mathfrak{q} \subseteq \mathfrak{r}$, we have to show that

$$(\mathfrak{p} \circ \mathfrak{q}) \cap \mathfrak{r} \subseteq (\mathfrak{p} \cap \mathfrak{r}) \circ \mathfrak{q}.$$

Let $(a,b) \in (\mathfrak{p} \circ \mathfrak{q}) \cap \mathfrak{r}$, say $a \equiv c \pmod{\mathfrak{p}}$ and $c \equiv b \pmod{\mathfrak{q}}$; then

$$a \equiv b \pmod{\mathfrak{r}}$$

and

$$b \equiv c \pmod{\mathfrak{r}},$$

because $\mathfrak{q} \subseteq \mathfrak{r}$; therefore $a \equiv c \pmod{\mathfrak{p} \cap \mathfrak{r}}$, and so $(a,b) \in (\mathfrak{p} \cap \mathfrak{r}) \circ \mathfrak{q}$. ∎

Using the commutativity of congruences, we are able to make the following deductions from the isomorphism theorems, constituting in effect a translation of the Jordan-Hölder and Schreier theorems to general algebras. We begin by proving an extension of the second isomorphism theorem.

Theorem 6.10 (Zassenhaus lemma)

Let A be an Ω-algebra, B and C subalgebras of A, and $\mathfrak{q}$ and $\mathfrak{r}$ congruences on B and C respectively. Assume that all congruences on $B \cap C$ commute; then $\mathfrak{q} \circ \mathfrak{r} \circ \mathfrak{q} \in \mathscr{C}_\Omega((B \cap C)^\mathfrak{q})$, $\mathfrak{r} \circ \mathfrak{q} \circ \mathfrak{r} \in \mathscr{C}_\Omega((B \cap C)^\mathfrak{r})$, and the identity mapping on A induces an isomorphism

$$(B \cap C)^\mathfrak{q}/\mathfrak{q} \circ \mathfrak{r} \circ \mathfrak{q} \cong (B \cap C)^\mathfrak{r}/\mathfrak{r} \circ \mathfrak{q} \circ \mathfrak{r}.$$

Proof:

Put $D = B \cap C$, $\mathfrak{q}' = \mathfrak{q} \cap C^2$, $\mathfrak{r}' = \mathfrak{r} \cap B^2$; then $\mathfrak{q} \circ \mathfrak{r} \circ \mathfrak{q} = \mathfrak{q} \circ \mathfrak{r}' \circ \mathfrak{q}$, and

$$\mathfrak{q} \circ \mathfrak{r} \circ \mathfrak{q} \circ \mathfrak{q} \circ \mathfrak{r} \circ \mathfrak{q} = \mathfrak{q} \circ \mathfrak{r}' \circ \mathfrak{q}' \circ \mathfrak{r}' \circ \mathfrak{q} = \mathfrak{q} \circ \mathfrak{r}' \circ \mathfrak{q};$$

hence $\mathfrak{q} \circ \mathfrak{r} \circ \mathfrak{q}$ is transitive; clearly, it is also reflexive and symmetric, and admits Ω; thus it is a congruence on $D^\mathfrak{q}$. Similarly, $\mathfrak{r} \circ \mathfrak{q} \circ \mathfrak{r}$ is a congruence on $D^\mathfrak{r}$, and $\mathfrak{q}' \circ \mathfrak{r}'$ is a congruence on D. Now

$$(\mathfrak{q} \circ \mathfrak{r} \circ \mathfrak{q}) \cap D^2 = \mathfrak{q}' \circ \mathfrak{r}';$$

hence by the second isomorphism theorem (Theorem 3.9),

$$D^\mathfrak{q}/\mathfrak{q} \circ \mathfrak{r} \circ \mathfrak{q} \cong D/\mathfrak{q}' \circ \mathfrak{r}'.$$

By symmetry we also have

$$D^\mathfrak{r}/\mathfrak{r} \circ \mathfrak{q} \circ \mathfrak{r} \cong D/\mathfrak{q}' \circ \mathfrak{r}',$$

and the result follows. ∎

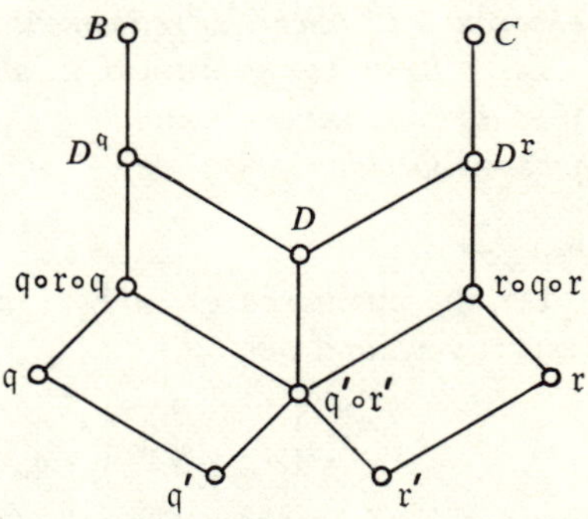

Figure 7

The factor $D^\mathfrak{q}/\mathfrak{q} \circ \mathfrak{r} \circ \mathfrak{q}$ is often called the *projection* of $C/\mathfrak{r}$ into $B/\mathfrak{q}$. Then Theorem 6.10 just states that under the given hypothesis, if $\bar{B}$ and $\bar{C}$ are any two factors of A, then the projection of $\bar{B}$ into $\bar{C}$ is isomorphic to the projection of $\bar{C}$ into $\bar{B}$.

Let A be an Ω-algebra with a subalgebra E; then by a *normal chain* from E to A is meant a finite chain of subalgebras of A:

$$E = A_0 \subseteq A_1 \subseteq \cdots \subseteq A_m = A, \tag{5}$$

together with a congruence $\mathfrak{q}_i$ on A_i ($i \geqslant 1$) such that A_{i-1} is precisely a $\mathfrak{q}_i$-class. Since A_{i-1} is a subalgebra, this means that the quotient $A_i/\mathfrak{q}_i$ has a trivial subalgebra, namely $A_{i-1}^{\mathfrak{q}_i}$. Qua set we have $A_{i-1} = A_{i-1}^{\mathfrak{q}_i} = E^{\mathfrak{q}_i}$, so that (5) may also be written

$$E = E^{\mathfrak{q}_1} \subseteq E^{\mathfrak{q}_2} \subseteq \cdots \subseteq E^{\mathfrak{q}_m} \subseteq A.$$

If in (5) each interval $[A_{i-1},A_i]$ is replaced by a normal chain from A_{i-1} to A_i such that the new subalgebras inserted, as well as the classes of congruences inserted, are unions of $\mathfrak{q}_i$-classes, we again obtain a normal chain from E to A, which is called a *refinement* of (5). Any such refinement may be obtained by taking a normal chain in each $A_i/\mathfrak{q}_i$ and applying the pullback along nat $\mathfrak{q}_i$.

A second normal chain

$$E = B_0 \subseteq B_1 \subseteq \cdots \subseteq B_n = A \tag{6}$$

with congruences $\mathfrak{r}_j$ is said to be *isomorphic* to (5), if $m = n$ and there is a permutation π of $1,2,\cdots,n$ such that

$$A_i/\mathfrak{q}_i \cong B_{i\pi}/\mathfrak{r}_{i\pi}.$$

Using these definitions we can establish the Schreier refinement theorem for Ω-algebras with commuting congruences:

Theorem 6.11 (Schreier refinement theorem)

Let A be an Ω-algebra with a subalgebra E, such that on any subalgebra of A all congruences commute. Then any two normal chains from E to A have isomorphic refinements.

Proof:

Let the chains be given by (5) and (6), with congruences $\mathfrak{q}_i$ and $\mathfrak{r}_j$ respectively, and put

$$\begin{aligned} F_{ij} &= (A_i \cap B_j)^{\mathfrak{q}_i}/\mathfrak{q}_i \circ \mathfrak{r}_j \circ \mathfrak{q}_i, && (\, i = 1,\cdots,m, \\ G_{ji} &= (A_i \cap B_j)^{\mathfrak{r}_j}/\mathfrak{r}_j \circ \mathfrak{q}_i \circ \mathfrak{r}_j; && \; j = 1,\cdots,n) \end{aligned}$$

then by the Zassenhaus lemma,

$$F_{ij} \cong G_{ji}. \tag{7}$$

If we put $D = A_i \cap B_j$, $\mathfrak{r}' = \mathfrak{r}_j \cap A_i^2$, $\mathfrak{q}' = \mathfrak{q}_i \cap B_j^2$, then since $E \subseteq A_i$ for $i = 1,\cdots,m$, we have

$$\begin{aligned} E^{\mathfrak{q}_i \circ \mathfrak{r}_j \circ \mathfrak{q}_i} = E^{\mathfrak{q}' \circ \mathfrak{r}' \circ \mathfrak{q}_i} &= E^{\mathfrak{r}' \circ \mathfrak{q}_i} \\ &= B_{j-1}^{\mathfrak{q}_i} \\ &= (A_i \cap B_{j-1})^{\mathfrak{q}_i}. \end{aligned}$$

Thus $F_{i1},\cdots,F_{in}$ form the factors of a normal chain in $A_i/\mathfrak{q}_i$; therefore (5) can be refined to a normal chain with factors $F_{11},F_{12},\cdots,F_{1n},F_{21},\cdots,F_{mn}$. By symmetry (6) can be refined to a normal chain with factors G_{ji}, and by (7) these refinements are isomorphic. ∎

A factor is said to be *trivial* if it consists of the trivial algebra. Clearly, if we have two isomorphic chains and omit the trivial factors from both, we again obtain two normal chains, isomorphic to each other but without repetitions. A normal chain without repetitions which has no proper refinements (i.e. no refinements without repetitions) is called a *composition series*. Thus a normal chain is a composition series if and only if all its factors are simple algebras.

Corollary 6.12 (Jordan-Hölder theorem)

If A and E are as in Theorem 6.11, *then any two composition series from E to A are isomorphic.* ∎

Corollary 6.13

If A and E are as in Theorem 6.11 *and if there exists a composition series from E to A, then any normal chain from E to A can be refined to a composition series.* ∎

An important special case is obtained by taking E to be a trivial subalgebra of A. In particular, this can always be done for groups; moreover, since congruences on groups commute, we obtain as a special case of Theorem 6.11 and its corollaries the usual Schreier theorem and Jordan-Hölder theorem for groups. As a second special case, let A be an Ω-algebra, with a normal chain (5). If $\mathfrak{q}_i = \mathfrak{q}_i^* \cap A_i^2$, where $\mathfrak{q}_i^*$ is a congruence on A, then (5) is said to be an *invariant chain*. It is easily seen that the refinements obtained in the proof of Theorem 6.11 will be invariant chains, provided we start with invariant chains. We thus obtain the refinement theorem for invariant chains, whose statement is left to the reader. An invariant chain without proper refinements or repetitions is called a *chief series*. We then have a corollary, analogous to Corollaries 6.12 and 6.13:

Corollary 6.14

If A is an Ω-algebra with commuting congruences, and E a subalgebra of A, then any two chief series from E to A are isomorphic; and when chief series exist, then any invariant chain from E to A may be refined to a chief series. ∎

The Jordan-Hölder theorem for general algebras has been proved under a number of different conditions; for an algebra with maximum and minimum condition for subalgebras, the following condition has been shown to be necessary and sufficient for the conclusion of Corollary 6.12 to hold (Goldie [52]): Given two congruences $\mathfrak{q}$ and $\mathfrak{r}$ occurring in different composition series, let B,C be the subalgebras on which $\mathfrak{q}$ and $\mathfrak{r}$ are defined and write $D = B \cap C$, $\mathfrak{q}' = \mathfrak{q} \cap D^2$, $\mathfrak{r}' = \mathfrak{r} \cap D^2$; then

$$E^{\mathfrak{q}' \circ \mathfrak{r}'} = E^{\mathfrak{r}' \circ \mathfrak{q}'}.$$

We turn now to the Krull-Schmidt theorem; to apply the lattice version proved in II.4, we consider representations of a given algebra A as a finite direct product:

(8) $$A \cong A_1 \times \cdots \times A_n.$$

If $\varepsilon_i: A \to A_i$ denotes the projection onto the factor A_i, then, clearly,

(i) for any $x,y \in A$, if $x\varepsilon_i = y\varepsilon_i$ $(i = 1,\cdots,n)$, then $x = y$,
(ii) given any family (a_i), $a_i \in A_i$ $(i = 1,\cdots,n)$, there exists an $x \in A$ such that $x\varepsilon_i = a_i$ (note that x is unique, by (i)).

Conversely, given a finite family of algebras and epimorphisms $\varepsilon_i: A \to A_i$ satisfying (i) and (ii), we can conclude that (8) holds. In fact, if $\theta: A \to \Pi A_i$ is the mapping defined by

$$x\theta = (x\varepsilon_i),$$

then θ is a homomorphism because each ε_i is a homomorphism, and θ is injective by (i) and surjective by (ii); hence it is an isomorphism. The conditions (i) and (ii) may also be expressed in terms of the kernels of the ε_i as follows:

Theorem 6.15

If an Ω-algebra A is written as a finite direct product

(8) $$A \cong A_1 \times \cdots \times A_n,$$

with projections $\varepsilon_i: A \to A_i$, and if $\mathfrak{q}_i = \ker \varepsilon_i$, then the $\mathfrak{q}_i$ satisfy

(9) $$\mathfrak{q}_1 \cap \cdots \cap \mathfrak{q}_n = \Delta,$$

(10) $$(\mathfrak{q}_1 \cap \cdots \cap \mathfrak{q}_{i-1}) \circ \mathfrak{q}_i = A^2 \qquad (i = 2,\cdots,n).$$

Conversely, any family of congruences $(\mathfrak{q}_i)$ satisfying (9) *and* (10) *gives rise to a direct product representation* (8), *where $A_i \cong A/\mathfrak{q}_i$. Moreover, the $\mathfrak{q}_i$ commute in pairs.*

Proof:

Given (8), let $\varepsilon^{(i)}: A \to A_1 \times \cdots \times A_i$ be the projection; then it is easily seen that $\ker \varepsilon^{(i)} = \mathfrak{q}_1 \cap \cdots \cap \mathfrak{q}_i$. In particular, since $\ker \varepsilon^{(n)} = \Delta$, we have (9). Further, for any elements $a_i \in A_i$ there exists $x \in A$ such that $x\varepsilon^{(n-1)} = (a_1,\cdots,a_{n-1})$, $x\varepsilon_n = a_n$. Since a_i was arbitrary in A_i, this means that given any $y,z \in A$, there exists $x \in A$ such that

$$x \equiv y \ (\mathrm{mod}\ \mathfrak{q}_1 \cap \cdots \cap \mathfrak{q}_{n-1}),$$
$$x \equiv z \ (\mathrm{mod}\ \mathfrak{q}_n).$$

Thus

$$(11) \qquad (\mathfrak{q}_1 \cap \cdots \cap \mathfrak{q}_{n-1}) \circ \mathfrak{q}_n = A^2.$$

By symmetry we have

$$(12) \qquad \left(\bigcap_{j \neq i} \mathfrak{q}_j\right) \circ \mathfrak{q}_i = A^2,$$

and a fortiori (10) holds. Further, by (12),

$$\mathfrak{q}_i \circ \mathfrak{q}_j = A^2 = \mathfrak{q}_j \circ \mathfrak{q}_i \qquad (i \neq j),$$

so that all the $\mathfrak{q}$'s commute.

Conversely, let $\mathfrak{q}_1,\cdots,\mathfrak{q}_n$ satisfying (9) and (10) be given, write $A_i = A/\mathfrak{q}_i$, and put $\varepsilon_i = \mathrm{nat}\ \mathfrak{q}_i$. By (9), if $x\varepsilon_i = y\varepsilon_i$ for $i = 1,\cdots,n$, then $x = y$. Now let $a_i \in A_i$ be given. By (10), if

$$(13) \qquad x^{(i)}\varepsilon_j = a_j (j = 1,\cdots,i) \text{ for some } x^{(i)} \in A,$$

then there exists $x^{(i+1)} \in A$ such that

$$x^{(i+1)}\varepsilon_j = a_j \qquad \text{for } j = 1,\cdots,i+1.$$

Now (13) holds for $i = 1$, and hence by induction, for all $i \leqslant n$. In particular, for $i = n$ we obtain (ii), and the conclusion follows. ∎

We note that (9) and (10) state just that $\mathfrak{q}_1,\cdots,\mathfrak{q}_n$ is an independent set in the lattice $\mathscr{C}_\Omega(A)$, regarding A^2 as least and Δ as greatest element. In other words, they are the conditions for Δ to be the direct join of $\mathfrak{q}_1,\cdots,\mathfrak{q}_n$ in the dual of $\mathscr{C}_\Omega(A)$.

We also recall that $\mathscr{C}_\Omega(A)$ is modular whenever all congruences on A commute (Proposition 6.9). Let A be an algebra for which this is the case and assume further that the lattice $\mathscr{C}_\Omega(A)$ is of finite length. Then the Krull-Schmidt theorem for lattices shows that in any two complete decompositions,

$$\mathfrak{q}_1 \cap \cdots \cap \mathfrak{q}_m = \mathfrak{r}_1 \cap \cdots \cap \mathfrak{r}_n = \Delta,$$

$m = n$ and each $\mathfrak{q}_i$ is related to some $\mathfrak{r}_j$. This means that there exists $\mathfrak{q}' \in \mathscr{C}_\Omega(A)$ such that

$$\mathfrak{q}_i \cap \mathfrak{q}' = \mathfrak{r}_j \cap \mathfrak{q}' = \Delta, \quad \mathfrak{q}_i \circ \mathfrak{q}' = \mathfrak{r}_j \circ \mathfrak{q}' = A^2.$$

(In fact, $\mathfrak{q}'$ may be taken as $\bigcap_{k \neq i} \mathfrak{q}_k$, from the proof of Theorem 4.11). Putting $A_i = A/\mathfrak{q}_i$, $B_j = A/\mathfrak{r}_j$, $A' = A/\mathfrak{q}'$, we thus have

(14) $$A \cong A_i \times A' \cong B_j \times A';$$

this establishes

Theorem 6.16 (exchange theorem for direct decompositions)

Let A be an Ω-algebra in which all congruences commute and whose lattice of congruences is of finite length. Then for any two direct decompositions

$$A \cong A_1 \times \cdots \times A_m \cong B_1 \times \cdots \times B_n$$

of A into indecomposable factors, $m = n$ and each A_i can be exchanged against some B_j. ∎

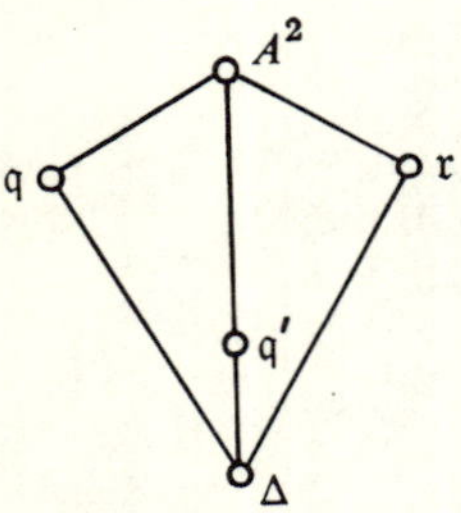

Figure 8

Usually one wants to be able to assert that the A_i and B_j which are related as in (14) are actually isomorphic, as Ω-algebras. This is not true in general (cf. Exercise 10), but the next theorem gives a simple condition for this to be the case. Let $\mathfrak{q},\mathfrak{r},\mathfrak{q}'$ be congruences on A such that

$$\mathfrak{q} \circ \mathfrak{q}' = \mathfrak{r} \circ \mathfrak{q}' = A^2, \quad \mathfrak{q} \cap \mathfrak{q}' = \mathfrak{r} \cap \mathfrak{q}' = \Delta.$$

These equations state that each $\mathfrak{q}'$-class is a common transversal for $A/\mathfrak{q}$ and $A/\mathfrak{r}$. If $A/\mathfrak{q}'$ has a trivial subalgebra, then the corresponding $\mathfrak{q}'$-class is itself a subalgebra which must then be isomorphic to both $A/\mathfrak{q}$ and $A/\mathfrak{r}$. Thus we obtain

Theorem 6.17 (Krull-Schmidt theorem for algebras)

Let A be an Ω-algebra in which all congruences commute, which has a chief series and a trivial subalgebra. Then in any two complete decompositions

$$A \cong A_1 \times \cdots \times A_m \cong B_1 \times \cdots \times B_n,$$

$m = n$, and for suitable numbering of the B's, $A_i \cong B_i$.

This follows from the remarks preceding the statement of the theorem, if we observe that any homomorphic image of A again has a trivial subalgebra. ∎

We have seen that congruences on any group commute; therefore the result may be applied to any group with a chief series, or, more generally, to any group with (unary or multiple) operators which has a chief series. In fact the result can be slightly strengthened in this case. We recall that the *centre* of a group G is defined as the set

$$Z = \{x \in G \mid xy = yx \quad \text{for all } y \in G\}.$$

Two subgroups H, K of G are said to be *centrally isomorphic* if there is an isomorphism $\alpha : H \to K$ such that $x^{-1} \cdot x\alpha \in Z$ for all $x \in H$. Then we have

Corollary 6.18

If G is any group with a chief series and

$$G \cong G_1 \times \cdots \times G_m \cong H_1 \times \cdots \times H_n$$

are any two complete decompositions, then $m = n$, and for a suitable numbering of the H's, G_i is centrally isomorphic to H_i. In particular, if G has trivial centre, then the factors G_i are uniquely determined.

For by Theorem 6.16 we have $G = G_1 \times K = H_1 \times K$ say, where direct factors of G have been identified with subgroups of G. Any element $x \in G$ has the form

$$x = x_1 z = y_1 z',$$

where $x_1 \in G_1$, $y_1 \in H_1$, and $z, z' \in K$. Now $x_1^{-1} y_1 = z z'^{-1} = t$ say, and K commutes elementwise with G_1 and with H_1; therefore, t commutes elementwise with G_1 and H_1; on the other hand, y_1 commutes with K and x_1 commutes with K, and so t also commutes with K, whence t commutes with every element of G, i.e. $t \in Z$. ∎

Many similar results, imposing different conditions, have been obtained. E.g. for modules over a ring a version of the Krull-Schmidt theorem which does not explicitly assume the existence of a chief series has been given by Jacobson ([56] ch. III); however, some finiteness condition has to be imposed, as examples show (Jónsson [57]). An analogous result for groupoids with neutral element, relative to representations as inner direct product, has been obtained by Jónsson & Tarski [47].

EXERCISES

1. If any two congruences on an Ω-algebra A commute, then the same is true of any quotient of A. (Hint: Use Corollary 3.12.)

2. By an *ordered semigroup* is meant a semigroup S which is also an ordered set, such that $a_i \leqslant b_i$ $(i = 1,2)$ implies $a_1a_2 \leqslant b_1b_2$. An element e of S is said to be *idempotent* if $e^2 = e$.

(a) Let S be an ordered semigroup with least element 1 which is also the unit element. Show that any element a such that $a^2 \leqslant a$ is an idempotent. If a and b are idempotents, then ab is an idempotent, provided that $ba \leqslant ab$.

(b) If S is as in (a) and has an order-preserving antiautomorphism $'$ of period two, i.e. a mapping of S into itself, $a \to a'$, such that (i) $a \leqslant b \Rightarrow a' \leqslant b'$, (ii) $a'' = a$, (iii) $(ab)' = b'a'$, show that for any idempotents a and b fixed by $'$, ab is an idempotent if and only if $ba = ab$. Deduce Proposition 6.6.

3. If A is an Ω-algebra and S any subset of A, write $a \sim b \pmod S$ if $a = b$ or $a,b \in S^\tau$, where τ is some translation of A. Further, write $a \approx b \pmod S$ if there is a finite sequence $x_0 = a, x_1, \cdots, x_n = b$ such that $x_{i-1} \sim x_i \pmod S$. Show that the relation '$a \approx b$' is a congruence on A, and that it is the least congruence which has a class containing S.

4. Let A and S be as in Exercise 3, and define

$$\mathfrak{q} = \{(x,y) \in A^2 \mid x^\tau \in S \Leftrightarrow y^\tau \in S \quad \text{for each translation } \tau\}.$$

Show that $\mathfrak{q}$ is the greatest congruence on A such that S is a union of $\mathfrak{q}$-classes.

5. (Malcev.) Let A and S be as in Exercise 3; show that S is the class of a unique congruence on A if and only if for all $x,y \in A$,

$$x^\tau \in S \Leftrightarrow y^\tau \in S \qquad \text{for all translations } \tau$$

holds precisely when $x \approx y$ (cf. definition Exercise 3).

6. (Malcev.) A translation on an Ω-algebra A is said to be *invertible* if it has an inverse which is also a translation. Show that the invertible translations form a group Γ on A, and if Γ acts transitively on A (i.e., if A as Γ-module consists of a single orbit), then all congruences on A commute. (Hint: If τ_{ab} is an invertible translation mapping a to b, verify that the mapping $x \to x\sigma$, where $x\sigma = a\tau_{xb}\tau_{bb}^{-1}$, is a translation.)

7. Let $\mathfrak{q}$ be a congruence on a lattice L. If $a \equiv b \pmod{\mathfrak{q}}$, show that $a \wedge b \equiv a \vee b \pmod{\mathfrak{q}}$.

8. (Funayama & Nakayama.) Show that the lattice of congruences on any lattice is distributive. (Hint: Use Exercise 7 to verify that $a \equiv b \pmod{\mathfrak{q} \cap (\mathfrak{r} \vee \mathfrak{s})}$ implies $a \equiv b \pmod{(\mathfrak{q} \cap \mathfrak{r}) \vee (\mathfrak{q} \cap \mathfrak{s})}$).

9. A congruence $\mathfrak{q}$ on an Ω-algebra A is said to be *fine*, if any transversal of $A/\mathfrak{q}$ generates A. If X is a minimal generating set of A and $\mathfrak{q}$ is a fine congruence on A, show that $\mathfrak{q}$ separates X.

10. (Jónsson.) If Ω consists of a single unary operator, construct two non-isomorphic Ω-algebras A, B, with two elements each, such that $A \times B \cong B \times B$. (Hint: Take the operation to be a permutation of the carrier in each case.)

7. LOCAL AND RESIDUAL PROPERTIES

In the last section we obtained a description of a finite direct product, in terms of the kernels of the projections onto the factors. For products with an arbitrary number of factors the description takes on a more complicated form. Let us therefore consider an Ω-algebra A with a family of epimorphisms

(1) $$\varepsilon_\lambda : A \to A_\lambda \qquad (\lambda \in \Lambda)$$

satisfying only the first of the conditions in II.6:

(i) for any $x,y \in A$, if $x\varepsilon_\lambda = y\varepsilon_\lambda$ for all $\lambda \in \Lambda$, then $x = y$.

The same argument then shows that A is isomorphic to a subalgebra of ΠA_λ; identifying A with this subalgebra of ΠA_λ, we see that the natural projection ε_λ, restricted to A, is still surjective. This suggests the following

Definition

If $(A_\lambda)_{\lambda\in\Lambda}$ is any family of Ω-algebras, then a subalgebra A of the direct product ΠA_λ which is such that $\varepsilon_\lambda | A$ is surjective (for each $\lambda \in \Lambda$) is said to be a *subdirect product* of the family $(A_\lambda)_{\lambda\in\Lambda}$.

Subdirect products usually arise in the following way:

Proposition 7.1

Let A be an Ω-algebra and $(\mathfrak{q}_\lambda)_{\lambda\in\Lambda}$ a family of congruences on A. Put $\mathfrak{q} = \bigcap \mathfrak{q}_\lambda$ and $A_\lambda = A/\mathfrak{q}_\lambda$; then $A/\mathfrak{q}$ is isomorphic to a subdirect product of the family (A_λ).

Proof:

Consider the mapping $\theta : A \to \Pi A_\lambda$ defined by

(2) $$(a\theta)\varepsilon_\lambda = a^{\mathfrak{q}_\lambda} \qquad (a \in A,\ \lambda \in \Lambda).$$

By definition, θ is a homomorphism, and its kernel is $\mathfrak{q}$; therefore, dividing by $\mathfrak{q}$, we obtain a monomorphism $A/\mathfrak{q} \to \Pi A_\lambda$. Thus $A/\mathfrak{q}$ may be embedded in ΠA_λ, and now (2) shows that $\theta\varepsilon_\lambda$ is an epimorphism. ∎

Corollary 7.2

If A is any Ω-algebra with a family of congruences $(\mathfrak{q}_\lambda)_{\lambda\in\Lambda}$ such that $\bigcap \mathfrak{q}_\lambda = \Delta$, then A is isomorphic to a subdirect product of the $A/\mathfrak{q}_\lambda$.

This is merely the case $\mathfrak{q} = \Delta$ and follows if we recall that $A/\Delta \cong A$. ∎

A family $(\mathfrak{q}_\lambda)_{\lambda \in \Lambda}$ of congruences on A such that $\bigcap \mathfrak{q}_\lambda = \Delta$ (i.e., $\bigcap \mathfrak{q}_\lambda$ separates A) is called a *separating* family of congruences. The corollary may then be taken as asserting that A can be expressed as a subdirect product of the A_λ whenever there is a separating family of congruences $(\mathfrak{q}_\lambda)_{\lambda \in \Lambda}$ with quotients A_λ. Let A be an Ω-algebra with a separating family of congruences $(\mathfrak{q}_\lambda)_{\lambda \in \Lambda}$ such that $\mathfrak{q}_\lambda \neq \Delta$ for all $\lambda \in \Lambda$; then A is said to be *subdirectly reducible*; otherwise, A is called *subdirectly irreducible*. Equivalently, A is subdirectly irreducible if and only if in every subdirect product representation (2) of A, at least one of the $\theta\varepsilon_\lambda$ is an isomorphism.

From this definition we obtain a useful representation theorem, due to Birkhoff [44].

Theorem 7.3

Every Ω-algebra A is a subdirect product of subdirectly irreducible Ω-algebras, which are homomorphic images of A.

Proof:

Let $\mathfrak{q}$ be any congruence on A; we shall say that $\mathfrak{q}$ is *meet-irreducible* if there exists no family of congruences $(\mathfrak{r}_\lambda)_{\lambda \in \Lambda}$ such that $\mathfrak{r}_\lambda \supset \mathfrak{q}$ for all $\lambda \in \Lambda$ and $\bigcap \mathfrak{r}_\lambda = \mathfrak{q}$. By Corollary 3.12 we see that $\mathfrak{q}$ is meet-irreducible if and only if $A/\mathfrak{q}$ is subdirectly irreducible. Now let $(\mathfrak{q}_\lambda)_{\lambda \in \Lambda}$ be the family of all meet-irreducible congruences on A; if we can show that,

$$\bigcap \mathfrak{q}_\lambda = \Delta, \tag{3}$$

the theorem will follow because, by Corollary 7.2, A is then a subdirect product of the subdirectly irreducible algebras $A/\mathfrak{q}_\lambda$. To establish (3), let $x,y \in A$, $x \neq y$ and let $\mathfrak{q}_0$ be a maximal congruence on A such that $(x,y) \notin \mathfrak{q}_0$ (which exists by Corollary 6.4). Then any congruence which properly contains $\mathfrak{q}_0$ also contains (x,y), and hence the intersection of all congruences which properly contain $\mathfrak{q}_0$ also contains (x,y). But $(x,y) \notin \mathfrak{q}_0$, and this shows that $\mathfrak{q}_0$ is meet-irreducible; since x,y was any pair of distinct elements of A, it follows that (3) holds. ∎

The somewhat imperfect duality which was already observed in set theory (in I.3) and which clearly extends (in an even less perfect form) to general algebras, suggests that we consider the following counterpart of a separating family of congruences. Let A be any Ω-algebra; then, by a *local system* of subalgebras of A, one understands a system $\mathscr{S}$ of nonempty subalgebras of A which is directed by inclusion and is such that $\bigcup \mathscr{S} = A$.

Trivial examples of local systems are (i) the system of all nonempty subalgebras of A and (ii) the system consisting of A alone (provided $A \neq \emptyset$). An important example of a local system is the system of all finitely generated subalgebras of A. As is easily seen, any local system in A which admits subalgebras (i.e. which contains, with any algebra, all its subalgebras) necessarily includes all finitely generated subalgebras of A.

We turn now to consider *properties* of Ω-algebras, sueh as being finite or being contained in a given algebra as subalgebra. We shall only be concerned with *abstract properties*, i.e. properties P such that if A has P, then every algebra isomorphic to A also has P. Thus, being finite is an abstract property, but being a subalgebra of a given algebra C, say, is not. If P is any property of algebras, then the algebra A is said to be *locally* P, if there is a local system of subalgebras of A, all having P. If every algebra which is locally P actually has P itself, P is said to be a *local property* of algebras. Thus, for example, the property of being abelian is a local property of groups; on the other hand, being finite is not a local property: the multiplicative group of all complex roots of unity is locally finite, but not finite.

Similarly we define the Ω-algebra A to be *residually* P if there is a separating family $(\mathfrak{q}_\lambda)_{\lambda \in \Lambda}$ of congruences on A such that each quotient $A/\mathfrak{q}_\lambda$ has P. By Corollary 7.2, A is residually P if and only if it can be expressed as a subdirect product of Ω-algebras having P. Now the property P is said to be a *residual property* if any Ω-algebra which is residually P actually has P itself. E.g., for groups, the property of being abelian is residual, for if G is residually abelian, it is a subdirect product of abelian groups, and hence is itself abelian, but being finite is not a residual property. We conclude this section with a result which establishes a connexion between residual and local properties.

Proposition 7.4

Any residual property of Ω-algebras which is preserved under homomorphic images is local.

Proof:

Let P be a property satisfying the hypothesis of the proposition and let A be an Ω-algebra with a local system $(A_\lambda)_{\lambda \in \Lambda}$ of subalgebras having P. For convenience we take the index set Λ to be preordered by the rule

$$\lambda \leqslant \mu \qquad \text{if and only if } A_\lambda \subseteq A_\mu.$$

Then it is clear that Λ is directed. Now form the direct product $D = \Pi A_\lambda$ with the projections $\varepsilon_\lambda : D \to A_\lambda$ and consider the subset

$$B = \{x \in D \mid \text{there exists } \lambda_0 = \lambda_0(x) \text{ such that } x\varepsilon_\lambda = x\varepsilon_{\lambda_0} \text{ for } \lambda \geqslant \lambda_0\}.$$

Thus B is the subset of elements of D whose coordinates are ultimately constant. The constant value of $x\varepsilon_\lambda$ will be denoted by $x\varepsilon$. We assert that B is a subalgebra of D and is in fact a subdirect product of the A_λ. For if $\omega \in \Omega(n)$, $x_1,\cdots,x_n \in B$, suppose that $x_i\varepsilon_\lambda = x_i\varepsilon$ for $\lambda \geqslant \lambda_i$; then since Λ is directed, there exists $\lambda_0 \in \Lambda$ such that $\lambda_0 \geqslant \lambda_i$ for $i = 1,\cdots,n$; hence $x_i\varepsilon_\lambda = x_i\varepsilon$ for $\lambda \geqslant \lambda_0$, and so

$$(x_1 \cdots x_n\omega)\varepsilon_\lambda = (x_1\varepsilon_\lambda)\cdots(x_n\varepsilon_\lambda)\omega = (x_1\varepsilon)\cdots(x_n\varepsilon)\omega;$$

this is constant for $\lambda \geqslant \lambda_0$, and so equals $(x_1, \ldots x_n\omega)\varepsilon$, by definition of ε. Given any $\lambda \in \Lambda$ and any $a \in A_\lambda$, we have $a \in A_\mu$ for all $\mu \geqslant \lambda$, hence the element x of D defined by

$$x\varepsilon_\mu = \begin{cases} a & \text{if } \mu \geqslant \lambda, \\ \text{arbitrary in } A_\mu & \text{otherwise,} \end{cases}$$

belongs to B and is such that $x\varepsilon = a$; this shows $\varepsilon_\lambda \mid B$ to be an epimorphism, and so B is a subdirect product of the A_λ.

For any $x \in B$, $x\varepsilon \in A$; thus ε is a mapping from B to A, and the above argument shows ε to be a homomorphism. Moreover, it is surjective, since every $a \in A$ is contained in A_λ for some $\lambda \in \Lambda$, and hence for all $\mu \geqslant \lambda$. Thus A is a homomorphic image of a subdirect product of the A_λ, and therefore has P. ▮

EXERCISES

1. For any hereditary property P, show that being locally P is a local property and being residually P is a residual property.

2. An abelian group is said to be *torsion-free* if there is no element of finite order, apart from the neutral element. Show that the property of being a torsion-free abelian group is both local and residual but is not preserved by homomorphic images.

3. A group G is said to be a p-group (where p is a prime) if the order of each element of G is a power of p. Show that being a p-group is a local property which is inherited by subgroups but which is not residual. (Hint: Every torsion-free abelian group is residually a p-group.)

4. A group G is said to be *ordered* if its carrier can be ordered in such a way that $a \leqslant b$, $a' \leqslant b'$ implies $aa' \leqslant bb'$. If G is an abstract group whose carrier can be ordered in such a way that G becomes a totally ordered group, we say that G can be totally ordered. Show that the property of groups: 'G can be totally ordered' is residual but is not preserved under homomorphic images. (Hint: The direct product of any family of totally ordered groups can be ordered by taking some well-ordering of the index set and then ordering the product lexicographically. Secondly, any free abelian group can be totally ordered, but not all its homomorphic images can be so ordered.)

5. (Dieudonné, Lorenzen.) Define the canonical ordering on the direct product of ordered groups ΠG_λ by the rule: $(x_\lambda) \leqslant (y_\lambda)$ if and only if $x_\lambda \leqslant y_\lambda$ for all $\lambda \in \Lambda$. Show that an abelian ordered group G (written additively) is a subdirect product of totally ordered groups (with the canonical ordering) if and only if, for any $x \in G$ and any positive integer n, $nx \geqslant 0$ implies $x \geqslant 0$.

6. Let R be a commutative ring with 1. If R has no nilptent elements $\neq 0$ (i.e. if $x^n = 0$ implies $x = 0$), show that R is subdirectly irreducible only if R is an integral domain. Deduce that every ring without nilpotent elements $\neq 0$ can be expressed as a subdirect product of integral domains.

7. A subdirect product A of a family $(A_\lambda)_{\lambda \in \Lambda}$ of algebras is said to be *irredundant*, if the natural projection onto a proper factor ΠA_μ ($\mu \in \Lambda'$, where $\Lambda' \subset \Lambda$) restricted to A, is not injective. If $\varepsilon_\lambda : A \to A_\lambda$ are the restrictions of the natural projections, show that the product is irredundant if and only if for each $\lambda \in \Lambda$, $\ker \varepsilon_\lambda \not\supseteq \bigcap_{\mu \neq \lambda} \ker \varepsilon_\mu$.

8. Show that any subdirect product of a finite number of factors can also be expressed as an irredundant subdirect product of some of these factors. Further, show that an irredundant subdirect product of a finite number of simple rings is necessarily their direct product.

9. Show that every distributive lattice with more than two elements is subdirectly reducible. (Hint: For any $a \in L$, consider the left and right segments generated by (a,a) in L^2, and show that the congruences generated by these segments intersect in the diagonal of L.)

For any set A, verify that the Boolean $\mathscr{B}(A)$ is a distributive lattice, and moreover, $\mathscr{B}(A) \cong 2^A$, where $2 = \{0,1\}$ is regarded as a lattice with the ordering $0 < 1$. Conversely, show that every distributive lattice L can be embedded in a lattice of the form $\mathscr{B}(A)$, for some set A, which may be taken to be finite if L is finite.

10. Show that the property of a group being simple is a local property. (Hint: G is simple if and only if, for any $x,y \in G$, $y \neq 1$, there is a product of conjugates of y and y^{-1} which equals x.) Using the fact that the alternating group of degree greater than four is simple, show that the group of all even permutations with a finite carrier on a countably infinite set is simple.

11. Show that being residually finite is a residual property of groups which is not local. (Use Exercises 1 and 10.)

12. Show that being a free abelian group is not a local property.

8. THE LATTICE OF CATEGORIES OF Ω-ALGEBRAS

We have seen that for a given operator domain Ω, all the Ω-algebras with all Ω-homomorphisms between them form a category, denoted by (Ω). In all that follows, the universe U will be arbitrary but fixed, so that we shall not refer to it explicitly, but in talking about Ω-algebras it will be understood that only algebras whose carrier belongs to U are considered.

A subcategory $\mathscr{K}$ of (Ω) is said to be *trivial* if it contains no algebra with more than one element; otherwise, it is *nontrivial*. If $\mathscr{K}$ contains, with any algebra A, all algebras isomorphic to A, and with any two isomorphic algebras A, A', it contains all the isomorphisms from A to A', then $\mathscr{K}$ is said to be *abstract*. The category $\mathscr{K}$ is said to be *regular* if every $\mathscr{K}$-homomorphism can be written in the form $\varepsilon\mu$, where ε is a $\mathscr{K}$-epimorphism and μ a $\mathscr{K}$-monomorphism. Clearly, an abstract category $\mathscr{K}$ is regular if and only if for every $\mathscr{K}$-homomorphism $\alpha: A \to B$, the image $A\alpha$ is a $\mathscr{K}$-algebra, the inclusion $i: A\alpha \to B$ is a $\mathscr{K}$-homomorphism, and α can be factored as $\alpha = \alpha_0 i$, where $\alpha_0: A \to A\alpha$ is a $\mathscr{K}$-epimorphism.

If $\mathscr{S}$ is any set of Ω-algebras, we may form a category from $\mathscr{S}$, e.g. by taking the full subcategory of (Ω) with $\mathscr{S}$ as class of objects; this will in general be neither abstract nor regular. However, we remark that if $\mathscr{K}$ is a subcategory of (Ω), there is a uniquely determined least abstract subcategory of (Ω) containing $\mathscr{K}$; we need only adjoin all isomorphic copies of $\mathscr{K}$-algebras and all possible isomorphisms. Given a set $\mathscr{S}$ of Ω-algebras, it is always possible to construct a regular category with $\mathscr{S}$ as object class, e.g. by allowing only isomorphisms as maps, but one often wishes to embed a given subcategory of (Ω) in a regular subcategory. To do this we remark that a set $\mathscr{S}$ of Ω-algebras together with certain monomorphisms and epimorphisms between them generates a regular subcategory of (Ω) if and only if, for every monomorphism μ and epimorphism ε such that $\mu\varepsilon$ is defined, there exist a monomorphism μ_1 and an epimorphism ε_1 in the set such that $\mu\varepsilon = \varepsilon_1\mu_1$.

Proposition 8.1

The class of abstract subcategories of (Ω) *forms a complete lattice, and so does the class of abstract regular subcategories of* (Ω).

Proof:

If $(\mathscr{K}_\lambda)_{\lambda\in\Lambda}$ is any family of abstract subcategories of (Ω), denote by $\mathscr{K} = \bigcap \mathscr{K}_\lambda$ the category whose object class is $\bigcap \mathrm{Ob}\, \mathscr{K}_\lambda$ and whose class of morphisms is $\bigcap \mathrm{Hom}\, \mathscr{K}_\lambda$. It is easily verified that this is again an abstract subcategory of (Ω), hence the abstract subcategories form a complete lattice by Proposition I.4.1. Now assume that in addition each $\mathscr{K}_\lambda$ is regular, and let $\alpha: A \to B$ be a $\mathscr{K}_\lambda$-homomorphism, for each $\lambda \in \Lambda$. Then $A\alpha$ is a $\mathscr{K}_\lambda$-algebra, the inclusion $i: A\alpha \to B$ a $\mathscr{K}_\lambda$-homomorphism, and $\alpha = \alpha_0 i$, where $\alpha_0: A \to A\alpha$ is a $\mathscr{K}_\lambda$-epimorphism, for each $\lambda \in \Lambda$; hence, i and α_0 are $\mathscr{K}$-homomorphisms. ∎

We shall denote by $\Gamma(\Omega)$ the lattice of all abstract regular subcategories of (Ω) and define certain closure operators on $\Gamma(\Omega)$, in the sense of II.1; this is possible because we are dealing with a complete lattice. Such closure operators (systematically introduced for classes of groups by P. Hall) will be denoted by small capitals; if A is a closure operator, we shall say that the category $\mathscr{K}$ is A-*closed* if $\mathrm{A}\mathscr{K} = \mathscr{K}$. The most important examples of closure operators on $\Gamma(\Omega)$ are the following:

(i) *Subalgebras*. For any $\mathscr{K} \in \Gamma(\Omega)$, denote by $\mathrm{S}\mathscr{K}$ the regular category generated by all subalgebras of $\mathscr{K}$-algebras and all restrictions of $\mathscr{K}$-homomorphisms cut down to subalgebras of $\mathscr{K}$-algebras. It is easily verified that S is a closure operator. If $\mathscr{K}$ is S-closed, we also say: $\mathscr{K}$ *admits subalgebras*, or: $\mathscr{K}$ is *hereditary*.

(ii) *Quotients*. Given $\mathscr{K} \in \Gamma(\Omega)$, we denote by $\mathrm{Q}\mathscr{K}$ the regular category generated by all homomorphic images of $\mathscr{K}$-algebras, together with all homomorphisms induced by $\mathscr{K}$-homomorphisms. The fact that Q is a closure operator follows from Corollary 3.12. A Q-closed category is said to *admit homomorphic images*.

(iii) *Direct products*. Given $\mathscr{K} \in \Gamma(\Omega)$, we denote by $\mathrm{P}\mathscr{K}$ the regular category generated by all direct products $P = \Pi A_\lambda$ of families of $\mathscr{K}$-algebras, together with the projections $\varepsilon_\lambda: P \to A_\lambda$ and the homomorphisms between products induced by $\mathscr{K}$-homomorphisms between the factors. A P-closed category is said to *admit direct products*. Again it is clear that P is a closure operator, if we bear in mind that the indexing set used is a member of the universe U.

(iv) *Local systems*. Given $\mathscr{K} \in \Gamma(\Omega)$, we denote by $\mathrm{L}\mathscr{K}$ the abstract category generated by Ω-algebras A which are locally $\mathscr{K}$-algebras, i.e.

which have a local system $(A_\lambda)_{\lambda \in \Lambda}$ of $\mathscr{K}$-algebras, with the inclusion mappings $i_\lambda : A_\lambda \to A$ and the homomorphisms induced by $\mathscr{K}$-homomorphisms between the algebras of appropriate local systems. When $\text{L}\mathscr{K} = \mathscr{K}$, $\mathscr{K}$ is called *local* (L is not in general a closure operator, cf. Kruse [67]).

(v) *Residual systems.* Let $\mathscr{K} \in \Gamma(\Omega)$ and denote by $\text{R}\mathscr{K}$ the regular category of all Ω-algebras A which are residually $\mathscr{K}$-algebras; this means that there is a separating family $(\mathfrak{q}_\lambda)$ of congruences such that $A/\mathfrak{q}_\lambda$ is a $\mathscr{K}$-algebra. Further, nat $\mathfrak{q}_\lambda$ is an $\text{R}\mathscr{K}$-homomorphism and a homomorphism $\alpha: B \to A$ is an $\text{R}\mathscr{K}$-homomorphism provided that $\alpha(\text{nat } \mathfrak{q}_\lambda)$ is a $\mathscr{K}$-homomorphism for all $\lambda \in \Lambda$. An R-closed category is said to be *residual.* R is again a closure operator, and moreover, every residual category admits direct products; more generally, for any category $\mathscr{K}$, $\text{P}\mathscr{K}$ is a subcategory of $\text{R}\mathscr{K}$.

If A and B are closure operators on $\Gamma(\Omega)$, we write $\text{A} \leqslant \text{B}$ to indicate that $\text{A}\mathscr{K}$ is a subcategory of $\text{B}\mathscr{K}$, for every $\mathscr{K} \in \Gamma(\Omega)$. Further, we define the operator AB by the equation

$$(\text{AB})\mathscr{K} = \text{A}(\text{B}\mathscr{K}).$$

In general AB need not be a closure operator, but it is not hard to see that there always exists a least closure operator containing A and B, which we denote by $\text{A} \vee \text{B}$. This follows because the collection of closure systems (corresponding to the closure operators) itself forms a complete lattice. It may also be directly verified as in the remarks preceding Proposition 6.6; as in that proposition, we can then deduce that AB is a closure operator provided that $\text{AB} \geqslant \text{BA}$. In fact the set of all operators (i.e. unary operations) on $\Gamma(\Omega)$ which satisfy J.1–2 of II.1 forms an ordered semigroup with 1 as least element, and a closure operator is just an element of this semigroup which is idempotent (cf. Exercise 6.2).

There is one relation between the operators defined above which is of importance in what follows.

Proposition 8.2

The operator $\text{U} = \text{SP}$ *is a closure operator and*

(1) $$\text{P} \leqslant \text{R} \leqslant \text{SP}.$$

Proof:

By the above remarks it is enough to verify that

(2) $$\text{PS} \leqslant \text{SP}$$

in order to show that SP is a closure operator. Let $A \in \text{PS}\mathscr{K}$; then $A = \Pi A_\lambda$, say, where A_λ is a subalgebra of a $\mathscr{K}$-algebra B_λ, say. Now ΠA_λ may be embedded in ΠB_λ, hence A is isomorphic to a subalgebra of ΠB_λ, i.e. $A \in \mathrm{SP}\mathscr{K}$; similarly, it follows that every $\mathrm{PS}\mathscr{K}$-homomorphism is an $\mathrm{SP}\mathscr{K}$-homomorphism, which proves (2). Now an Ω-algebra is residually $\mathscr{K}$ if and only if it is a subdirect product of $\mathscr{K}$-algebras. In particular, every direct product of $\mathscr{K}$-algebras is residually $\mathscr{K}$, whence $\mathrm{P} \leqslant \mathrm{R}$, and a subdirect product is clearly a subalgebra of a direct product, thus $\mathrm{R} \leqslant \mathrm{U}$, which establishes (1). ▮

EXERCISES

1. Show that $\mathrm{P}\mathscr{K}$ and $\mathrm{R}\mathscr{K}$ always contain trivial algebras, and if ∅ has an Ω-algebra structure (i.e. if $\Omega(0) = \emptyset$), then $\mathrm{L}\mathscr{K}$ always contains ∅ as an algebra.

2. Show that $\mathrm{SQ} \leqslant \mathrm{QS}$, but that equality does not necessarily hold. (Use Corollary 3.12 to establish the inequality, and verify that it is strict for the category of groups and homomorphisms.)

3. Show that in general $\mathrm{SP} \neq \mathrm{PS}$.

4. Show that $\mathrm{QL} \leqslant \mathrm{LQ}$.

5. Show that $\mathrm{SR} = \mathrm{RS} = \mathrm{U}$.

6. Interpret Proposition 7.4 in the terminology of this section.

Chapter III

Free Algebras

Many important classes of algebras, among them groups, rings, and lattices, consist of all the homomorphic images of certain 'free' algebras in the class, which are essentially determined by the cardinal of a free generating set. These are the *varieties* of algebras, which form the subject of Chapter IV, but free algebras are also of importance in more general situations, and we therefore devote a chapter to the study of properties of free algebras which are independent of the notion of a variety. A free algebra is itself a special case of the notion of a universal functor in category theory, and so we shall first describe universal functors in general categories (cf. Samuel [48], MacLane [63]).

1. UNIVERSAL FUNCTORS

Let $\mathscr{K}$ be a category and F any functor from $\mathscr{K}$ to St. Thus F associates with each $a \in \text{Ob } \mathscr{K}$ a set $F(a)$ and with each $\mathscr{K}$-morphism $\alpha: a \to b$ a mapping $F(\alpha): F(a) \to F(b)$. Given $\xi \in F(a)$, we shall write $\xi\alpha$ instead of $\xi F(\alpha)$; then the fact that F is a functor is expressed by the equations

$$(\xi\alpha)\beta = \xi(\alpha\beta), \quad \xi\varepsilon_a = \xi \qquad (\xi \in F(a),\ \alpha, \beta \in \text{Hom}\,\mathscr{K}),$$

whenever both sides are defined.

If there exist a $\mathscr{K}$-object u and an element ρ of $F(u)$ with the property that to each $\xi \in F(a)$ there corresponds precisely one $\xi' \in \mathrm{Hom}(u,a)$ such that

$$\xi = \rho\xi',$$

then u is said to be a *universal $\mathscr{K}$-object*, with *universal morphism* ρ, for the functor F. Thus the class $\bigcup F(a)$ is generated by ρ, under right multiplication by $\mathscr{K}$-morphisms; this is often referred to as the *universal property* of the pair (u,ρ). Of course a universal object need not exist, but when it does exist, it is essentially unique:

Proposition 1.1

Let F be a functor from a category $\mathscr{K}$ to St *and suppose that (u_1,ρ_1) and (u_2,ρ_2) are both universal for this functor. Then u_1 and u_2 are equivalent, and in fact there exists a unique equivalence $\theta: u_1 \to u_2$ such that $\rho_1\theta = \rho_2$.*

Proof:

By the universal property of u_1, there exists a unique morphism $\theta: u_1 \to u_2$ such that

(1) $$\rho_1\theta = \rho_2,$$

and by the universal property of u_2, there exists a morphism $\phi: u_2 \to u_1$ such that $\rho_2\phi = \rho_1$. Hence $\rho_1\theta\phi = \rho_1 = \rho_1\varepsilon_{u_1}$; by uniqueness we find that

$$\theta\phi = \varepsilon_{u_1},$$

and similarly,

$$\phi\theta = \varepsilon_{u_2}.$$

This shows θ to be a $\mathscr{K}$-equivalence. Since it is the only morphism satisfying (1), a fortiori it is the only equivalence. ∎

If $\mathscr{K}$ and $\mathscr{L}$ are any categories we shall say that $\mathscr{L}$ is *represented* in $\mathscr{K}$ if there is a covariant functor F from $\mathscr{L}^\circ \times \mathscr{K}$ to St. Thus to any pair $A \in \mathrm{Ob}\ \mathscr{L}$, $a \in \mathrm{Ob}\ \mathscr{K}$, there corresponds a set $F(A,a)$ and the correspondence $(A,a) \to F(A,a)$ is contravariant in A and covariant in a. Given any morphisms $\phi: B \to A$ and $\alpha: a \to b$ in $\mathscr{L}$ and $\mathscr{K}$ respectively, there is a mapping $F(\phi,\alpha): F(A,a) \to F(B,b)$. We denote the effect of this mapping on $\xi \in F(A,a)$ by $\phi\xi\alpha$. Then the functorial character of F is expressed by the equations

$$(\psi\phi)\xi = \psi(\phi\xi),\ \varepsilon_A\xi = \xi \qquad (\phi,\psi \in \mathrm{Hom}\ \mathscr{L}),$$

$$\xi(\alpha\beta) = (\xi\alpha)\beta,\ \xi\varepsilon_a = \xi \qquad (\alpha,\beta \in \mathrm{Hom}\ \mathscr{K}),$$

which hold whenever both sides are defined.

In most applications the objects of $\mathscr{L}$ and $\mathscr{K}$ will be sets with a certain structure, and $F(A,a)$ consists of certain mappings from A to a (qua sets). For this reason we generally refer to the elements of $F(A,a)$ as the *admissible morphisms* of the representation. Instead of the single universal $\mathscr{K}$-object we now have one universal $\mathscr{K}$-object for each $\mathscr{L}$-object, which will soon be shown to be a functor from $\mathscr{L}$ to $\mathscr{K}$. For example, if $\mathscr{L}$ is a subcategory of $\mathscr{K}$, $\mathscr{L}$ is represented in $\mathscr{K}$ and each $\mathscr{L}$-object is its own universal object for this representation. A less trivial example is the case where $\mathscr{K}$ is a subcategory of $\mathscr{L}$; more generally we shall say that $\mathscr{K}$ is *subordinate* to $\mathscr{L}$, in symbols $\mathscr{K} \prec \mathscr{L}$, if there is a functor ι from $\mathscr{K}$ to $\mathscr{L}$ such that for any $\mathscr{K}$-morphisms $\alpha,\alpha' : a \to b$ $(a,b \in \mathrm{Ob}\ \mathscr{K})$, $\alpha\iota = \alpha'\iota$ implies $\alpha = \alpha'$[1]. The $\mathscr{L}$-object $a\iota$ corresponding to a will be called the *$\mathscr{L}$-carrier* of a and the morphism $\alpha\iota$ the *$\mathscr{L}$-morphism* defined by α. In particular, if $\mathscr{K}$ and $\mathscr{L}$ consist of sets with some structure and mappings preserving that structure, then $\mathscr{K} \prec \mathscr{L}$ if the $\mathscr{L}$-objects have less structure than the $\mathscr{K}$-objects and the functor ι has the effect of ignoring or 'forgetting' the $\mathscr{K}$-structure, i.e. is a *forgetful functor*, in MacLane's terminology. A typical example is the functor from groups to sets, which assigns to each group its carrier and regards a homomorphism as a mapping between the carriers. This shows the category Gp of groups and homomorphisms to be subordinate to the category St of sets and mappings; of course it is not a subcategory because distinct groups may have the same carrier.

When $\mathscr{K} \prec \mathscr{L}$ and ι is the corresponding functor, we can always represent $\mathscr{L}$ in $\mathscr{K}$ by putting $F(A,a) = \mathrm{Hom}\,(A,a\iota)$, $F(\phi,\alpha) : \xi \to \phi\xi(\alpha\iota)$. We remark that in all cases of interest to us the subordinate categories are obtained by a forgetful functor. Thus e.g., the category Gp is subordinate to St; the corresponding representation is obtained by associating with each set X and each group G, all the mappings from X to the carrier of G. As we shall see later, there is a universal object for X, namely the free group on X. To take another example, the category of ordered sets and order-homomorphisms is subordinate to St, and this gives rise to a representation of sets in ordered sets, but there is no universal object in general. Hereafter we shall usually omit explicit mention of the functor ι; this is consistent with our practice of not using a special symbol for the carrier of an algebra or distinguishing between a homomorphism and the corresponding mapping of the underlying carriers.

We return now to the general case of categories $\mathscr{K}$ and $\mathscr{L}$, where $\mathscr{L}$ is represented in $\mathscr{K}$, and suppose that for each $\mathscr{L}$-object A there is a universal $\mathscr{K}$-object $U(A)$ with morphism $\rho(A) : A \to U(A)$ having the universal

[1]Such a functor $\mathscr{L}$ is said to be *faithful*.

property. Then the couple $(U(A), \rho(A))$, which by Proposition 1.1 is determined up to $\mathscr{K}$-equivalence by the representation, is called the *universal functor* for the given representation. This name is justified by

Theorem 1.2

Let $\mathscr{K}$ and $\mathscr{L}$ be given categories with a representation of $\mathscr{L}$ in $\mathscr{K}$ for which a universal functor U exists. Then U is indeed a functor from $\mathscr{L}$ to $\mathscr{K}$ and is determined up to $\mathscr{K}$-equivalence, and the associated mapping ρ is a natural transformation from I to U.

Proof:

Let $\alpha : A \to B$ be an $\mathscr{L}$-morphism; by composition we obtain a morphism $\alpha\rho : A \to U(B)$, and hence a unique $\mathscr{K}$-morphism $U(\alpha) : U(A) \to U(B)$ such that the diagram

$$\begin{array}{ccc} A & \xrightarrow{\alpha} & B \\ {\scriptstyle\rho}\downarrow & & \downarrow{\scriptstyle\rho} \\ U(A) & \xrightarrow{U(\alpha)} & U(B) \end{array}$$

commutes. This shows ρ to be natural. If $\beta : B \to C$ is another $\mathscr{L}$-morphism, we have the commutative diagram

$$\begin{array}{ccccc} A & \xrightarrow{\alpha} & B & \xrightarrow{\beta} & C \\ {\scriptstyle\rho}\downarrow & & {\scriptstyle\rho}\downarrow & & \downarrow{\scriptstyle\rho} \\ U(A) & \xrightarrow{U(\alpha)} & U(B) & \xrightarrow{U(\beta)} & U(C) \end{array}$$

By definition of $U(\alpha\beta)$, we have $\alpha\beta\rho = \rho U(\alpha\beta)$, as well as $\alpha\beta\rho = \rho U(\alpha)U(\beta)$; hence from the uniqueness we conclude that

$$U(\alpha\beta) = U(\alpha)U(\beta).$$

Further, if $\varepsilon_A : A \to A$ is the identity morphism, then

$$\varepsilon_A \rho = \rho = \rho U(\varepsilon_A),$$

whence $U(\varepsilon_A) = \varepsilon_{U(A)}$. This shows U to be a functor, and the uniqueness follows from Proposition 1.1. ∎

If $\mathscr{K}$ is subordinate to $\mathscr{L}$, we also have a representation of $\mathscr{K}$ in $\mathscr{L}$, but this trivially has a universal functor, namely

$$\eta_a : a \to a\iota;$$

going over to opposite categories, we have a representation of $\mathscr{L}^\circ$ in $\mathscr{K}^\circ$, which may or may not possess a universal functor. Generally, if $\mathscr{K}$ is represented in $\mathscr{L}$, then $\mathscr{L}^\circ$ is represented in $\mathscr{K}^\circ$, and if the universal functor for this representation exists, we obtain a covariant functor from $\mathscr{L}^\circ$ to $\mathscr{K}^\circ$, or equivalently, a covariant functor from $\mathscr{L}$ to $\mathscr{K}$. This functor is called the *couniversal functor* for the representation of $\mathscr{K}$ in $\mathscr{L}$. Clearly it is again determined up to $\mathscr{K}$-equivalence.

An important example for the representation of categories is the following. Let $\mathscr{K}$ be any category and Λ a preordered set. Then a Λ-system in $\mathscr{K}$ is a family $(a_\lambda,\alpha_{\lambda\mu})$ of $\mathscr{K}$-objects a_λ indexed by Λ, together with $\mathscr{K}$-morphisms $\alpha_{\lambda\mu}:a_\lambda \to a_\mu$ for all $\lambda,\mu \in \Lambda$ such that $\lambda \leqslant \mu$, subject to the conditions

$$(2) \qquad \alpha_{\lambda\lambda} = 1, \qquad \alpha_{\lambda\mu}\alpha_{\mu\nu} = \alpha_{\lambda\nu} \text{ whenever } \lambda \leqslant \mu \leqslant \nu.$$

We denote by $\mathscr{F}(\mathscr{K})$ the category whose objects are Λ-systems in $\mathscr{K}$, for arbitrary preordered sets Λ, and whose morphisms are defined thus: Given $\mathscr{F}(\mathscr{K})$-objects $(a_\lambda,\alpha_{\lambda\mu})_{\lambda,\mu\in\Lambda}$ and $(b_\xi,\beta_{\xi\eta})_{\xi,\eta\in \mathrm{M}}$, an $\mathscr{F}(\mathscr{K})$-morphism between these objects is defined by an order-homomorphism $\lambda \to \lambda'$ from Λ to M together with a family of $\mathscr{K}$-morphisms

$$\phi_\lambda : a_\lambda \to b_{\lambda'}$$

such that

$$\alpha_{\lambda\mu}\phi_\mu = \phi_\lambda\beta_{\lambda'\mu'} \quad (\lambda,\mu \in \Lambda).$$

If the number 1 is regarded as an ordered set consisting of a single element, then the 1-systems in $\mathscr{K}$ and their $\mathscr{F}(\mathscr{K})$-morphisms form a full subcategory of $\mathscr{F}(\mathscr{K})$ which is isomorphic to $\mathscr{K}$. We shall therefore take $\mathscr{K}$ to be embedded in $\mathscr{F}(\mathscr{K})$ by means of this isomorphism, so that henceforth $\mathscr{K}$ may be regarded as a subcategory of $\mathscr{F}(\mathscr{K})$. This gives rise to a representation of $\mathscr{F}(\mathscr{K})$ in $\mathscr{K}$, and we may enquire whether there exists a universal functor or a couniversal functor for this representation. In answering this question, it is advantageous to consider not the whole of $\mathscr{F}(\mathscr{K})$ but a subcategory (containing $\mathscr{K}$, of course) which is obtained by limiting the class of preordered sets in some way. E.g. we may admit only totally ordered sets, or only totally unordered sets, or directed sets. We give a few examples which will be used later.

(i) *The universal functor for $\mathscr{F}(\mathscr{K})$, using totally unordered sets.* In the case of totally unordered sets, there are no morphisms to make up an $\mathscr{F}(\mathscr{K})$-object, and (2) is vacuous. By definition, the universal functor associates with every family $(a_\lambda)_{\lambda\in\Lambda}$ of $\mathscr{K}$-objects a $\mathscr{K}$-object $p = \bigsqcup a_\lambda$, which is unique up to $\mathscr{K}$-equivalence, and a family of $\mathscr{K}$-morphisms,

$$\rho_\lambda : a_\lambda \to p, \tag{3}$$

such that for every family of $\mathscr{K}$-morphisms $\phi_\lambda : a_\lambda \to b$ ($b \in \text{Ob}\,\mathscr{K}$) there exists a unique $\mathscr{K}$-morphism $\phi : p \to b$ such that $\phi_\lambda = \rho_\lambda \phi$. The $\mathscr{K}$-object $\bigsqcup a_\lambda$ is called the *free composition* of the family (a_λ), with the *canonical morphisms* (3). Such a free composition exists in particular for the category St; it associates with any family of sets (A_λ) a set $P = \bigsqcup A_\lambda$, which is the union of pairwise disjoint sets A'_λ such that A'_λ is equipotent with A_λ. In this case, $\bigsqcup A_\lambda$ is also called the *disjoint sum* of the A_λ. In III.6 we shall show that the free composition exists in (Ω), and we shall consider other categories of Ω-algebras with this property.

(ii) *The couniversal functor for* $\mathscr{F}(\mathscr{K})$, *using totally unordered sets.* This functor associates with every family $(a_\lambda)_{\lambda\in\Lambda}$ of $\mathscr{K}$-objects a $\mathscr{K}$-object $c = \prod a_\lambda$, unique up to $\mathscr{K}$-equivalence, and a family of $\mathscr{K}$-morphisms

$$\sigma_\lambda : c \to a_\lambda \tag{4}$$

such that for every family of morphisms $\phi_\lambda : b \to a_\lambda$ ($b \in \text{Ob}\,\mathscr{K}$), there is a unique $\mathscr{K}$-morphism $\phi : b \to c$ such that $\phi_\lambda = \phi\sigma_\lambda$. The $\mathscr{K}$-object $\prod a_\lambda$ is called the *direct composition* of the family (a_λ), with the *canonical morphisms* (4). E.g., in the case of St the direct composition is just the Cartesian product; for (Ω) it is the direct product, and we shall see in III.6 that the direct composition also exists in certain subcategories of (Ω), even when these categories do not admit direct products.

(iii) *The universal functor for* $\mathscr{F}(\mathscr{K})$, *using directed sets.* With each Λ-system $(a_\lambda, \alpha_{\lambda\mu})$ in $\mathscr{K}$ the universal functor associates a $\mathscr{K}$-object d and a family $\rho_\lambda : a_\lambda \to d$ satisfying

$$\alpha_{\lambda\mu}\rho_\mu = \rho_\lambda \qquad (\lambda \leqslant \mu),$$

and such that, for every family of morphisms $\phi_\lambda : a_\lambda \to b$ satisfying $\alpha_{\lambda\mu}\phi_\mu = \phi_\lambda$, there exists a morphism $\phi : d \to b$ such that $\phi_\lambda = \rho_\lambda\phi$. The object d is called the *direct limit* of the given system and is written

$$d = \varinjlim\, (a_\lambda, \alpha_{\lambda\mu}).$$

A Λ-system in $\mathscr{K}$, where Λ is directed, is also called a *directed system.*

(iv) *The couniversal functor for* $\mathscr{F}(\mathscr{K})$, *using sets directed downwards.*

This functor associates with each Λ-system $(a_\lambda, \alpha_{\lambda\mu})$ a $\mathscr{K}$-object c and a family $\sigma_\lambda : c \to a_\lambda$ satisfying

$$\sigma_\lambda \alpha_{\lambda\mu} = \sigma_\mu \qquad (\lambda \leqslant \mu),$$

and such that, for every family $\phi_\lambda : b \to a_\lambda$ satisfying $\phi_\lambda \alpha_{\lambda\mu} = \phi_\mu$, there exists a morphism $\phi : b \to c$ such that $\phi_\lambda = \phi \sigma_\lambda$. The object c is called the *inverse limit* of the system given and is written

$$c = \varprojlim (a_\lambda, \alpha_{\lambda\mu}).$$

It is not hard to prove that direct and inverse limits exist for the category (Ω); more precisely, direct limits exist in any local category of Ω-algebras admitting homomorphic images, and inverse limits exist in any hereditary residual category of Ω-algebras. We shall prove only the first of these statements, as this is all we require.

Proposition 1.3

Any local category of Ω-algebras admitting homomorphic images admits direct limits.

The proof is similar to that of Proposition II.7.4. Let $\mathscr{K}$ be a local category of Ω-algebras admitting homomorphic images; given any Λ-system $(A_\lambda, \alpha_{\lambda\mu})$ of $\mathscr{K}$-algebras and homomorphisms, where Λ is directed, we form $P = \prod A_\lambda$, where the product is taken over all algebras A_λ with nonempty carrier, and consider the set T of *threads* in P, i.e. elements $x = (x_\lambda)$ such that for some $\lambda_0 \in \Lambda$ (depending on x) we have

$$x_\lambda \alpha_{\lambda\mu} = x_\mu \qquad \text{for all } \mu \geqslant \lambda \geqslant \lambda_0.$$

Two threads x and y are said to be *equivalent* if there is a $\lambda_1 \in \Lambda$ such that $x_\lambda = y_\lambda$ for all $\lambda \geqslant \lambda_1$. This relation is clearly reflexive and symmetric, and because Λ is directed, it is also transitive, so that we have an equivalence relation, $\mathfrak{q}$ say, on T. As in the proof of Proposition II.7.4 we see that T is a subalgebra of P and that $\mathfrak{q}$ is a congruence on T. Putting $D = T/\mathfrak{q}$, we have, for each $\lambda \in \Lambda$, a homomorphism

$$\rho_\lambda : A_\lambda \to D \tag{5}$$

defined as follows: Given $a \in A_\lambda$, consider the thread x given by

$$x_\mu = \begin{cases} a\alpha_{\lambda\mu} & \text{if } \mu \geqslant \lambda \\ \text{any element of } A_\mu & \text{otherwise.} \end{cases}$$

Clearly, the $\mathfrak{q}$-class containing x depends only on a and not on the choice of the coordinates x_μ with $\mu \not\geqslant \lambda$. We may therefore write $x^{\mathfrak{q}} = a\rho_\lambda$, and note that for any $\mu \geqslant \lambda$, the element $a\alpha_{\lambda\mu}$ determines the same $\mathfrak{q}$-class as a, i.e.

$$\alpha_{\lambda\mu}\rho_\mu = \rho_\lambda \qquad \text{for } \lambda \leqslant \mu.$$

We assert that D is the direct limit, with canonical homomorphisms ρ_λ. To establish the universal property, let $\phi_\lambda : A_\lambda \to B$ be a family of homomorphisms satisfying $\alpha_{\lambda\mu}\phi_\mu = \phi_\lambda$. This means in effect that the coordinates of any thread x are mapped by the ϕ_λ to a family (b_λ) of elements of B such that $b_\lambda = b_\mu$ for all $\lambda,\mu \geqslant \lambda_0$ (for some $\lambda_0 \in \Lambda$); moreover, this constant value b_λ is the same for equivalent threads, and we therefore obtain a mapping $\phi : D \to B$, which is easily seen to satisfy

(6) $$\phi_\lambda = \rho_\lambda \phi.$$

Now the images of the canonical mappings $\rho_\lambda : A_\lambda \to D$ form a local system of $\mathscr{K}$-algebras for D, hence D is itself a $\mathscr{K}$-algebra and each ρ_λ is a $\mathscr{K}$-homomorphism; further, since the ϕ_λ are $\mathscr{K}$-homomorphisms, it follows by (6) and the regularity of $\mathscr{K}$ that ϕ is a $\mathscr{K}$-homomorphism. ∎

EXERCISES

1. Verify that the embedding of $\mathscr{K}$ in $\mathscr{F}(\mathscr{K})$ defines in fact a representation of $\mathscr{F}(\mathscr{K})$ in $\mathscr{K}$ and show that for the case $\mathscr{K} = \mathrm{St}$ this representation has a universal functor and a couniversal functor.

2. Let $\mathscr{K}, \mathscr{L}, \mathscr{M}$ be any categories such that $\mathscr{K}$ is subordinate to $\mathscr{L}$ and $\mathscr{L}$ subordinate to $\mathscr{M}$; verify that $\mathscr{K}$ is subordinate to $\mathscr{M}$. If, further, the respresentations of $\mathscr{M}$ in $\mathscr{L}$ and of $\mathscr{L}$ in $\mathscr{K}$ have universal functors U, V respectively, show that the representation of $\mathscr{M}$ in $\mathscr{K}$ has the universal functor $A \to V(U(A))$. Conversely, if $\mathscr{M}$ has a universal functor in $\mathscr{K}$, then so does $\mathscr{L}$, but $\mathscr{M}$ may not have a universal functor in $\mathscr{L}$. (For the last part, take $\mathscr{M} = \mathscr{F}(\mathscr{K})$, with totally unordered index sets, and $\mathscr{L} = \mathscr{F}_0(\mathscr{K})$ the subcategory of finite families of $\mathscr{K}$-objects.)

3. Let $\mathscr{K}$ be subordinate to $\mathscr{L}$ and assume that the representation of $\mathscr{L}$ in $\mathscr{K}$ has a universal functor U. If, further, $\mathscr{L}$ admits free composition and $a = \bigsqcup a_\lambda$, where (a_λ) is a family of $\mathscr{L}$-objects, then $U(a)$ is the free $\mathscr{K}$-composition of the family $(U(a_\lambda))$.

4. Prove that any residual category of Ω-algebras admitting subalgebras admits inverse limits.

5. If A is the direct limit of a directed system of Ω-algebras, show that A can be defined as an Ω-algebra such that the canonical mappings are homomorphisms. If the algebras of the system belong to a category $\mathcal{K}$ admitting homomorphic images, show that A is locally $\mathcal{K}$.

2. Ω-WORD ALGEBRAS

We shall now apply the results of the preceding section to the category (Ω). Clearly, (Ω) is subordinate to St, by the functor which associates with each algebra its carrier. Thus we have a representation of St in (Ω), and our object is to show that this representation has a universal functor. Such a functor associates with any set X an Ω-algebra which we shall call the Ω-*word algebra on* X and denote by $W_\Omega(X)$. It is of some interest to have a constructive existence proof of this algebra; we shall therefore begin with the construction of $W_\Omega(X)$ and verify its universal property later on.

Let Ω be any operator domain and X any set, and define an Ω-algebra $W(\Omega;X)$, the *algebra of* Ω-*rows* in X, as follows: by an Ω-*row in* X we understand a finite sequence (i.e. an n-tuple for $n \geqslant 1$) of elements of the disjoint sum $\Omega \sqcup X$. On the set $W(\Omega;X)$ of all Ω-rows in X we define an Ω-algebra structure by juxtaposition; thus, if $\omega \in \Omega(n)$ and $a_i \in W(\Omega;X)$ $(i = 1, \cdots, n)$, say

$$a_i = (a_{i1}, \cdots, a_{ik_i}) \qquad (a_{ij} \in \Omega \sqcup X),$$

then

$$a_1 \cdots a_n\omega = (a_{11}, \cdots, a_{1k_1}, a_{21}, \cdots, a_{nk_n}, \omega). \tag{1}$$

When X is disjoint from Ω, we may replace the disjoint sum by the ordinary union, and if we identify Ω-rows consisting of a single term with the corresponding element of $\Omega \cup X$, we can regard Ω and X as subsets of $W(\Omega;X)$; then both sides of (1) may be denoted by

$$a_{11} \cdots a_{1k_1} a_{21} \cdots a_{nk_n}\omega. \tag{2}$$

For simplicity of notation we often assume that X is disjoint from Ω; this is no loss of generality, as will soon become clear.

Definition

The subalgebra of $W(\Omega;X)$ generated by X is called the *Ω-word algebra on X* and is denoted by $W_\Omega(X)$. Its elements are called *Ω-words in X*, and X is called its *alphabet*.

The first point to notice is that $W_\Omega(X)$ is determined essentially by the cardinal of the set X.

Proposition 2.1

If X, Y are any sets, then the Ω-word algebras on X and Y are isomorphic:

$$W_\Omega(X) \cong W_\Omega(Y),$$

if and only if X is equipotent to Y.

Proof:

If $\theta: X \to Y$ is a bijection, then we obtain an isomorphism between $W(\Omega;X)$ and $W(\Omega;Y)$ by replacing, in each Ω-row on X, each $x \in X$ by $x\theta$. Restricting this isomorphism to $W_\Omega(X)$, we get a mapping whose image is $W_\Omega(Y)$, it is clear that this is in fact an isomorphism between $W_\Omega(X)$ and $W_\Omega(Y)$.

Conversely, assume that

$$W_\Omega(X) \cong W_\Omega(Y). \tag{3}$$

On any Ω-algebra A we may define a congruence $\mathfrak{q}$ by putting $a \equiv b (\text{mod } \mathfrak{q})$ if and only if $a = b$ or $a = x\omega$, $b = y\bar{\omega}$ for some $x \in A^n$, $y \in A^m$, $\omega \in \Omega(n)$, $\bar{\omega} \in \Omega(m)$. The congruence properties are easily verified; we denote by A^0 the quotient algebra $A/\mathfrak{q}$ so defined. Then for any Ω-word algebra $W = W_\Omega(X)$, the algebra W^0 consists of the singletons of elements of X and a further class which contains all the other elements. Thus $|W_\Omega(X)^0| = |X| + 1$. If (3) holds, we therefore have

$$|X| + 1 = |Y| + 1,$$

whence $|X| = |Y|$. ∎

We remark that the isomorphism between $W_\Omega(X)$ and $W_\Omega(Y)$ obtained in the first part of the proof is uniquely determined by θ. This follows from a quite general lemma:

Lemma 2.2

Let A be an Ω-algebra and X a generating set of A. Then any homomorphism of A into another Ω-algebra is completely determined by its restriction to X.

For if θ_1 and θ_2 are two homomorphisms from A to B which agree on X,

(4) $$x\theta_1 = x\theta_2$$

for all $x \in X$, let A' be the subset of all elements $x \in A$ for which (4) holds; then A' is a subalgebra of A, for, if $a_i\theta_1 = a_i\theta_2$ $(i = 1,\cdots,n)$ and $\omega \in \Omega(n)$, then

$$\begin{aligned}(a_1\cdots a_n\omega)\theta_1 &= (a_1\theta_1)\cdots(a_n\theta_1)\omega \\ &= (a_1\theta_2)\cdots(a_n\theta_2)\omega = (a_1\cdots a_n\omega)\theta_2.\end{aligned}$$

Since A' contains the generating set X of A by hypothesis, it follows that $A' = A$, i.e., (4) holds throughout A. ∎

To obtain a more explicit description of Ω-words we introduce the notions of length and valency. Given any Ω-row in X,

$$w = c_1\cdots c_N \qquad (c_i \in \Omega \sqcup X),$$

we define the *length* of w as the integer N and denote it by $l(w)$. The *valency* of w, denoted by $v(w)$, is defined as

$$v(w) = \sum v(c_i),$$

where

$$v(c) = \begin{cases} 1 & \text{if } c \in X, \\ -n+1 & \text{if } c \in \Omega(n). \end{cases}$$

Thus the elements of X have the same valency as constant operators. With these definitions we have the following criterion for an Ω-row to be an Ω-word (cf. P. Hall [58]; for a history of the theorem see also Rosenbloom [50]).

Theorem 2.3

An Ω-row $w = c_1\cdots c_N$ in X is an Ω-word if and only if for every left segment $w_i = c_1\cdots c_i$ of w,

(5) $$v(w_i) > 0 \qquad (i = 1,\cdots,N),$$

and further,

(6) $$v(w) = 1.$$

We remark that intuitively this is obvious, for $v(c)$ essentially represents the 'element-balance', for $c \in \Omega \sqcup X$. Thus, if $c \in \Omega(n)$, then c requires an input of n elements and has an output of one, so that

$$v(c) = \text{output} - \text{input}.$$

Now (5) ensures that at any given stage there are enough elements to operate on, and (6) states that the final output is one element.

To prove the theorem, we show more generally, by induction on the length $l(w)$, that w is a sequence of r Ω-words if and only if (5) holds and

$$(6') \qquad v(w) = r.$$

This includes the assertion of the theorem for $r = 1$. The result clearly holds for words of length 1. Now let w be an Ω-word, say

$$w = a_1 \cdots a_n \omega \qquad (a_i \in W_\Omega(X),\ \omega \in \Omega(n)).$$

By the induction hypothesis, we have $v(a_i) = 1$ and $v(\omega) = 1 - n$; hence

$$v(w) = n + 1 - n = 1.$$

Moreover, every left segment of each a_i has positive valency, so the same holds for w. Now, if w is a sequence of r words, then (5) again holds and (6′) follows by addition.

Conversely, let w be an Ω-row satisfying (5) and (6′). If $l(w) = 1$, then $v(w) = 1$ and $w \in X \sqcup \Omega(0)$, so w is then an Ω-word. If $l(w) > 1$, let $w = w'c$, where $c \in \Omega \sqcup X$ and $v(w') = r' > 0$ by (5). By the induction hypothesis w' is then a sequence of r' Ω-words. Further, we have

$$v(c) = v(w) - v(w') = r - r'.$$

Now either $r - r' = 1$ (then c is an Ω-word and hence w is a sequence of $r' + 1 = r$ words), or $r - r' = 1 - s \leqslant 0$, in which case c is an s-ary operator. Now

$$s = r' - (r - 1) \leqslant r',$$

therefore we can form from c and the s preceding words just one new Ω-word and obtain a sequence of $r' - s + 1 = r$ Ω-words, as asserted. ∎

From the proof of this theorem we obtain

Corollary 2.4

An Ω-row satisfying (5) *and* (6′) *can be written as a sequence of r Ω-words*

$$w = w_1 w_2 \cdots w_r$$

in exactly one way. ∎

Corollary 2.5

If $w = c_1 \cdots c_N$ is any Ω-word of the form

$$(7) \qquad w = a_1 \cdots a_n \omega \qquad (a_i \in W_\Omega(X)),$$

then any proper subsequence $w' = c_i c_{i+1} \cdots c_j$ $(j - i < N - 1)$, *which is itself an* Ω*-word, occurs within a single factor* a_k *of the expression* (7).

For, if not, let $w' = uv$, where u is a right segment of a_h. By the valency condition for w', $v(u) > 0$, while the valency condition for a_h shows that $v(u) \leqslant 0$ unless $u = a_h$. Thus w' is of the form $w' = a_h a_{h+1} \cdots a_k w''$, where $v(w'') = -(k - h) \leqslant 0$. This is impossible if w'' is a left segment of a_{k+1}, hence $k = n$ and $w'' = \omega$; but this is possible only if $h = 1$ and $w' = w$, which contradicts the hypothesis. ∎

Let A be any Ω-algebra; from Corollary 2.4 it is easy to see that if in an Ω-word w in X we replace each element of X by an element of A, then we obtain a uniquely determined element of A. This is clear if $l(w) = 1$, so let us assume that $l(w) > 1$ and use induction. We have

$$w = w_1 \cdots w_n \omega,$$

where $\omega \in \Omega(n)$ and the w_i are uniquely determined, by Corollary 2.4, and are Ω-words of shorter length than w. When we replace the elements of X by elements of A, each w_i becomes an element of A, by induction, and hence w does too. Moreover, the element of A obtained in this way is unique. This remark will now be used to establish the universal property for Ω-word algebras.

Theorem 2.6

Let A *be an* Ω*-algebra and* X *an arbitrary set. Then any mapping* $\theta : X \to A$ *extends in just one way to a homomorphism* $\bar{\theta} : W_\Omega(X) \to A$.

Proof:

We define a mapping $\bar{\theta} : W_\Omega(X) \to A$ to extend θ as follows: Every Ω-word w is unique of the form

$$w = c_1 \cdots c_N \qquad (c_i \in \Omega \sqcup X);$$

we write

$$w\bar{\theta} = c'_1 \cdots c'_N,$$

where

$$c' = \begin{cases} c & \text{if } c \in \Omega, \\ c\theta & \text{if } c \in X. \end{cases}$$

Thus $w\bar{\theta}$ is just the unique element of A obtained by replacing $x \in X$ by $x\theta$. The remark preceding the theorem shows that $w\bar{\theta}$ is well-defined, and it is easily verified that $\bar{\theta}$ is a homomorphism, which is unique by Lemma 2.2. ∎

If w is any Ω-word in X, and $x_1,\cdots,x_n$ are the actual elements of X occurring in w, then w may be regarded as a function of $x_1,\cdots,x_n$, and we shall often indicate its dependence on $x_1,\cdots,x_n$ by writing

$$w = w(x_1,\cdots,x_n).$$

If $\theta: X \to A$ is any mapping into an Ω-algebra A and $x_i\theta = a_i$, then the image of w under the induced homomorphism $\bar{\theta}$ is naturally denoted by

$$w\bar{\theta} = w(a_1,\cdots,a_n).$$

The assertion of Theorem 2.6, that $w(a_1,\cdots,a_n)$ is uniquely determined by $w(x_1,\cdots,x_n)$ and the mapping $\theta: x_i \to a_i$, depended essentially on the fact that the operators are always written on one side of the row of elements on which they act. Thus e.g. if '$+$' is a binary operator and the result of operating on (a,b) is denoted by '$a + b$', it is necessary to distinguish between

$$a + (b + c) \quad \text{and} \quad (a + b) + c \tag{8}$$

by placing parentheses (as we have done) or by some other means. However, if in accordance with the notations used in Chapter II we write '$ab+$' for the result of operating by '$+$', then the two expressions (8) become

$$abc++ \quad \text{and} \quad ab+c+.$$

Now parentheses are no longer necessary; this observation is due to Łukasiewicz. In particular cases we shall usually keep to the accepted notation for operators.

An important consequence of Theorem 2.6 is the following result, asserting the existence of presentations of Ω-algebras (cf. III.8 below).

Theorem 2.7

Any Ω-algebra A can be expressed as a homomorphic image of an Ω-word algebra $W_\Omega(X)$, for a suitable set X.

For if X is any generating set of A, then the identity mapping on X may be extended to a homomorphism $\phi: W_\Omega(X) \to A$. The image under this homomorphism is a subalgebra of A containing X, and hence must be A itself. Thus A is a homomorphic image of $W_\Omega(X)$. ∎

Informally, the theorem may be taken to state that if A is an Ω-algebra generated by a set X, then any element $a \in A$ can be expressed as an Ω-word in X, possibly in different ways. The minimum of the lengths of

Ω-words representing a will be called the *length* of a relative to X and denoted by $l_X(a)$, or more briefly by $l(a)$. It is clear that $l(a) = 1$ if and only if $a \in X \cup \Omega(0)$; if $l(a) > 1$, then by taking an Ω-word of length $l(a)$ representing a, we obtain a representation

(9) $$a = b_1 \cdots b_n \omega, \text{ where } \omega \in \Omega \text{ and } l(a) = \sum l(b_i) + 1.$$

Every element of A has finite length, relative to a given generating set of A, and this makes it possible to prove results about A by induction on the length. As a case in point we shall prove here a theorem due to Higman [52], on ordered algebras, which is needed later (in Chapter VII). More precisely, the result concerns preordered algebras, and we begin by generalizing the notion of a partial well-ordering (I.4) to preordered sets.

A preordered set A is said to be *partly well-ordered* if every strictly descending sequence[1]

$$a_1 > a_2 > \cdots$$

terminates and every subset of pairwise incomparable elements in A is finite. For ordered sets, this reduces to the previous definition; more generally, if A is preordered and $A/\mathfrak{q}$ is the associated ordered set (cf. Exercise I.3.6), then A is partly well-ordered if and only if the ordered set $A/\mathfrak{q}$ is partly well-ordered. Other ways of expressing this condition are given in

Lemma 2.8

For any preordered set A the following three conditions are equivalent:

(i) *Every infinite sequence in A contains an ascending subsequence, i.e., given (a_i) in A, there is an infinite sequence (n') of integers such that $m' < n'$ implies $a_{m'} \leqslant a_{n'}$.*
(ii) *For every infinite sequence (a_i) in A there exists a pair of integers m,n such that $m < n$ and $a_m \leqslant a_n$.*
(iii) *A is partly well-ordered.*

Proof:

Evidently (i) implies (ii), and (ii) implies (iii), from the definition of a partly well-ordered set; so it remains to show that (iii) implies (i). Thus assume that A is partly well-ordered and take an infinite sequence (a_i) in A. For the moment let us call an element a_i *strictly minimal* if $a_n < a_i$ for

[1] Recall that '$a > b$' means '$a \geqslant b$ but not $b \geqslant a$'.

no n. Then the number of strictly minimal elements in the sequence must be finite, and we can therefore choose one, say, $a_{1'}$ such that

(10) $$a_{1'} \leqslant a_n \qquad \text{for infinitely many } n.$$

Omitting all terms a_n which do not satisfy (10), we obtain an infinite sequence, (b_j) say, such that $b_1 = a_{1'} \leqslant b_n$ for all n. Repeating the argument with the sequence $(b_j)(j \geqslant 2)$, we obtain by induction on n an infinite ascending subsequence of (a_i), which shows that (i) holds. ∎

If A and B are any preordered sets, then their Cartesian product $A \times B$ is preordered by the rule

$$(a,b) \leqslant (a',b') \qquad \text{if and only if } a \leqslant a' \text{ and } b \leqslant b'.$$

The Cartesian product of two partly well-ordered sets is again partly well-ordered. For any infinite sequence of elements of the product contains an infinite subsequence in which the first factors are in ascending order, and this contains an infinite subsequence in which the second factors are in ascending order. A corresponding definition and proof show that the product of any finite number of partly well-ordered sets is again partly well-ordered.

We now come to Ω-algebras; an Ω-algebra A is said to have a *divisibility preordering* if its carrier is preordered in such a way that for any $\omega \in \Omega$,

(a) $a_1 \cdots a_n \omega \leqslant b_1 \cdots b_n \omega$ whenever $a_i \leqslant b_i$ $(i = 1, \cdots, n)$,
(b) $a \leqslant b_1 \cdots b_n \omega$ whenever $a \leqslant b_i$ for some i.

Now Higman's theorem may be stated as follows:

Theorem 2.9

Let A be an Ω-algebra with a divisibility preordering, where Ω is finite. If A is generated by a set X which is partly well-ordered, then A is itself partly well-ordered.

Proof:

If A is not partly well-ordered then there exists a sequence (a_i) in A such that

(11) $$a_i \nleqslant a_j \qquad \text{for all } i,j \text{ such that } i < j.$$

Let us call a sequence (a_i) *nonascending* if it satisfies (11). Among all the nonascending sequences (a_i) in A we choose one in which a_1 is of minimal length. Among the nonascending sequences beginning with this a_1 we choose one with a_2 of minimal length, and so on. Then the infinite sequence $a_1,a_2,\cdots$ so constructed is itself nonascending. Now, for each $i = 1,2,\cdots$, either $l(a_i) = 1$, or

$$(12) \qquad a_i = b_{i1}\cdots b_{in_i}\omega_i, \qquad b_{ik} \in A, \qquad \sum l(b_{ik}) + 1 = l(a_i).$$

We note that when $l(a_i) = 1$, then $a_i \in X \cup \Omega(0)$, and since $\Omega(0)$ is finite, the set $X \cup \Omega(0)$ is partly well-ordered. It follows that there can be at most finitely many a_i of length 1, since they form a nonascending sequence, by (11). Thus the a_i in (12) still form an infinite sequence.

We assert that the set, B say, of all elements b_{ik} occurring in (12) is partly well-ordered. For if this were not so, then we could find a nonascending sequence $(b_{i'i^*})$ $(i = 1,2,\cdots)$ in B, i.e.

$$(13) \qquad b_{i'i^*} \nleqslant b_{j'j^*} \qquad \text{for all } i,j \text{ such that } i < j.$$

Let i_0 be the least value of i' occurring in the sequence; then by omitting a finite number of terms from the sequence, we may assume that $i_0 = 1'$. Now consider the sequence

$$(14) \qquad a_1,\cdots,a_{1'-1}, \quad b_{1'1^*}, b_{2'2^*},\cdots.$$

If $a_h \leqslant b_{i'i^*}$ for some $h < 1'$ and some i, then $a_h \leqslant a_{i'}$, and this contradicts (11). Thus $a_h \nleqslant b_{i'i^*}$, and together with (11) and (13) this shows that (14) is nonascending. But by (12), $l(b_{1'1^*}) < l(a_{1'})$, and this contradicts the choice of $a_{1'}$. Thus, B contains no nonascending sequence, i.e., B is partly well-ordered. Now Ω is finite, so some $\omega \in \Omega$ occurs infinitely often in (12); hence by going over to a subsequence of (a_i) we may assume that $\omega = \omega_i$ for all i. Let $a(\omega) = n$; then the Cartesian power B^n is again partly well-ordered; hence the sequence $a_i = b_{i1}\cdots b_{in}\omega$ contains an ascending subsequence; but this contradicts (11), and the result follows. ∎

The theorem may also be stated in terms of graphs, and in fact the above proof is based on the proof of a more general graph-theoretical result given by Nash-Williams [63]; see also Kruskal [60] and Exercise 10 below.

EXERCISES

1. If every operator is written on one side of the row of elements on which it acts, and for each n all the n-ary operators are written on the same side (i.e.

either on the right or on the left), which may vary for different n, are parentheses needed?

2. (P. Hall.) Show that the number of Ω-words of length k in which k_n occurrences are n-ary operators (regarding elements of X as 0-ary operators) is

$$\frac{(k-1)!}{k_0!k_1!\cdots}.$$

3. Show that an Ω-algebra A is an Ω-word algebra if there is a generating set X such that for any ω, $\bar{\omega} \in \Omega$ and any $a_i, b_j \in A$, (i) $a_1 \cdots a_n\omega \notin X$ and (ii) $a_1 \cdots a_n\omega = b_1 \cdots b_m\bar{\omega}$ implies that $m = n$, $\omega = \bar{\omega}$, and $a_i = b_i (i = 1, \cdots, n)$.

4. Show that A is locally an Ω-word algebra if and only if it satisfies (ii) of Exercise 3. Give an example of an Ω-algebra which is locally an Ω-word algebra without being itself an Ω-word algebra. (Take Ω to consist of a single unary operator.)

5. Show that any subalgebra of $W_\Omega(X)$ is of the form $W_\Omega(Y)$, for some Y. (Take Y to be a minimal generating set of the subalgebra and apply Exercise 3).

6. Show that the automorphism group of the Ω-word algebra on X is isomorphic to the symmetric group on X.

7. Let A and B be Ω-algebras on the same set X as the generating set and denote by H the subalgebra of $A \times B$ generated by the elements (x,x) $(x \in X)$. If ε is the projection $A \times B \to A$, restricted to H, show that ε is always surjective, and that it is an isomorphism for all choices of B (generated by X) if and only if A is the Ω-word algebra on X.

8. (P. Erdös.) Show that if a set X of positive integers is such that any infinite subset contains two integers one of which divides the other, then the same holds for the set of all products of elements of X.

9. If X is any preordered set, show that the Ω-word algebra $W_\Omega(X)$ may be given a divisibility preordering which induces the given preordering on X.

10. Let Ω be a partly weli-ordered operator domain; then the Ω-algebra A is said to have a *divisibility preordering* if it is preordered in such a way that for any ω, $\bar{\omega} \in \Omega$,

(a′) $a_1 \cdots a_r\omega \leqslant b_1 \cdots b_s\bar{\omega}$ whenever $a_i \leqslant b_{i'}$ for some suffixes $1' < 2' < \cdots < r'$ in the range $1, \cdots, s$ and $\omega \leqslant \bar{\omega}$,

(b) $a \leqslant b_1 \cdots b_r\omega$ whenever $a \leqslant b_i$ for some i.

Verify that this reduces to the definition given in the text when Ω is taken to be finite and totally unordered. Show that with the new definition, Theorem 2.9

holds for any partly well-ordered operator domain. (Show first that the finite sequences of elements of B, defined as in the proof of Theorem 2.9, are partly well-ordered, by regarding them as elements of the free semigroup on B and applying Theorem 2.9; now use this semigroup in place of B^n in the last part of the proof.)

3. CLONES OF OPERATIONS

Let A be a nonempty set and denote by $\mathcal{O}(A)$ the set of all finitary operations on A. As we have seen, an Ω-algebra structure on A is specified by a certain subset of $\mathcal{O}(A)$; from the operations in this subset, we can form others by composition, and the decisive role is played not by the set of operations on A defined by Ω, but by the set of all operations obtainable from them by composition. The way in which the operations on A are combined may be regarded as the effect of certain operators acting on $\mathcal{O}(A)$, and they provide $\mathcal{O}(A)$ with a certain algebraic structure, which we shall now describe more closely.

We remark that $\mathcal{O}(A)$ has the form of a disjoint sum

$$\mathcal{O}(A) = \coprod \mathcal{O}_n(A),$$

where $\mathcal{O}_n(A)$ is the set of all n-ary operations on A. Given any m elements $\alpha_1,\cdots,\alpha_m \in \mathcal{O}_n$ and $\beta \in \mathcal{O}_m$, there exists a unique n-ary operation γ defined by

$$(1) \qquad c\gamma = (c\alpha_1)\cdots(c\alpha_m)\beta \qquad \text{for all } c \in A^n.$$

This operation γ will be denoted by $\alpha_1\cdots\alpha_m\beta$ and called the *composition of* $\alpha_1,\cdots,\alpha_m$ *with* β, so that we have

$$(2) \qquad c(\alpha_1\cdots\alpha_m\beta) = (c\alpha_1)\cdots(c\alpha_m)\beta.$$

Further, for each $n > 0$, there are n elements $\delta_n^{(i)} \in \mathcal{O}_n$ defined by

$$(3) \qquad c\delta_n^{(i)} = c_i \qquad \text{where } c = (c_1,\cdots,c_n) \in A^n.$$

Thus the effect of $\delta_n^{(i)}$ is to pick out the ith coordinate; $\delta_n^{(i)}$ is called a *unit operator*. Now $\mathcal{O}(A)$ may be regarded as a partial algebra with composition as $(m+1)$-ary operation (at least in the instance (1)) and the unit operators as 0-ary operations, whose values (in the instance (3)) are n-ary operations on A. For each type of composition it is necessary to specify its domain of definition; in fact, whether the composition of $\alpha_1,\cdots,\alpha_m$ with β is defined depends only on the arities of $\alpha_1,\cdots,\alpha_m,\beta$. Any set of operations on A admitting the operations (1) and (3), i.e. closed under composition and containing the unit operators, is called a *closed set of operations on* A, or more briefly, a *clone* on A (P. Hall). In particular, $\mathcal{O}(A)$ is a clone, and a general clone on A is also referred to as a *subclone* of $\mathcal{O}(A)$. If $\mathcal{F}$, $\mathcal{G}$ are any clones, a *clone-homomorphism* $\mathcal{F} \rightarrow \mathcal{G}$ is a mapping which preserves the arity of each element and is compatible with the clone operations. Thus clones form an example of graded algebras in the sense of Higgins [63].

Let A be an Ω-algebra; then the action of Ω on A defines certain operations on A; the clone generated by these operators is called the *clone of action* of Ω on A. Denoting this clone by $\mathcal{F}$, we see that the elements of $\mathcal{F}$ are precisely those operations on A which can be obtained by repeated composition from Ω and the unit operators. It follows that the Ω-subalgebras of A are just those subsets of A which admit all the operations of $\mathcal{F}$. Generally, if X is any subset of A, then the set $X\mathcal{F}$ consisting of all the values of operations in $\mathcal{F}$ as the arguments range over X, is a subset admitting $\mathcal{F}$, and hence a subalgebra of A. It is in fact the subalgebra generated by X, as is easily seen. More precisely, we have

Proposition 3.1

Let A be an Ω-algebra and $\mathcal{F}$ the clone of action of Ω on A. If $c = (c_1,\cdots,c_n) \in A^n$, then $c\mathcal{F}$ is the Ω-subalgebra of A generated by $c_1,\cdots,c_n$.

The proof is an immediate consequence of the definition of $\mathcal{F}$. ∎

In order to study the clone of action of Ω more closely we need the notion of a centralizer in $\mathcal{O}(A)$. Let $\alpha,\beta \in \mathcal{O}(A)$, where A is any set, and denote the arities of α, β by m, n respectively. If C is any $m \times n$ matrix of elements of A, we can operate on each row of C with β and thus obtain a column of m elements of A, which may be written as $C\beta$. For any column b of m elements of A, denote by αb the result of applying α to b. Then by applying α to the n columns of C, we obtain a row of n elements of A, which is written αC. With these conventions both $(\alpha C)\beta$ and $\alpha(C\beta)$ are defined as elements of A. If

$$(4) \qquad (\alpha C)\beta = \alpha(C\beta) \qquad \text{for all } m \times n \text{ matrices } C \text{ over } A,$$

then we say that α and β *commute*. Thus e.g., in the case of an Ω-algebra, an Ω-endomorphism is a unary operation commuting with all the operations defined by the elements of Ω.

Let A be any set and U any subset of $\mathcal{O}(A)$; an element α of $\mathcal{O}(A)$ is said to *centralize* U if it commutes with each element of U. The set of all elements centralizing U is called the *centralizer* of U in $\mathcal{O}(A)$ and is denoted by U^*, while the centralizer of $\mathcal{O}(A)$ itself is called the *centre* of $\mathcal{O}(A)$. For ease of notation it is convenient to write the elements of U^* on the opposite side from the elements of U; thus if U acts from the right, U^* acts from the left, and the commutative law then takes the form of the associative law (4).

Proposition 3.2

Let A be an arbitrary set; then the centralizer of any subset of $\mathcal{O}(A)$ is a subclone of $\mathcal{O}(A)$. In particular the centre of $\mathcal{O}(A)$ is the clone consisting of all unit operators provided that A has more than one element.

Proof:

Clearly a unit operator commutes with every operation. Now let U be any subset of $\mathcal{O}(A)$; choose $\alpha_1,\cdots,\alpha_m, \beta \in U^*$, where the α_i are n-ary and β is m-ary, and let η be any k-ary operation in U. Then for any $k \times n$ matrix C over A,

$$\begin{aligned}(\eta C)(\alpha_1\cdots\alpha_m\beta) &= [(\eta C)\alpha_1]\cdots[(\eta C)\alpha_m]\beta \\ &= [\eta(C\alpha_1)]\cdots[\eta(C\alpha_m)]\beta \\ &= \{\eta[(C\alpha_1)\cdots(C\alpha_m)]\}\beta \\ &= \eta\{[(C\alpha_1)\cdots(C\alpha_m)]\beta\} \\ &= \eta[C(\alpha_1\cdots\alpha_m\beta)],\end{aligned}$$

which shows that U^* also admits composition, and is therefore a clone.

Assume now that A has more than one element and let α belong to the centre of $\mathcal{O}(A)$. If α is 0-ary its value must be fixed under every permutation of A, which is false because A has more than one element. Therefore $a(\alpha) = n \geqslant 1$; now suppose that $\alpha \neq \delta_n^{(i)}(i = 1,\cdots,n)$; then the value of $c\alpha$, as $c = (c_1,\cdots,c_n)$ ranges over A^n, is different from c_i. Thus we can find an $n \times n$ matrix C over A such that the column $C\alpha$ is different from all the columns of C. Define $\beta \in (\mathcal{O}_n(A))^n$ as follows (acting from the left); $\beta x = x$ for all columns x such that $x \neq C\alpha$, while $\beta(C\alpha)$ is the first column of C. Then

$$(\beta C)\alpha = C\alpha \neq \beta(C\alpha),$$

which is a contradiction; hence α must be a unit operator. ∎

Let A be an Ω-algebra; denote the set of operations defined by the action of Ω by U and the clone of action of Ω by $\mathcal{F}$. Thus $\mathcal{F}$ is the clone generated by U, and it is clear from Proposition 3.2 that the centralizer U^* of U also centralizes $\mathcal{F}$. It follows that $\mathcal{F} \subseteq U^{**}$, but here equality need not hold. We shall call U^{**} the *bicentralizer* of Ω; if, moreover,

(5) $$\mathcal{F} = U^{**},$$

we say that Ω acts *bicentrally*. This requirement (5) is very strong, and in practice it is more useful to have weaker conditions. We shall say that Ω acts *fully* on A if for each $\phi \in U^{**}$ and each $c \in A^n$, where $n = a(\phi)$, there exists $\phi' \in \mathscr{F}$ such that

$$c\phi = c\phi'. \tag{6}$$

If for every finite set of n-tuples $c_1,\cdots,c_r \in A^n$ there exists $\phi' \in \mathscr{F}$ such that

$$c_i\phi = c_i\phi' \qquad (i = 1,\cdots,r), \tag{7}$$

then Ω is said to act *densely* on A. Clearly, if Ω acts bicentrally, it acts densely, and if Ω acts densely, it acts fully. If A is regarded as a discrete topological space, then $\mathcal{O}_n(A) = A^{A^n}$ is a topological space with respect to the product topology, and hence $\mathcal{O}(A) = \bigsqcup \mathcal{O}_n(A)$ becomes a topological space. The resulting topology is called the topology of *pointwise convergence*, and to say that Ω acts densely on A amounts to saying that $\mathscr{F}$, the clone of action of Ω, is dense in the bicentralizer of Ω. When A is finite, this is the same as acting bicentrally. More generally, we have

Proposition 3.3

Let A be an Ω-algebra and denote the centralizer of Ω by Θ. If for all n, A^n *qua Θ-algebra, is finitely generated, then Ω acts densely if and only if Ω acts bicentrally.*

Proof:

Assume first that Ω acts densely. Let ϕ be any element in the bicentralizer of Ω, say $a(\phi) = n$. Any $c \in A^n$ is of the form θY, where $\theta \in \Theta$ and the rows of Y come from a finite generating set of A^n. Since Y is finite, the clone of action of Ω contains ϕ' such that $y\phi = y\phi'$ for each row y of Y. Hence

$$c\phi = (\theta Y)\phi = \theta(Y\phi) = \theta(Y\phi') = (\theta Y)\phi' = c\phi'.$$

Thus (6) holds for all $c \in A^n$, which means that Ω acts bicentrally. Conversely, any Ω acting bicentrally acts densely. ∎

We now go on to ask under what conditions Ω acts densely. Our main objective is the density theorem, which states that Ω acts densely on A if and only if it acts fully on A^r for all r. First we need a lemma on bicentralizers.

Lemma 3.4

Let A be an Ω-algebra with bicentralizer Φ. Then for any $r \geqslant 1$, the bicentralizer of Ω acting on A^r is also Φ.

Proof:

Let Θ be the centralizer of Ω acting on A and Θ′ the centralizer of Ω acting on A^r. Given any n-ary operator $\omega \in \Omega$, an m-ary operation θ on A^r commutes with ω if for every $m \times n$ matrix C over A^r,

$$(\theta C)\omega = \theta(C\omega). \tag{8}$$

Since the values of θ are elements of A^r, θ may be regarded as a row of r (mr)-ary operations on A: $\theta_1,\cdots,\theta_r$ say, and (8) just states that each θ_i commutes with ω, as operator on A. Thus $\theta \in \Theta'$ if and only if $\theta_i \in \Theta$ $(i = 1,\cdots,r)$. Now let Φ and Φ′ be the centralizers of Θ and Θ′ respectively; we have to show that $\Phi' = \Phi$. Given $\phi \in \Phi'$, where ϕ is n-ary say, then

$$\phi = (\phi_1,\cdots,\phi_r)$$

consists of r (nr)-ary operations on A such that

$$(\theta C)\phi = \theta(C\phi) \qquad \text{for each } \theta \in \Theta' \text{ and } C \in A^{m\times n\times r}. \tag{9}$$

Here the notation indicates that C is an $m \times n \times r$ array of elements of A. Each θ_i acts on an $m \times r$ slab of C; we choose θ such that θ_i picks out the (1,1)-element of the $m \times r$ slab. Such an operation is made up of unit operators and therefore belongs to Θ′. Now (9), with this choice of θ, shows that ϕ_1, acting on an $n \times r$ slab, does not depend on the (h, j)-element of the slab, except for $j = 1$. In other words, ϕ_1, and likewise each ϕ_i, may be regarded as an n-ary operation. The same choice of θ shows further that $\phi_1 = \cdots = \phi_r$. Thus ϕ acts as n-ary operation on A. Since it centralizes Θ, it must belong to Φ, whence $\Phi' \subseteq \Phi$. Conversely, if $\phi \in \Phi$, then ϕ centralizes Θ′ because Ω and Φ acting on A^r have the same centralizer, and so $\phi \in \Phi'$. Thus $\Phi' = \Phi$. ∎

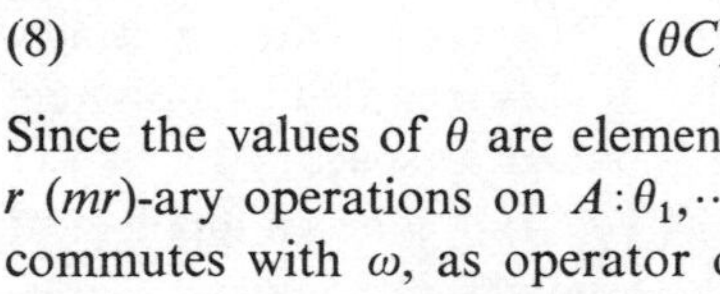

Figure 9

Consider now an Ω-algebra A; denote the clone of action by $\mathscr{F}$ and the bicentralizer by Φ. To say that Ω acts fully on A means: for each $c \in A^n$ and each n-ary $\phi \in \Phi$ there exists $\phi' \in \mathscr{F}$ such that $c\phi = c\phi'$. If $c = (c_1,\cdots,c_n)$, then $c\phi'$ is an element of the Ω-subalgebra generated by

$c_1, \cdots, c_n$, and conversely, every element of this subalgebra has the form $c\phi'$ ($\phi' \in \mathscr{F}$). Thus Ω acts fully if and only if, for each $\phi \in \Phi$ and each $c \in A^n$, where $a(\phi) = n$, $c\phi \in c\mathscr{F}$. Similarly, Ω acts densely if and only if for each $\phi \in \Phi$ and each $r \times n$ matrix C over A, $C\phi$ belongs to the subalgebra of A^r generated by the n columns of C. Since Φ is also the bicentralizer of Ω acting on A^r, this just states that Ω acts fully on A^r and we obtain

Theorem 3.5 (density theorem)

Let A be an Ω-algebra. Then Ω acts densely on A if and only if it acts fully on A^r for $r = 1,2,\cdots$. ∎

To obtain useful conditions for the density theorem to hold it is necessary to specialize the algebras somewhat. We shall consider only one condition, which in particular may be applied in the case of modules over a ring to obtain the usual form of the density theorem. An Ω-algebra A is said to be *fully retractable* if, for each integer $n = 1,2,\cdots$ and each subalgebra B of A^n, there exists an Ω-endomorphism of A^n with B as image. Then we have

Proposition 3.6

In any fully retractable Ω-algebra A, Ω acts densely on A.

Proof:

Fix r and let Φ be the bicentralizer of Ω acting on A^r. Given any n-ary operation $\phi \in \Phi$, and any $c_1, \ldots, c_r \in A^n$, denote by B the subalgebra of A^r generated by the n columns of $C = \begin{pmatrix} c_1 \\ \vdots \\ c_r \end{pmatrix}$. If π is an endomorphism of A^r with B as image, then $C = \pi X$, where X is an $r \times n$ matrix with elements in A. Hence

$$C\phi = (\pi X)\phi = \pi(X\phi) \in B.$$

This shows that $C\phi = C\phi'$ for some $\phi' \in \mathscr{F}$, because B is generated by the columns of C; in other words, Ω acts fully on A^r, and hence, by the density theorem, it acts densely on A. ∎

If for our Ω-algebras we take R-modules, where R is a given ring, then any completely reducible module (i.e., any module which can be written as a sum of irreducible modules) is fully retractable in the above sense, because if M is completely reducible, then so is M^r, and any submodule is then a direct summand. In this way we obtain the Chevalley-Jacobson density theorem for modules (Jacobson [56], ch. VI): If M is a completely reducible unital R-module, where R is a ring with unit element, then R acts densely on M.

EXERCISES

1. Show that when A has one element, $\mathcal{O}(A)$ coincides with its centre.

2. For any set A, find the centralizer of $\mathcal{O}_1(A)$ and, more generally, of $\mathcal{O}_n(A)$.

3. (P. Hall.) An abstract clone is a partial algebra A defined as follows: With each $c \in A$ a nonnegative integer $\alpha(c)$ is associated; A has constant operators $d_n^{(i)}$ $(i = 1,\cdots,n;\ n = 1,2,\cdots)$ such that $\alpha(d_n^{(i)}) = n$ and for each integer r there is an $(r+1)$-ary operation

$$(a_1,\cdots,a_r,b) \to a_1\cdots a_r b\mu$$

which is defined whenever $\alpha(a_1) = \cdots = \alpha(a_r)$, $\alpha(b) = r$; moreover, in this case $\alpha(a_1\cdots a_r b\mu) = \alpha(a_1)$. These operations are subject to the laws

(i) $$(ab_1\mu)\cdots(ab_s\mu)c\mu = abc\mu\mu,$$

where $a = (a_1,\cdots,a_r)$, $b = (b_1,\cdots,b_s)$, and $\alpha(a_i) = n$, $\alpha(b_j) = r$, $\alpha(c) = s$,

(ii) $$ad_r^{(i)}\mu = a_i,$$

where a is as in (i).

Show that every clone of operations is an abstract clone.

4. Define $W_\Omega(X)$ as an abstract clone and show that for any Ω-algebra A there is a natural clone homomorphism $W_\Omega(X) \to \mathcal{O}(A)$. Conversely, if A is any set, show that any clone homomorphism $W_\Omega(X) \to \mathcal{O}(A)$ defines an Ω-algebra structure on A.

5. Let R be a ring, not necessarily with a unit element, and define an R-module M to be completely reducible if any submodule of M is a direct summand and $xR = 0$ for $x \in M$ implies $x = 0$. Show that for any $x \in M$, $x \in xR$, and hence verify that the density theorem still holds in this case.

4. REPRESENTATIONS IN CATEGORIES OF Ω-ALGEBRAS

We have seen that the Ω-word algebras may be described as the universal functor of the representation of the category St in (Ω). This situation may be generalized by replacing St by a category $\mathscr{L}$ subordinate to St, and replacing (Ω) by a category $\mathscr{K}$ of Ω-algebras subordinate to $\mathscr{L}$. This gives rise to a representation of $\mathscr{L}$ in $\mathscr{K}$ (called the natural representation), and our object will be to find conditions for this representation to have a universal functor.

In what follows, every subcategory $\mathscr{K}$ of (Ω) is assumed to be abstract and regular, unless the contrary is stated. For such categories $\mathscr{K}$ the notion of a representation will be restricted as follows: If $\mathscr{L} \prec$ St, we shall say that $\mathscr{L}$ is *represented* in $\mathscr{K}$ if there is a representation of $\mathscr{L}$ in $\mathscr{K}$ in the sense of III.1 such that (i) any admissible morphism $\rho: X \to A$ ($X \in \text{Ob}\, \mathscr{L}$, $A \in \text{Ob}\, \mathscr{K}$) determines a mapping (again denoted by ρ) from the carrier of X to the carrier of A, (ii) the subalgebra A_0 of A generated by the image under ρ is a $\mathscr{K}$-algebra, (iii) the inclusion $i: A_0 \to A$ is a $\mathscr{K}$-homomorphism, and (iv) if ρ_0 is the result of cutting down ρ to A_0, so that $\rho = \rho_0 i$, then ρ_0 is admissible. Thus, e.g., if $\mathscr{K}$ is a hereditary subcategory of (Ω), these supplementary conditions merely state that every admissible morphism $\rho: X \to A$ can be cut down to the subalgebra generated by the image under ρ. Generally, if $\mathscr{K} \prec \mathscr{L} \prec$ St, where $\mathscr{K} \subseteq (\Omega)$, then by the *natural representation* of $\mathscr{L}$ in $\mathscr{K}$ we understand the representation whose admissible morphisms are all the $\mathscr{L}$-morphisms

$$\rho: X \to A$$

from an $\mathscr{L}$-object X to the $\mathscr{L}$-carrier of a $\mathscr{K}$-algebra A such that the subalgebra A_0 generated by the image of any $\mathscr{L}$-subobject of X under the set mapping corresponding to ρ is a $\mathscr{K}$-algebra and the inclusion i: $A_0 \to A$ is a $\mathscr{K}$-homomorphism. It is easily verified that this is in fact a representation.

If $\alpha: X \to Y$ is any set mapping, we shall also write im α in place of $X\alpha$. We note the following consequence of the definitions.

Proposition 4.1

If a category $\mathscr{L}$ which is subordinate to St *is represented in an abstract regular category $\mathscr{K}$ of Ω-algebras, and a universal functor (U,u) exists for this representation, then $U(X)$ is generated by* im u.

Proof:

By definition, the universal mapping

$$u: X \to U(X)$$

is an admissible morphism. Let U_0 be the subalgebra of $U = U(X)$ generated by im u; then $u = u_0 i$, where $u_0: X \to U_0$ is admissible and

$i: U_0 \to U$ is the inclusion mapping. Hence there exists a unique $\mathscr{K}$-homomorphism $\alpha: U \to U_0$ such that $u_0 = u\alpha$. It follows that

$$u = u_0 i = u\alpha i.$$

Here both $1: U \to U$ and $\alpha i: U \to U$ are $\mathscr{K}$-homomorphisms; hence, by uniqueness, $\alpha i = 1$, which shows i to be surjective, i.e., $U_0 = U$. ∎

Consider now a representation of a category $\mathscr{L}$ in a residual category $\mathscr{K}$ of Ω-algebras. This representation is said to be *residual* if the following condition holds: If A is a $\mathscr{K}$-algebra which is a subdirect product of a family $(A_\lambda)_{\lambda \in \Lambda}$ of $\mathscr{K}$-algebras with projections $\varepsilon_\lambda: A \to A_\lambda$, then for any mapping ρ from the carrier of an $\mathscr{L}$-object X to the carrier of A such that $\rho\varepsilon_\lambda$ corresponds to an admissible morphism $X \to A_\lambda$, the mapping ρ itself corresponds to an admissible morphism $X \to A$. With these definitions we can state the main result on the existence of a universal functor.

Theorem 4.2

Let $\mathscr{K}$ be a residual category of non-empty Ω-algebras and let $\mathscr{L}$ be any category subordinate to St *which has a residual representation in $\mathscr{K}$. Then the representation has a universal functor.*

Proof:

Let X be an $\mathscr{L}$-object and $W_\Omega(X)$ the Ω-word algebra on X, regarded as a set, with the canonical injection

(1) $$\rho: X \to W.$$

Denote by $(\mathfrak{q}_\lambda)_{\lambda \in \Lambda}$ the family of all congruences on W such that $W/\mathfrak{q}_\lambda \in \mathrm{Ob}\,\mathscr{K}$ and the mapping

(2) $$\rho(\mathrm{nat}\ \mathfrak{q}_\lambda): X \to W/\mathfrak{q}_\lambda$$

is admissible. Putting $\rho_\lambda = \rho(\mathrm{nat}\ \mathfrak{q}_\lambda)$ and $A_\lambda = W/\mathfrak{q}_\lambda$, we have a family of admissible mappings $\rho_\lambda: X \to A_\lambda$, and composing these we obtain a mapping

(3) $$\sigma: X \to \prod A_\lambda.$$

Let $U(X)$ be the subalgebra of $A = \prod A_\lambda$ generated by im σ; cutting down σ to $U(X)$ we obtain from (3) a mapping

$$u: X \to U(X)$$

such that $ui = \sigma$, where $i: U(X) \to A$ is the inclusion mapping. We assert that (U,u) is the required functor. In the first place, if ε_λ denotes the projection of $\prod A_\mu$ onto A_λ, then for any $\lambda \in \Lambda$,

$$\sigma\varepsilon_\lambda = \rho_\lambda.$$

Thus $\sigma\varepsilon_\lambda$ is admissible for all $\lambda \in \Lambda$, and hence so is σ. Since $\sigma = ui$, it follows that $U(X)$ is a $\mathscr{K}$-algebra, i is a $\mathscr{K}$-homomorphism, and u is admissible. We note incidentally that since im ρ generates W, im ρ_λ generates A_λ and so does im $i\varepsilon_\lambda$; but $i\varepsilon_\lambda$ is a homomorphism, whence im $i\varepsilon_\lambda = A_\lambda$, i.e. $i\varepsilon_\lambda$ is an epimorphism. Therefore $U(X)$ is a subdirect product of the A_λ.

Now let B be any $\mathscr{K}$-algebra and

$$\alpha : X \to B$$

an admissible mapping. Cutting down α to the subalgebra generated by im α, we may assume that B is generated by im α. The mapping α may be extended to an epimorphism

$$\beta : W \to B;$$

if ker $\beta = \mathfrak{q}$, then $\beta = (\text{nat } \mathfrak{q})\beta^*$, where

$$\beta^* : W/\mathfrak{q} \to B$$

is an isomorphism. Now α is admissible, and by construction,

$$\alpha = \rho\beta = \rho(\text{nat } \mathfrak{q})\beta^*;$$

since β^* is an isomorphism, it follows that $\alpha\beta^{*-1} = \rho(\text{nat } \mathfrak{q})$ is admissible. Hence $\mathfrak{q} = \mathfrak{q}_\lambda$ for some $\lambda \in \Lambda$, and we have the following diagram:

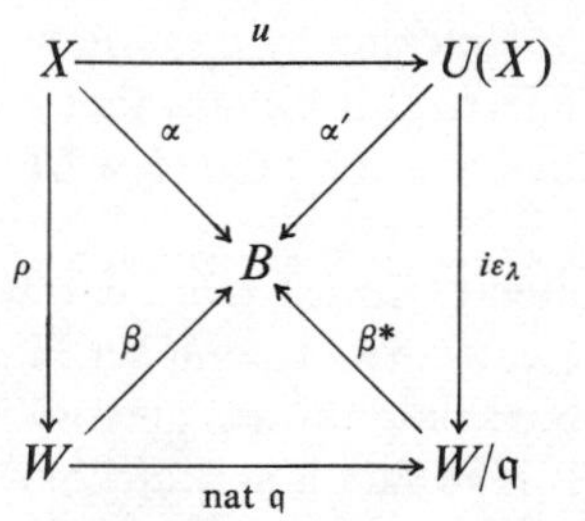

where $\alpha' = i\varepsilon_\lambda\beta^*$, and the square and all the triangles except possibly the topmost one are commutative. By going round the other triangles and the square, we find

$$\alpha = \rho\beta = \rho(\text{nat } \mathfrak{q})\beta^* = ui\varepsilon_\lambda\beta^* = u\alpha',$$

so the topmost triangle commutes also, and α' is the required mapping. Clearly it is a $\mathscr{K}$-homomorphism (since $i\varepsilon_\lambda$ and β^* are), and it is unique, for if $\alpha = u\alpha_1 = u\alpha_2$, then α_1 and α_2 agree on im u and hence (by Lemma 2.2) on $U(X)$, i.e., $\alpha_1 = \alpha_2$. ∎

In particular, if $\mathscr{K} \prec \mathscr{L}$, then the natural representation of $\mathscr{L}$ in $\mathscr{K}$ is residual and we obtain

Corollary 4.3

If $\mathscr{K}$ is a residual subcategory of (Ω) *and* $\mathscr{K} \prec \mathscr{L} \prec \mathrm{St}$, *then the natural representation of $\mathscr{L}$ in $\mathscr{K}$ has a universal functor.* ∎

Corollary 4.4

If $\mathscr{K}$ is a hereditary subcategory of (Ω) *admitting direct products and* $\mathscr{K} \prec \mathscr{L} \prec \mathrm{St}$, *then the natural representation of $\mathscr{L}$ in $\mathscr{K}$ has a universal functor.* ∎

Let $\mathscr{L} \prec \mathrm{St}$ be represented in a category $\mathscr{K}$ of Ω-algebras and suppose that the universal functor U exists for this representation. If for every $\mathscr{L}$-object X, the universal mapping

$$u : X \to U(X)$$

is injective, the functor U is said to be *injective*. It is easily seen that the universal functor U is injective if and only if each $\mathscr{L}$-object X can be represented by an injection in some $\mathscr{K}$-algebra. For if $\rho : X \to A$ is an admissible injection, then factoring by u we obtain $\rho' : U(X) \to A$ such that $\rho = u\rho'$, and since ρ is injective, it follows that u must be injective. Conversely, when u is injective, we can take $A = U(X)$ to obtain an injective representation of X.

A universal functor often exists for representations other than algebraic ones. For example, the category of compact Hausdorff spaces and continuous mappings is subordinate to the category of Hausdorff spaces and continuous mappings (it is in fact a subcategory). The natural representation has a universal functor, which associates with every Hausdorff space X a compact Hausdorff space $U(X)$, called the *Stone-Čech-compactification* of X (cf. Samuel [48] or Kelley [55]). This functor is injective on the subcategory of completely regular Hausdorff spaces (a space X is *completely regular* if for every point $p \in X$ and every neighbourhood N of p there is a real-valued continuous function f on X such that $f(p) = 1$, $f(x) = 0$ for $x \notin N$).

EXERCISES

1. If $\mathcal{K}$ and $\mathcal{L}$ are abstract regular categories of Ω-algebras, show that any abstract regular category containing both $\mathcal{K}$ and $\mathcal{L}$ as subcategories defines a representation of $\mathcal{L}$ in $\mathcal{K}$.

2. If $\mathcal{K} \subseteq (\Omega)$, $\mathcal{K} \prec \mathcal{L} \prec \mathrm{St}$ and $\mathcal{K}$ is regular and residual, verify that the natural representation of $\mathcal{L}$ in $\mathcal{K}$ is residual.

5. FREE ALGEBRAS IN CATEGORIES OF Ω-ALGEBRAS

We now specialize the results of the previous section to the case $\mathcal{L} = \mathrm{St}$. If the universal functor for the natural representation of St in the category $\mathcal{K}$ of Ω-algebras exists, then for any set X, the algebra $U(X)$ is called the *universal $\mathcal{K}$-algebra on X.* When the universal functor is injective, we may identify X with its image in $U(X)$; in this case $U(X)$ is called the *free* $\mathcal{K}$-algebra on X, while X is called a *$\mathcal{K}$-free* generating set of $U(X)$, and we also say that $\mathcal{K}$ is a *category with free algebras.* By the functorial property, the free $\mathcal{K}$-algebra on X, when it exists, is determined up to isomorphism by X. Thus e.g., (Ω) is a category with free algebras, by Theorem 2.6.

Clearly a trivial category cannot have free algebras on any set with more than one element; in all other cases the existence of free algebras follows from the existence of the universal functor:

Proposition 5.1

If $\mathcal{K}$ is a nontrivial category of Ω-algebras and the universal functor exists for the natural representation of St *in $\mathcal{K}$, then there is a cardinal number α such that for every set X of cardinal at least α, the free $\mathcal{K}$-algebra on X exists. Moreover, every $\mathcal{K}$-algebra is a homomorphic image of a free $\mathcal{K}$-algebra.*

Proof:

By hypothesis $\mathcal{K}$ contains an algebra A with more than one element. Let X_0 be a generating set of A which has more than one element; then the inclusion mapping $X_0 \to A$ is admissible (in the natural representation), hence the universal mapping

$$u_0 : X_0 \to U(X_0)$$

is injective. Put $\alpha = |X_0|$; then for any set X of cardinal at least α, there is a surjection $\rho: X \to X_0$, and for any such ρ the mapping $\rho u_0: X \to U(X_0)$ is admissible. Moreover, given $x,y \in X$, $x \neq y$, we may choose ρ such that $x\rho \neq y\rho$; therefore $x\rho u_0 \neq y\rho u_0$, and so $xu \neq yu$, where $u: X \to U(X)$ is the universal mapping. Since this holds for any pair of elements x,y of X, u is injective.

In the proof we obtained A as the homomorphic image of the free $\mathscr{K}$-algebra $U(X_0)$; this applies if A is any $\mathscr{K}$-algebra, and it establishes the last assertion. ∎

Under the hypotheses of Proposition 5.1, $\mathscr{K}$ need not have free algebras on all sets X; for example, the category consisting of all uncountable groups and the trivial group (and all homomorphisms between them) has a universal functor: $U(X)$ is the free group on X if X is uncountable and is the trivial group otherwise. Thus for finite nonempty sets X, $U(X)$ is not the free group on X. However, if $\mathscr{K}$ is hereditary, we can take A in the proof of Proposition 5.1 to be generated by a two-element set X_0, and we thus obtain

Corollary 5.2

If $\mathscr{K}$ is a nontrivial hereditary category of Ω-algebras and the universal functor for the natural representation of St *in $\mathscr{K}$ exists, then $\mathscr{K}$ is a category with free algebras.* ∎

To find conditions for the existence of the universal functor, we use Corollary 4.3. Combined with Proposition 5.1 this yields

Theorem 5.3

Let $\mathscr{K}$ be a subcategory of (Ω) which is nontrivial and residual. Then there exists a cardinal number α such that for every set X of cardinal at least α, the free $\mathscr{K}$-algebra on X exists, and every $\mathscr{K}$-algebra is a homomorphic image of a free $\mathscr{K}$-algebra. ∎

In particular, the conditions of this theorem hold for nontrivial hereditary subcategories $\mathscr{K}$ of (Ω) admitting direct products.

Consider now a category $\mathscr{K}$ of Ω-algebras with free algebras. In this case we shall write $F(X)$ or $F(X; \mathscr{K})$ in place of $U(X)$. From the uniqueness of free algebras we obtain a number of useful consequences.

Proposition 5.4

Let $\mathscr{K}$ be a subcategory of (Ω) *with free algebras, X any set, and Y a subset of X. If $i: Y \to X$ is the inclusion mapping, then the induced homomorphism $F(i): F(Y) \to F(X)$ is injective.*

For the inclusion mapping $i: Y \to X$ has a right inverse $\tau: X \to Y$ such that $i\tau = 1_Y$, hence $F(i)$ has a right inverse and so is injective. ∎

This proposition allows $F(Y)$ to be identified with a subalgebra of $F(X)$, clearly, this subalgebra is proper whenever Y is a proper subset of X, hence X is a minimal generating set of $F(X)$.

Proposition 5.5

Let $\mathscr{K}$ be a subcategory of (Ω) *with free algebras, X a set, and $\mathfrak{q}$ an equivalence on X. Identify X with a subset of $F(X)$ (by means of the universal mapping) and let $\bar{\mathfrak{q}}$ be the congruence on $F(X)$ generated by $\mathfrak{q}$. Then*

(1) $$\bar{\mathfrak{q}} \cap X^2 = \mathfrak{q},$$

and

$$F(X)/\bar{\mathfrak{q}} \cong F(X/\mathfrak{q}).$$

Proof:

Put $X' = X/\mathfrak{q}$ and write $F = F(X)$, $F' = F(X')$ for short. The natural mapping $\theta: X \to X'$ gives rise to a homomorphism $F(\theta): F \to F'$, whose kernel $\mathfrak{k}$ contains $\mathfrak{q}$ and therefore also $\bar{\mathfrak{q}}$. We assert that

(2) $$\mathfrak{k} \cap X^2 = \mathfrak{q}.$$

For if $(x,y) \in \mathfrak{k} \cap X^2$, then $xF(\theta) = yF(\theta)$, and hence $(x,y) \in \mathfrak{q}$; thus $\mathfrak{k} \cap X^2 \subseteq \mathfrak{q}$ and the reverse inequality is clear. Since $\mathfrak{k} \supseteq \bar{\mathfrak{q}}$, we can factor $F(\theta): F \to F'$ by nat $\bar{\mathfrak{q}}$ and obtain a homomorphism

(3) $$\phi: F/\bar{\mathfrak{q}} \to F'.$$

Clearly this reduces to the identity on $X' = X/\mathfrak{q}$. Therefore $\phi \mid X'$ has an inverse; this inverse extends to a homomorphism $F' \to F/\bar{\mathfrak{q}}$ because F' is free and so ϕ itself has an inverse, i.e. (3) is an isomorphism. This means that $\mathfrak{k} = \bar{\mathfrak{q}}$ and now (1) follows from (2). ∎

As an illustration (which will be used later), let A be the free $\mathscr{K}$-algebra on a,b,c. If we add the relation $b = c$ we obtain the free $\mathscr{K}$-algebra on a and b.

Since two sets are equivalent (in the category St) if and only if they have the same cardinal number, it follows that in a category $\mathscr{K}$ with free algebras there is just one free $\mathscr{K}$-algebra F_α, up to isomorphism, for each cardinal α. This algebra is called the free $\mathscr{K}$-algebra of *rank* α. The question now arises whether free $\mathscr{K}$-algebras of different ranks can be isomorphic. From the remark following Proposition 5.4, it follows that the set X is a minimal generating set of the free $\mathscr{K}$-algebra $F(X)$; therefore, by Proposition II.5.5, two free $\mathscr{K}$-algebras of different ranks are nonisomorphic if at least one of the ranks is infinite. This no longer always holds when both ranks are finite, but we have the following sufficient conditions, due to Jónsson & Tarski [61].

Theorem 5.6

Let $\mathscr{K}$ be a subcategory of (Ω) *with free algebras and suppose that $\mathscr{K}$ has at least one finite algebra with more than one element. Then free $\mathscr{K}$-algebras of different ranks are nonisomorphic.*

Proof:

By the remarks preceding the theorem, we need only consider the case where both ranks are finite. We shall prove the stronger result that for finite n, every generating set of F_n has at least n elements. Thus n is characterized by the isomorphism type of F_n as the least cardinal of a generating set of F_n, from which the theorem follows.

Let X be a free generating set of F_n consisting of n elements, Y any other generating set, and B a finite $\mathscr{K}$-algebra with more than one element. By definition of F_n, each mapping $X \to B$ can be extended to a unique $\mathscr{K}$-homomorphism $F_n \to B$; therefore, the set H of all $\mathscr{K}$-homomorphisms $F_n \to B$ is equipotent with B^X. Let $\theta \in H$; then the restriction $\theta \mid Y$ defines a mapping $Y \to B$, i.e. an element of B^Y, and distinct homomorphisms define distinct mappings (by Lemma 2.2). Thus we have an injection $B^X \to B^Y$. It follows that $|B^X| \leqslant |B^Y|$, and since B,X are both finite, $n = |X| \leqslant |Y|$. ∎

A further property of free algebras which for certain categories characterizes the free algebras in a manner which is independent of the choice of generating set, is the following

Proposition 5.7

Let $\mathscr{K}$ be a category of Ω-algebras with free algebras and let F be the free $\mathscr{K}$-algebra on a set X. If A, B are any $\mathscr{K}$-algebras, $\beta: F \to B$ is a $\mathscr{K}$-homomorphism and $\alpha: A \to B$ is a $\mathscr{K}$-epimorphism, then there exists a $\mathscr{K}$-homomorphism $\gamma: F \to A$ such that $\gamma\alpha = \beta$ (of course γ may not be unique).

Proof:

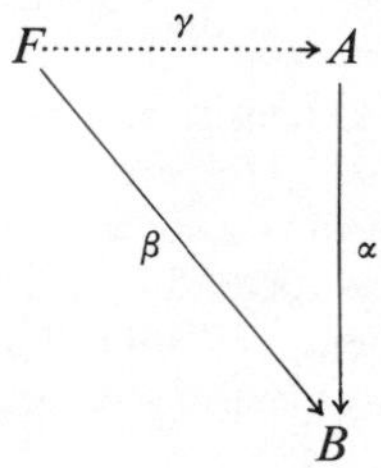

We define γ on X as follows: Let $x \in X$; then $x\beta \in B$ and so $x\beta = a\alpha$ for some $a \in A$. Choose such an a corresponding to each $x \in X$ and put $x\gamma = a$; then, by definition,

$$x\gamma\alpha = x\beta \qquad (x \in X).$$

Now γ may be extended to a homomorphism $\gamma: F \to A$; then $\gamma\alpha$ and β are homomorphisms which agree on X and hence are equal. ∎

A $\mathscr{K}$-algebra F with the property stated in Proposition 5.7 is said to be *$\mathscr{K}$-projective.* Thus, the proposition states that every free $\mathscr{K}$-algebra is $\mathscr{K}$-projective. For the category of groups and homomorphisms the converse is true: every projective group is free. A similar result holds for abelian groups, but the corresponding assertion for R-modules is false, when R is a semisimple ring with minimum condition which is not a skew field (cf. e.g. MacLane [63]).

EXERCISES

1. Show that the category of all nonabelian groups and the trivial group (with all homomorphisms) has a universal functor but has no free groups on sets of a single element.

2. Verify that for any category $\mathscr{K}$ with free algebras and any set X, the free $\mathscr{K}$-algebra on X has X as a minimal generating set.

3. Generalize Exercise 2.7 to categories with free algebras.

4. (Jónsson & Tarski.) Let $\mathscr{K}$ be a category with free algebras and denote by F_n the free $\mathscr{K}$-algebra of rank n. If every n-element generating set of F_n is a free generating set (for all finite n), show that every generating set of F_n has at least n elements, and deduce that the conclusion of Theorem 5.6 holds in this case.

5. (Jónsson & Tarski.) Let Ω consist of a binary and two unary operators, whose effects are written $x \cdot y$, $x\lambda$, $x\rho$ respectively, and denote by $\mathscr{K}$ the category of all Ω-algebras such that $x\lambda \cdot x\rho = x$, $(x \cdot y)\lambda = x$, $(x \cdot y)\rho = y$. Given that $\mathscr{K}$ is a category with free algebras (proved in IV. 3.3 below), show that all free $\mathscr{K}$-algebras of nonzero finite rank are isomorphic. (If A is free on $X \cup \{a, b\}$ and $X \cap \{a, b\} = \emptyset$, show that A is also free on X.)

6. (Malcev.) Let $\mathscr{K}$ be a category with free algebras and F_n the free $\mathscr{K}$-algebra of rank n. If no proper quotient of F_n is isomorphic to F_n, then any generating set of F_n consisting of n elements is free.

7. Show that every unital module over a skew field is free, and deduce that the lattice of submodules of such a module is complemented. (Use Proposition 5.7.)

8. (R. S. Pierce.) If $\mathscr{K}$ is an abstract hereditary category with free algebras and $\mathrm{Q}\mathscr{K}$ the category of homomorphic images (II.8), show that an algebra A is $\mathrm{Q}\mathscr{K}$-free if and only if it is $\mathscr{K}$-free.

6. FREE AND DIRECT COMPOSITION OF Ω-ALGEBRAS

Two universal functors which exist for many categories of Ω-algebras are the free and direct composition introduced in III.1. If $\mathscr{K}$ is any regular category of Ω-algebras, then $\mathscr{K}$ is subordinate to the category $\mathscr{F}(\mathscr{K})$ of families of $\mathscr{K}$-algebras (over totally unordered sets), and so we may apply Corollary 4.3:

Theorem 6.1

Let $\mathscr{K}$ be any regular residual category of Ω-algebras. Then the free $\mathscr{K}$-composition of any family of $\mathscr{K}$-algebras exists. ∎

In a category $\mathscr{K}$ with free composition, a $\mathscr{K}$-algebra A is said to be the *free product* of the $\mathscr{K}$-algebras $A_\lambda (\lambda \in \Lambda)$ if the A_λ are subalgebras of A, with free composition $P = \bigsqcup A_\lambda$, such that

(i) there is an isomorphism $\theta : A \to P$ whose restriction to A_λ is the canonical mapping,
(ii) $A_\lambda \cap A_\mu$ is the minimal subalgebra of A for $\lambda \neq \mu$.

When a family (A_λ) of $\mathscr{K}$-algebras is given, there may be no $\mathscr{K}$-algebra A which is their free product (for instance, it is necessary for the minimal subalgebras of all the A_λ to be isomorphic), but when such a free product exists, it is by definition determined (up to isomorphism) by the free composition of the A_λ. Let us say that a $\mathscr{K}$-algebra A is *retractable* if there is a $\mathscr{K}$-homomorphism

$$\theta : A \to C$$

onto the minimal subalgebra C of A such that $\theta | C = 1$. Then we can state the following condition for free products to exist.

Proposition 6.2

Given a category $\mathscr{K}$ of Ω-algebras, let (A_λ) be a family of $\mathscr{K}$-algebras whose free composition P exists. If each A_λ is retractable and for any $\lambda,\mu \in \Lambda$ the minimal subalgebras of A_λ and A_μ are isomorphic, then P is the free product of the algebras A_λ.

Proof:

The assertion holds trivially when Λ consists of a single element, so we may assume that Λ has more than one element.

We have to show that the canonical mappings

(1) $$\rho_\lambda : A_\lambda \to P$$

are injective, and identifying A_λ with its image in P, we must then show that the A_λ intersect minimally in P. Let C_λ be the minimal subalgebra of A_λ; by hypothesis, there is a retraction

(2) $$\varepsilon_\lambda : A_\lambda \to C_\lambda,$$

and since all the C_λ are isomorphic, we can take an isomorphism

(3) $$\theta_\lambda : C_\lambda \to C$$

with a fixed algebra C. Now take any pair of distinct indices of Λ, say $\lambda = 1,2$. We define a family of mappings $A_\lambda \to A_1$ by

$$\phi_\lambda = \begin{cases} 1 & \lambda = 1, \\ \varepsilon_\lambda \theta_\lambda \theta_1^{-1} & \lambda \neq 1. \end{cases}$$

By the universal property, there exists a homomorphism $\phi : P \to A_1$ such that $\phi_\lambda = \rho_\lambda \phi$. Taking $\lambda = 1$, we find that $\rho_\lambda \phi = 1$, whence ρ_1 is injective; moreover, if $c \in A_1\rho_1 \cap A_2\rho_2$, say $c = a\rho_1 = b\rho_2$, then $c\phi = a\phi_1 = a$ and $c\phi = b\phi_2 = b\varepsilon_2\theta_2\theta_1^{-1} \in C_1$, whence $a \in C_1$, $c = a\rho_1 \in C_1\rho_1$. Thus $A_1\rho_1 \cap A_2\rho_2$ is contained in $C_1\rho_1$, and hence is the minimal subalgebra of P. Since 1,2 were arbitrary in Λ, this shows P to be the free product. ∎

If every minimal $\mathscr{K}$-subalgebra is trivial, the minimal subalgebras are all isomorphic and every $\mathscr{K}$-algebra is retractable; more generally, this still holds if we only know that every $\mathscr{K}$-algebra has a trivial subalgebra. We need only introduce a 0-ary operator which picks out a specific trivial subalgebra in each $\mathscr{K}$-algebra. Thus, as a corollary of Proposition 6.2, we obtain the following result, due to Sikorski [53].

Corollary 6.3

If every $\mathscr{K}$-algebra has a trivial subalgebra and $\mathscr{K}$ admits free composition amalgamating these subalgebras, then the free product of any family of $\mathscr{K}$-algebras exists. ∎

It will be shown in Chapter IV and may easily be verified by Theorem 6.1 that the category Gp of all groups admits free composition. Therefore, by Corollary 6.3, Gp admits free products. A similar result holds for semigroups and for rings, but not for rings with unit element (cf. Cohn [59]).

In conclusion we note a simple sufficient condition for the existence of direct compositions.

Theorem 6.4

Let $\mathscr{K}$ be any category of Ω-algebras admitting direct products. Then the direct $\mathscr{K}$-composition of any family of $\mathscr{K}$-algebras exists and coincides with the direct product.

To prove this result it is enough to show that the direct product $P = \Pi A_\lambda$ with the projections ε_λ as canonical homomorphisms satisfies the universal property. Thus, given a family $\phi_\lambda : B \to A_\lambda$ of homomorphisms, we require a homomorphism $\phi : B \to P$ such that $\phi_\lambda = \phi\varepsilon_\lambda$, i.e., for any $b \in B$,

$$b\phi\varepsilon_\lambda = b\phi_\lambda. \tag{4}$$

Such a mapping ϕ is obtained by composing the ϕ_λ; it is uniquely determined by (4) and it is a $\mathscr{K}$-homomorphism because the ϕ_λ are. ∎

Examples to show that this condition is not necessary will be given in the exercises.

EXERCISES

1. Show that for any category $\mathscr{K}$ with free algebras, which admits homomorphic images, the free composition of any family of $\mathscr{K}$-algebras exists.

2. Show that the category of finite abelian groups and homomorphisms admits homomorphic images but has no free algebras. Further show that the family $(C_n)_{n \geqslant 1}$, where C_n is cyclic of order n, has no free composition in the class. (Use Exercise 1 for the second part.)

3. Show that the category of commutative rings without zero-divisors has free algebras, but does not admit homomorphic images. Give an example to show that free compositions do not exist in general.

4. Show that the category of all finite abelian groups (and homomorphisms) does not admit free composition.

5. (J. D. Reid.) Let $\mathscr{K}$ be a category of Ω-algebras admitting free composition and let (A_λ) be a family of $\mathscr{K}$-algebras which has a representation in $A_1 (1 \in \Lambda)$. Show that the canonical mapping $\rho_1 : A_1 \to \bigsqcup A_\lambda$ is injective. Deduce Corollary 6.3.

6. Show that the category of abelian torsion groups does not admit direct products, but does possess a direct composition. What happens for arbitrary torsion groups?

7. DERIVED OPERATORS

Let Ω be an operator domain; we have seen that on each Ω-algebra A, a given element $\omega \in \Omega$ defines a certain operation. In addition to these operations given explicitly there are the operations obtained by composing the operators of Ω, which constitute the clone of action of Ω on A. They may also be expressed in terms of the clone structure on $W_\Omega(X)$, as indicated in Exercise 3.4, but this will not be needed, as we shall not be concerned with the process of composition. To obtain the operation defined by an arbitrary Ω-word $w = w(x_1, \cdots, x_n)$ in $x_1, \cdots, x_n$, let a be the image in A under the homomorphism defined by $x_i \to a_i$, so that

$$a = w(a_1, \cdots, a_n).$$

We may thus regard the word w as defining an n-ary operator $\bar{w}$ by the rule

$$a_1 \cdots a_n \bar{w} = w(a_1, \cdots, a_n).$$

The operator $\bar{w}$ is said to be *derived* from Ω. In this sense, every Ω-word gives rise to a derived operator; for a sufficiently large alphabet X, this includes in particular the operators of Ω itself; in fact this is true for any infinite set X. For, if $\omega \in \Omega(n)$ and $x_1,\cdots,x_n$ are any distinct elements of X, then ω is obtained from $x_1x_2\cdots x_n\omega$, or also from $x_n\cdots x_2x_1\omega$, but not from $x_1x_1\cdots x_1\omega$, unless $n = 1$. We remark that from each n-ary derived operator, we obtain k-ary operations on A by specializing $n - k$ of the arguments to be elements of A. The unary derived operators and unary operations obtained by specializing the derived operators are again called *translations*, or more precisely, *derived translations*, to stress the fact that they may not be obtainable by specializing the operations of Ω; thus in a group G the operation $x \to x^2$ cannot always be expressed in terms of the operations axb and $ax^{-1}b$.

The set of all derived operators of Ω is denoted by $\overline{\Omega}$. It is clear that any Ω-algebra A also admits $\overline{\Omega}$; more generally, the Ω-subalgebras and $\overline{\Omega}$-subalgebras of A are the same. Now let Ω' be a subset of $\overline{\Omega}$; then Ω' is called a *restriction* of Ω. Every Ω-algebra A has an Ω'-algebra structure induced by restriction, and any subalgebra of A admits Ω', but not necessarily conversely. Likewise, every Ω-homomorphism is also an Ω'-homomorphism, but not necessarily conversely. In fact, the category (Ω) is subordinate to (Ω'), under the forgetful functor, which forgets the part of $\overline{\Omega}$ not in Ω'.

As an example, consider the category Gp of groups and consider the commutator as a derived operator:

$$[x,y] = x^{-1}y^{-1}xy. \tag{1}$$

By limiting ourselves to this operator, we may regard Gp as subordinate to the category of groupoids, Γ say. Then a Γ-subalgebra of a group is a subset of the group which is closed under commutation. Such a subset need not admit inversion (e.g. a non-torsion abelian group); the image admits inversion $x \to x^{-1}$, and if nonempty, it admits also the constant operator e, but it may not admit multiplication, and so is not always a subgroup (cf. Carmichael [37], p. 39).

This construction leads to a representation of groupoids in groups; since Gp is residual and regular, the universal functor for this representation exists, by Theorem 4.2. Other examples will be considered in Chapter VII.

If Ω is an operator domain containing only unary and 0-ary operators, then every derived operator has arity at most one. This is intuitively obvious, and may also be deduced from Theorem II.5.6, which gave a criterion of the operators to be at most unary, in terms of the lattice of subalgebras. Similarly, it is possible to express in terms of the lattice of subalgebras the condition that Ω contain 0-ary operators only or unary operators only (Exercise II.5.5). Once binary operators are admitted, this is no longer possible; on the contrary, in most cases one can express all the operators in terms of binary ones.

***Theorem** 7.1*

Let Ω be any operator domain; then there exists an operator domain Ω^* consisting of at most binary operators, such that Ω may be embedded in the domain of derived operators of Ω^*, and such that every Ω-algebra A may be embedded in an Ω^*-algebra A^*. Moreover, if A is infinite, A^* may be taken to have the same carrier as A, whereas if A is finite, A^* may be taken to be countable.

In the proof we have to use the fact that an infinite set is equipotent with its square. We shall assume this for now, so as not to interrupt the thread; as we need a more general statement later on, we refer the reader to Lemma VI.6.1 for the proof.

We shall take Ω^* to consist of one unary operator ω' for each operator $\omega \in \Omega$, and an additional binary operator μ. Suppose first that A is infinite; then A is equipotent with A^2 and so there is a bijection

$$\nu : A \to A^2.$$

More generally, we define the action of ν on an n-tuple of A by the rule

$$(x_1,\cdots,x_n)\nu = (x_1,\cdots,x_n\nu).$$

Clearly, this provides a bijection between A^n and A^{n+1}. It follows that ν^k is a bijection between A and A^{1+k}. Now, for any $\omega \in \Omega(n)$ define the action of the corresponding unary operator $\omega' \in \Omega^*$ by

$$x\omega' = x\nu^{n-1}\omega \qquad (n \geqslant 1); \tag{2}$$

if $n = 0$, we put $x\omega' = \omega$. Denote ν^{-1} by μ; then μ is a binary operator on A, and by (2), when $n \geqslant 2$,

$$x_1\cdots x_n\omega = x_1\cdots x_n\mu^{n-1}\omega'.$$

Hence ω has been expressed as a derived Ω^*-operator. If A is finite, we can embed it in a countable set A^* and define the operators of Ω arbitrarily on $A^*\backslash A$; now we can apply the first part to define A^* as an Ω^*-algebra. ∎

Of course the construction in this proof is not of any practical interest because it depends essentially on the choice of the bijection $\nu : A \to A^2$. Another more practical method of reducing operators to binary ones is given in Chapter IV.

It looks at first sight as if not every operation on a finite set can be expressed in terms of binary ones. Thus on a set of k elements there are k^{k^n} n-ary operations, so that the number of ternary operations is k^{k^3}. By combining two binary operations α, β we can form four ternary operations, namely $xy\alpha z\beta$, $xy\beta z\alpha$, $xyz\alpha\beta$, and $xyz\beta\alpha$; so the number of these operations is at most k^{4k^2}, and this is less than k^{k^3} for $k \geqslant 5$. We note that for $k = 3$, the number of unary operations is 27, the number of binary operations is 19,683, and the number of ternary operations is nearly 10^{13}. Despite these appearances to the contrary it can be shown that the clone of any finite set is generated by binary operations, cf. IX. 2, p. 338.

Let $\mathscr{K}$ be any category of Ω-algebras and Ω' a restriction of Ω. If we regard the objects of $\mathscr{K}$ as Ω'-algebras, then with the $\mathscr{K}$-homomorphisms, regarded as Ω'-homomorphisms, they form a subcategory $\mathscr{L}$ of (Ω'). The category s$\mathscr{L}$ will be denoted by $\mathscr{K}'$ and called the *category derived from* $\mathscr{K}$ *by restricting* Ω to Ω'. For simplicity, let us assume that $\mathscr{K}$ is hereditary; then by applying Corollary 4.4 we obtain

Theorem 7.2

Let $\mathscr{K}$ be a hereditary category of Ω-algebras admitting direct products and let $\mathscr{K}'$ be the category derived by restricting Ω to Ω'. Then there is a universal functor for the natural representation of $\mathscr{K}'$ in $\mathscr{K}$.

This follows once it is shown that $\mathscr{K}$ is subordinate to $\mathscr{K}'$, but this is immediate from the definitions. ∎

In Chapter IV we shall give an explicit construction of this universal functor for a wide class of categories, as well as a criterion for the functor to be injective.

We conclude this section with a criterion for the commutativity of congruences, expressed in terms of derived operators, which is due to Malcev [54].

Theorem 7.3

Let $\mathscr{K}$ be a category of Ω-algebras with free algebras. Then the congruences on every $\mathscr{K}$-algebra are permutable if and only if there exists a derived ternary operator ω such that

$$xxz\omega = z, \qquad xzz\omega = x. \tag{3}$$

Proof:

If ω exists, satisfying (3), then the translation $axb\omega$ interchanges a and b; hence the congruences on any $\mathscr{K}$-algebra A commute, by Proposition II.6.8. To prove the converse, take the free $\mathscr{K}$-algebra F on a,b,c and denote by $\mathfrak{q}$ the congruence generated by (a,b), and by $\mathfrak{r}$ the congruence generated by (b,c). Then $(a,c) \in \mathfrak{q} \circ \mathfrak{r}$, and by hypothesis, $\mathfrak{q} \circ \mathfrak{r} = \mathfrak{r} \circ \mathfrak{q}$; hence $(a,c) \in \mathfrak{r} \circ \mathfrak{q}$, i.e. there exists $d \in F$ such that $(a,d) \in \mathfrak{r}$ and $(d,c) \in \mathfrak{q}$. Since $d \in F$, we can express d as an Ω-word in a,b,c say

$$d = abc\omega,$$

where ω is a ternary derived operator. Now by Proposition 5.5, $F/\mathfrak{q}$ is the free $\mathscr{K}$-algebra on a,c and $F/\mathfrak{r}$ is the free $\mathscr{K}$-algebra on a,c. But in $F/\mathfrak{q}$, $d = c$, whence

$$aac\omega = c, \tag{4}$$

while in $F/\mathfrak{r}$, $d = a$, and so

$$acc\omega = a. \tag{5}$$

Both (4) and (5) are equations between the elements of the free $\mathscr{K}$-algebra on a,c, and therefore these equations hold identically in every $\mathscr{K}$-algebra. ∎

EXERCISES

1. Let Ω consist of a ternary operator ω and Ω' consist of σ,τ where $xy\sigma = xxy\omega$, $xy\tau = xyy\omega$. If W, W' are the Ω-word algebra and the Ω'-word algebra, respectively, on a set X, show that the canonical homomorphism $W' \to W$ (obtained by extending the identity mapping $1: X \to X$) is not injective.

2. Show that there is no derived ternary operator on lattices to satisfy the conditions of Theorem 7.3. (Take a totally ordered set.)

3. Show that on a relatively complemented lattice all congruences commute. (Take $xyz\omega$ to be the complement of y in $[x \wedge z, x \vee z]$ and apply Theorem 7.3.)

4. Show that on a free groupoid with at least three free generators not all congruences commute.

5. (Malcev.) A quasigroup may be defined as an algebra with three binary operators, satisfying certain identities, namely the product $ab\mu$ and the solutions of $xb\mu = c$, $ay\mu = c$, denoted by $cb\rho$, $ca\lambda$ respectively (cf. IV.2). Show that on a nonempty quasigroup, regarded as an algebra with respect to μ, λ, and ρ, all congruences commute. (Take $xyz\omega = xay\lambda\mu az\lambda\rho$, where a is any fixed element.)

6. Let Ω be any operator domain, and suppose that with each $\omega \in \Omega(n)$, a derived n-ary operator w_ω is associated. If A is any Ω-algebra and $W = W_\Omega(X)$, show that any mapping $\theta: X \to A$ can be extended to a unique mapping $\bar{\theta}: W \to A$ such that for any $\omega \in \Omega(n)$ and $a \in A^n$,

$$(a\omega)\bar{\theta} = (a\bar{\theta})w_{\omega}.$$

Does this still hold if W is replaced by the free algebra on X in a category $\mathcal{K}$ with free algebras and A by any $\mathcal{K}$-algebra? (Try the category of algebras with a unary operator $x \to x'$ satisfying $x'' = x$.)

8. PRESENTATIONS OF Ω-ALGEBRAS

In practice, algebraic structures often arise in a natural way as sets of homomorphisms of a given structure. For instance, groups arise as sets of automorphisms, semigroups as sets of endomorphisms, rings as sets of endomorphisms of abelian groups. But it may also happen that an Ω-algebra is defined abstractly, by its carrier together with the effect of the operators—in the form of 'multiplication tables', giving for each n-ary operator its effect on each n-tuple. This description can often be given in a more economical form. Suppose for definiteness that we are dealing with an algebra A in a given category $\mathcal{K}$ of Ω-algebras. Instead of prescribing the whole carrier, it is enough to give a generating set X of A, and instead of complete multiplication tables, we need only give enough entries to determine the remaining ones completely. Each entry in the multiplication table for $\omega \in \Omega(n)$ is of the form

$$a_1 \cdots a_n \omega = b \qquad (a_1, \cdots, a_n, b \in A).$$

If we express $a_1, \cdots, a_n, b$ in terms of the generating set X, we obtain a relation

$$f(x) = g(x),$$

where f,g are certain Ω-words in $x_1, \cdots, x_r \in X$. Any set of relations which suffices to determine the effect of all the operators in A is called a *set of defining relations* for A in terms of the generating set X.

If Φ is such a set of defining relations, the definition of A in terms of X and Φ is called a *presentation* of A and this is denoted by

$$A = \mathcal{K}\{X \mid \Phi\}. \tag{1}$$

For a given algebra A there are of course many presentations, since there are usually many ways of choosing a generating set X, and depending on this choice, and on $\mathcal{K}$, many ways of choosing a set of defining relations.

In practice, one tries to keep the category $\mathscr{K}$ as small as possible so as to simplify the defining relations. Thus, e.g., a presentation of a particular group which happens to be abelian can usually be simplified if we restrict ourselves to the category of abelian groups, since any relations expressing commutativity can then be omitted.

It will be convenient to extend the notion of presentation defined above as follows: Instead of taking X to be a generating set of A, we allow X to be any set of symbols each of which is identified with a certain element of A in such a way that the elements of A so obtained form a generating set. This is more general insofar as distinct elements of X may represent the same element of A. It also corresponds more closely to the practical use that is made of presentations, for when the category $\mathscr{K}$ is suitably restricted, any set X with any set of defining relations will give a $\mathscr{K}$-algebra (cf. Theorem 8.2 below), but it is usually a nontrivial problem to decide when two given elements of X represent the same element of A (this is in fact a special case of the word problem, cf. III.9).

Thus a presentation of a $\mathscr{K}$-algebra A is a set X together with a set Φ of relations between Ω-words in X. We remark that when the category $\mathscr{K}$ of Ω-algebras is unrestricted, the presentation $\mathscr{K}\{X \mid \Phi\}$ may define no algebra at all in $\mathscr{K}$. For example, let $\mathscr{K}$ be the category of infinite groups and $\{X \mid \Phi\}$ the presentation of a finite group, say $\mathscr{K}\{x \mid x^2 = 1\}$; clearly, there is no infinite group with this presentation. To obtain a sufficient condition which is convenient for the applications, we need a lemma, which will be used again later.

Lemma 8.1

Let A and B be any Ω-algebras. Given a set X and mappings $\alpha: X \to A$, $\beta: X \to B$ such that (i) *A is generated by* im α, (ii) *any relation in A between the elements $x\alpha$ ($x \in X$) also holds between the corresponding elements $x\beta$ in B; then there exists a unique homomorphism $\phi: A \to B$ such that $\alpha\phi = \beta$.*

Proof:

Let $W = W_\Omega(X)$ be the Ω-word algebra on X; then the mappings α, β may be extended to homomorphisms $\bar{\alpha}: W \to A$, $\bar{\beta}: W \to B$ respectively. Since im $\bar{\alpha}$ is a subalgebra of A containing im α, it follows by (i) that im $\bar{\alpha} = A$, i.e. $\bar{\alpha}$ is an epimorphism. Let $\mathfrak{q} = \ker \bar{\alpha}$; then there is an isomorphism

$$\alpha^*: W/\mathfrak{q} \to A,$$

such that $(\text{nat}\ \mathfrak{q})\alpha^* = \bar{\alpha}$. Now by (ii), $\mathfrak{q} \subseteq \ker \bar{\beta}$, hence there exists a homomorphism $\beta^* : W/\mathfrak{q} \to B$ such that $(\text{nat}\ \mathfrak{q})\beta^* = \bar{\beta}$. If we put $\phi = (\alpha^*)^{-1}\beta^*$, then

$$\alpha\phi = i\bar{\alpha}\phi = i(\text{nat}\ \mathfrak{q})\alpha^*\phi = i(\text{nat}\ \mathfrak{q})\beta^* = i\bar{\beta} = \beta,$$

where $i : X \to W$ is the inclusion mapping. Thus ϕ is a mapping of the required form, and it is unique because any two homomorphisms satisfying the conditions of the lemma must agree on im α and hence on A. ∎

Theorem 8.2

Let $\mathscr{K}$ be a residual category of Ω-algebras. Then every presentation

$$\mathscr{K}\{X \mid \Phi\} \tag{2}$$

defines a $\mathscr{K}$-algebra, unique up to isomorphism.

Proof:

Let $W = W_\Omega(X)$ and let $(\phi_\lambda)_{\lambda \in \Lambda}$ be the family of all epimorphisms from W to some $\mathscr{K}$-algebra A_λ such that

$$f(x\phi_\lambda) = g(x\phi_\lambda) \qquad \text{for all relations } (f,g) \text{ in } \Phi.$$

If we put $\mathfrak{q}_\lambda = \ker \phi_\lambda$, then $W/\mathfrak{q}_\lambda \cong A_\lambda$, and hence if $\mathfrak{q} = \bigcap \mathfrak{q}_\lambda$, $W/\mathfrak{q}$ is a subdirect product of the A_λ, and so is a $\mathscr{K}$-algebra. We assert that

$$W/\mathfrak{q} \cong \mathscr{K}\{X \mid \Phi\}. \tag{3}$$

In the first place, if $\phi = \text{nat}\ \mathfrak{q}$, then $X\phi$ generates $W/\mathfrak{q}$, and clearly, all the relations Φ hold in $W/\mathfrak{q}$; thus (3) will follow if we can show that Φ is a set of defining relations for $W/\mathfrak{q}$, as $\mathscr{K}$-algebra. This amounts to showing that every $\mathscr{K}$-algebra generated by X with relations Φ is a homomorphic image of $W/\mathfrak{q}$. Let B be a $\mathscr{K}$-algebra generated by X with the relations Φ holding; then by the lemma there is a homomorphism $\theta : W/\mathfrak{q} \to B$ which is induced by the identity mapping on X, which is what we had to show. It is also clear from this that the algebra given by the presentation (2) is unique up to isomorphism. ∎

Later we shall meet important classes of categories in which the condition of Theorem 8.2 is always satisfied. We have seen already that the category of infinite groups does not satisfy this condition; it is not residual because it does not contain the trivial group. As another corollary of the lemma we have a theorem first proved by Dyck [1882] in the case of groups.

Theorem 8.3

Let $\mathscr{K}$ be any category of Ω-algebras, and A any $\mathscr{K}$-algebra, with the presentation

$$A = \mathscr{K}\{X \mid \Phi\}.$$

If B is a $\mathscr{K}$-algebra with generating set X, such that all the relations Φ hold in B, then B is a homomorphic image of A. ▮

This theorem expresses the fact, which for residual categories also follows from Theorem 8.2, that the algebra A with the presentation (1) is universal for the representation of X in $\mathscr{K}$-algebras by mappings such that the relations Φ hold for the images of the elements of X.

The presentation (1) of an algebra A is said to be *finite*, if both X and Φ are finite. For an algebra A which is finitely presented in this way, there is a simple method of obtaining all finite presentations from a given one, due (for groups) to Tietze (cf. Shoda [49]).

Theorem 8.4

Let $\mathscr{K}$ be a residual category of Ω-algebras, and let A be a $\mathscr{K}$-algebra with a finite presentation

(1) $$A = \mathscr{K}\{X \mid \Phi\}.$$

Then any other finite presentation of A is obtained from (1) *by the following operations and their inverses:*

(i) *If (u,v) is a consequence of Φ, replace Φ by $\Phi \cup \{(u,v)\}$.*
(ii) *If u is any word in X and y is any letter not occurring in X, replace X by $X \cup \{y\}$ and Φ by $\Phi \cup \{(y,u)\}$.*

Proof:

Clearly (i) and (ii) when applied to any finite presentation of A yield another finite presentation. Now let A be given by the finite presentation (1) and let

(4) $$A = \mathscr{K}\{Y \mid \Psi\}$$

be another finite presentation. Since (4) determines A only up to isomorphism, we may assume that $X \cap Y = \emptyset$. Now each $y \in Y$ can be expressed in terms of X by (1), say $y = f_y(x)$, and we thus obtain the presentation

(5) $$A = \mathscr{K}\{X \cup Y \mid \Phi \cup \Phi'\}, \quad \Phi' = \{(y, f_y(x)) \mid y \in Y\}$$

from (1) by operations of the form (ii). But each element of Ψ is a consequence of $\Phi \cup \Phi'$, so by applying (i) we get

(6) $$A = \mathscr{K}\{X \cup Y \mid \Phi \cup \Phi' \cup \Psi\}.$$

Finally, each $x \in X$ has the form $x = g_x(y)$, and if

$$\Psi' = \{(x, g_x(y)) \mid x \in X\},$$

we have, by another application of (ii),

(7) $$A = K\{X \cup Y \mid \Phi \cup \Phi' \cup \Psi \cup \Psi'\}.$$

Clearly (7) is symmetric in the presentations (1) and (4); by reversing the steps, with the roles of (1) and (4) interchanged, we therefore reach (4). ∎

The main application of this result is in the following situation: The algebra A is given by a finite presentation (1) and Y is another generating set. Then $y = f_y(x)$ for each $y \in Y$, and as in the above proof we reach a presentation (5). This may now be simplified by applying the inverses of (i), (ii) to eliminate as many as possible of the elements of the old generating set X. In using Theorem 8.4 it is important to note that we are given that (1) and (4) are presentations of the same algebra. If we are given merely two finite presentations, there is in general no method for deciding when the corresponding algebras are isomorphic (the existence of such a method in the general case would imply a positive solution of the word problem, cf. III.9).

EXERCISES

1. Let G be the group with the presentation $G = \mathrm{Gp}\{x, y \mid x^3 = y^2 = 1\}$; if $z = xy^{-1}$, show that in terms of y and z, G has the presentation $G = \mathrm{Gp}\{y, z \mid y^2 = (yz)^3 = 1\}$.

2. Prove Theorems 8.2 and 8.3 by using Theorem 4.2. (See the remark following Theorem 8.3.)

9. THE WORD PROBLEM

Let $\mathscr{K}$ be a residual category of Ω-algebras, and let A be a $\mathscr{K}$-algebra, given by a presentation

(1) $$A = \mathscr{K}\{X \mid \Phi\}.$$

The elements of A are represented by Ω-words in X, where two words f, g represent the same element of A if and only if one can be obtained from the other by applying the relations Φ and the rules in $\mathscr{K}$. Here the role of $\mathscr{K}$ is simply that of providing certain identical relations or laws (cf. IV.1) which must hold in $\mathscr{K}$. Thus for the definition of A by means of a presentation (1) to be of any practical value, we require an algorithm, i.e. a rule for deciding in a finite number of steps when two Ω-words in X represent the same element of A. The problem of finding such an algorithm is called the *word problem* for A (in the presentation (1)). For certain classes of algebras, the word problem has been completely solved; thus e.g. the basis theorem for finitely generated abelian groups implies that every finitely generated abelian group has a presentation for which the word problem can be solved by inspection. The word problem has also been solved for certain classes of nonabelian groups, but for groups in general it has been shown to be insoluble. More precisely, there exist groups with a presentation for which the word problem is insoluble (Novikov [55], Boone [57], Britton [58]). Even if the word problem in a group G is insoluble, in the sense described, there may be an algorithm for enumerating all the pairs of words which represent equal elements in G (though for a given pair, no means is at hand for deciding in a finite number of steps whether they are equal). Such a group is said to have a *recursively soluble* word problem. An example of such a group is any finitely presented group, or more generally, any finitely generated subgroup of a finitely presented group. Now Higman [61] has proved that conversely, every finitely generated group with a recursively soluble word problem can be embedded in a finitely presented group. Since not every finitely generated group can be so embedded, this shows the existence of groups whose word problem is recursively insoluble.

To demonstrate the insolubility, one has of course to define much more precisely what constitutes an algorithm. On the other hand, in proving the solubility of word problems (which is all we shall do), the above vague formulation is sufficient. We shall discuss one general method of solving the word problem which can often be applied in practical cases. For simplicity we take for $\mathscr{K}$ the category (Ω) of all Ω-algebras and homomorphisms, and suppose that the Ω-algebra A has the presentation

$$A = \Omega\{X \mid \Phi\}, \tag{2}$$

where Φ consists of the relations

$$u_\lambda(x) = v_\lambda(x) \qquad (\lambda \in \Lambda). \tag{3}$$

Thus A consists of equivalence classes of Ω-words in X, where two Ω-words f, g are equivalent if and only if there exist words $w_0, w_1, \cdots, w_n$ such that $w_0 = f$, $w_n = g$, and in any successive pair of words w_{k-1}, w_k, one is obtained from the other by replacing an occurrence of $u_\lambda(x)$ by $v_\lambda(x)$ (or vice versa). These equivalence classes of words are just the $\mathfrak{q}$-classes of the Ω-word algebra $W = W_\Omega(X)$, where $\mathfrak{q}$ is the congruence on W generated by the pairs (u_λ, v_λ), and $W/\mathfrak{q} \cong A$. Let

$$\theta : W \to A \tag{4}$$

be an epimorphism with kernel $\mathfrak{q}$; we recall that a transversal of A in W is a subset T of W meeting each $\mathfrak{q}$-class in exactly one element. T may also be characterized by the property that the restriction $\theta' = \theta \mid T$ is a bijection. In general θ' will not be a homomorphism, because T is not necessarily a subalgebra of W. However, we can use θ to define T as Ω-algebra isomorphic to A. Let $\tau : W \to W$ be the mapping which associates with each $w \in W$ the unique representative in T of the class $w^{\mathfrak{q}}$; this mapping τ is called the *retraction*[1] associated with T. Then each $\omega \in \Omega(n)$ defines an n-ary operation ω_T on T by the rule

$$a_1 \cdots a_n \omega_T = (a_1 \cdots a_n \omega)\tau.$$

This determines A (up to isomorphism) as soon as we know a transversal T and the corresponding retraction τ. Thus the word problem for A is reduced to the determination of a transversal for A in W.

When A is given by the presentation (2), we have a homomorphism (4), e.g. by taking $A = W/\mathfrak{q}$ and $\theta = \operatorname{nat} \mathfrak{q}$. Now if S is an arbitrary subset of W, it is usually easy to determine whether the restriction $\theta \mid S$ is surjective. If it is also injective it will follow that S is a transversal, but this is often difficult to check directly. Instead we may proceed as follows: Let S be any subset of W such that $S^{\mathfrak{q}} = W$ (i.e. $\theta \mid S$ is surjective), and for each $w \in W$ choose an element $w\sigma \in S \cap w^{\mathfrak{q}}$ in such a way that $w\sigma = w$ whenever $w \in S$. The resulting mapping $\sigma : W \to S$ is no longer uniquely determined by S (unless S is a transversal); let us call it an *idempotent mapping compatible with* $\mathfrak{q}$. We now define an Ω-algebra structure on S by the rule

$$a_1 \cdots a_n \omega_S = (a_1 \cdots a_n \omega)\sigma \qquad (\omega \in \Omega(n),\ a_i \in S). \tag{5}$$

Suppose that the relations (3) hold in S, i.e.

$$u_\lambda(x\sigma) = v_\lambda(x\sigma). \tag{6}$$

[1] In the special case when τ is a homomorphism this agrees with the definition in III.6.

Then we assert that S is a transversal for A. For since the defining relations for A hold in the algebra S, it follows that S is a homomorphic image of A, under the mapping which takes $x\theta$ to $x\sigma$ $(x \in X)$. Hence for any $s,t \in S$, if $s \equiv t \pmod{\mathfrak{q}}$, then $s = s\sigma = t\sigma = t$. Therefore each class $s^{\mathfrak{q}}$ meets S in a single element, and so S is a transversal.

If $\mathscr{K}$ is any residual category with free algebras, then the same reasoning applies if in place of W we take F, the free $\mathscr{K}$-algebra on X, though this is of practical use only if the word problem can be solved for F.[2] Thus we obtain

Theorem 9.1

Let $\mathscr{K}$ be a residual category with free algebras, and A a $\mathscr{K}$-algebra given by the presentation

$$A = \mathscr{K}\{X \mid \Phi\}, \tag{1}$$

where Φ consists of the relations

$$u_\lambda(x) = v_\lambda(x) \qquad (\lambda \in \Lambda).$$

Further, let F be the free $\mathscr{K}$-algebra on X and $\mathfrak{q}$ the congruence generated by all pairs $(u_\lambda, v_\lambda)(\lambda \in \Lambda)$. Then

$$A \cong F/\mathfrak{q}.$$

Moreover, if S is a subset of F such that $S^{\mathfrak{q}} = F$ and σ is an idempotent mapping of F into S compatible with $\mathfrak{q}$, then S may be defined as an Ω-algebra by the rule

$$a_1 \cdots a_n \omega_S = (a_1 \cdots a_n \omega)\sigma \qquad (\omega \in \Omega(n),\ a_i \in S).$$

If the resulting Ω-algebra is a $\mathscr{K}$-algebra and satisfies the relations

$$u_\lambda(x\sigma) = v_\lambda(x\sigma),$$

then S is a transversal for A and the algebra S is isomorphic to A. ∎

As an example in the use of this method, consider the definition of the symmetric group of degree three by generators and defining relations. We have the presentation

$$G = \mathrm{Gp}\{x,y \mid x^2 = y^3 = 1,\ x^{-1}yx = y^2\}. \tag{7}$$

[2] For many categories $\mathscr{K}$ this is the case; for (Ω) itself the word problem has, of course, the trivial solution that two words are equal if and only if they are identical.

Using the given relations, we can reduce every group word in x and y to the form

(8) $$x^r y^s \qquad (r = 0,1;\ s = 0,1,2).$$

Thus there are at most six $\mathfrak{q}$-classes, i.e., the order of G is at most six. To show that it is exactly six, we must show that the six elements (8) represent distinct elements of G. This may be done by taking a concrete realization of G, i.e. an example of a group on two generators x and y satisfying the defining relations of G. Since the symmetric group on three letters is such a group, this proves that G has six elements. But we may also establish the result without recourse to a realization of G, by applying Theorem 9.1. Thus we take the six elements (8) in F, compute their products in F, and reduce them to one of the elements (8). This gives the following multiplication table:

	1	x	y	xy	y^2	xy^2
1	1	x	y	xy	y^2	xy^2
x	x	1	xy	y	xy^2	y^2
y	y	xy^2	y^2	x	1	xy
xy	xy	y^2	xy^2	1	x	y
y^2	y^2	xy	1	xy^2	y	x
xy^2	xy^2	y	x	y^2	xy	1

We note incidentally that it is not necessary to define the mapping σ for the whole of F, but only for the products of the generators. To complete the construction, it is now only necessary to show that the above table defines a group for which the given relations between x and y are satisfied. That the relations hold is immediate, from an inspection of the table, and of the laws defining groups only the associative law is not immediately obvious; but the process of checking it is purely mechanical, although tedious. We note that once the number of possible $\mathfrak{q}$-classes has been shown to be finite, as in this example, the remaining part of the problem can always be carried out in a finite number of steps. We therefore obtain the following corollary to Theorem 9.1:

Corollary 9.2

If F and $\mathfrak{q}$ are as in Theorem 9.1 *and S is a finite subset of F such that $S^{\mathfrak{q}} = F$ and there is an algorithm associating with any element a of F an element of $S \cap a^{\mathfrak{q}}$, then the word problem for $A = F/\mathfrak{q}$ is soluble.* ∎

As we saw in the above example, in the case of groups the main task is the verification of the associative law. This work can be cut down if we take as our category not the category of all groups but the category of permutation groups on the set S.

There is a graphical representation of the process described in Theorem 9.1 which is sometimes helpful. We represent the elements of F as vertices of a graph; let $w \in F$ be an element in which u_λ occurs, for some $\lambda \in \Lambda$, say

$$w = f(x, u_\lambda),$$

where f is an Ω-word in the x's and a single occurrence of u_λ. If w' is the word defined by

$$w' = f(x, v_\lambda),$$

then we draw a segment from w to w' in our graph and say: w' may be reached by a *direct move* from w, and w is reached by an *inverse move* from w'. In this way we obtain an oriented graph Γ on the elements of F as vertices. Clearly, the different $\mathfrak{q}$-classes are just the connected components of Γ. Now apply Theorem I.4.9; let us call an element of F *reduced* if no direct moves can be applied to it (i.e., if it is minimal in the preordering on Γ); then we have

Theorem 9.3

Let $\mathscr{K}$ be a residual category with free algebras and let A be a $\mathscr{K}$-algebra with the presentation

$$A = \mathscr{K}\{X \mid \Phi\}.$$

Assume further that:

(i) *For each $w \in F$ there exists an integer k such that the number of successive direct moves which can be applied to w cannot exceed k.*
(ii) *If w_1 and w_2 are obtained by a direct move from the same element $u \in F$, then there is an element $v \in F$ which can be obtained by direct moves from w_1 and w_2.*

Then each element $w \in F$ can be transformed by a finite number of direct moves into a reduced word $\bar{w}$, which depends only on w and not on the moves chosen. Moreover, the reduced words form a transversal of A in F.

For the elements of A correspond to $\mathfrak{q}$-classes, i.e., connected components of the graph defined by the presentation. Therefore the minimal elements form a transversal of A in F, by Theorem I.4.9, and the result follows from Theorem 9.1. ∎

The reduced word $\bar{w}$ is often called the *normal form* of w. This theorem again provides a solution for the word problem for A as presented in (1): to see whether w_1 and w_2 represent the same element of A, we need only pass to the corresponding reduced words $\bar{w}_1$, $\bar{w}_2$ and see whether they are equal.

We note that when $\mathscr{K} = (\Omega)$, condition (i) of the theorem is always fulfilled if $l(u_\lambda) > l(v_\lambda)$ for all $\lambda \in \Lambda$, for then each move decreases the length of w and so no more than $l(w)$ can be applied in succession.

But in other cases care may be needed in applying the theorem. Suppose e.g. that we have a ring, in which a direct move consists in replacing an element u by u^*. Then we have the infinite chain of reductions

$$0 = u - u = u^* - u = u^* - u^* = 0 = \cdots$$

so (i) is not satisfied. However, in any concrete instance it is usually clear what has to be done to make the result applicable. In the case of rings a precise formulation, under quite general conditions, has been given by G. M. Bergman; this is sketched in X. 4.

EXERCISE

1. Let $\mathscr{K}$ be a residual category of Ω-algebras which has free algebras. If there is an algorithm for deciding when two presentations define isomorphic $\mathscr{K}$-algebras, show that the word problem for presentations of $\mathscr{K}$-algebras is soluble.

Chapter IV

Varieties

Many important classes of algebras occurring in practice, such as groups, rings, lattices, etc., may be completely described by identical relations. Such equational classes or varieties have many useful properties; in particular, they always possess free algebras, and the members of the variety may be characterized as homomorphic images of the free algebras. Any variety $\mathscr{V}$ of Ω-algebras gives rise to a subcategory of (Ω), namely the full subcategory whose objects are all the members of $\mathscr{V}$. The resulting category is always local, residual, and hereditary, and admits homomorphic images and direct products. Conversely, the algebras of any residual category admitting homomorphic images form a variety. Thus most of the results of Chapter III apply to varieties; in addition, there are a number of features special to varieties, which also are discussed briefly.

1. DEFINITION AND BASIC PROPERTIES

Let an operator domain Ω and a set X be given; as before, we write $W = W_\Omega(X)$ for the Ω-word algebra on X. If A is any Ω-algebra and $\alpha: W \to A$ a homomorphism, then α maps any Ω-word w to an element $w\alpha$ of A, which is a value of w in A. If we regard w as a derived operator, then its values in A just constitute the image of the operation in A defined by w.

Definition

A *law* or *identity* over Ω in the alphabet X is a pair $(w_1,w_2) \in W^2$, or sometimes the equation

$$(1) \qquad w_1 = w_2$$

formed from this pair. We say that the law (1) *holds* in A, or that A *satisfies* (1), if under every homomorphism $W \to A$ the values of w_1 and w_2 coincide. In other words: the derived operators w_1 and w_2 define the same operation on A.

This relation between laws and algebras establishes a Galois connexion between the class of all Ω-algebras (within a given universe U) and the set of all laws in the given alphabet X. If Σ is any set of laws, then $\mathscr{V}_\Omega(\Sigma)$, the *variety* defined by Σ, is the class of all Ω-algebras which satisfy all the laws in Σ. Thus, by a variety of Ω-algebras, we understand the class of all Ω-algebras satisfying some given set of laws. Other terms used instead of 'variety' are 'equationally definable class' (Tarski) and 'primitive class' (Malcev).

Of course varieties as defined above still depend on the alphabet X, but we shall now show that all varieties may be obtained by using any fixed alphabet which is infinite, but otherwise arbitrary. Usually, we shall take as our standard alphabet a countable set

$$X_0 = \{x_1,x_2,\cdots\}$$

which is indexed by the positive integers.

Theorem 1.1

Let Σ be any set of laws over Ω in an alphabet X. Then the variety defined by Σ may also be defined by a set Σ_0 of laws in the standard alphabet X_0 (or more generally, in any infinite alphabet).

Proof:

We shall say that two laws (not necessarily in the same alphabet) are *equivalent* if in every Ω-algebra, either both hold or neither holds. For example, by renaming the variables, we pass from any law to an equivalent law. Since any law in X depends only on finitely many elements in X, we can always replace it by an equivalent law in X_0, and thus obtain a set Σ_0 of laws in X_0 which is equivalent to Σ. ∎

Henceforth we shall assume every variety to be defined by a set of laws in the standard alphabet. We now look more closely at the Galois connexion between the class of all Ω-algebras and W_0^2, where $W_0 = W_\Omega(X_0)$. This Galois connexion defines two closure systems (cf. II.1): a closed set of Ω-algebras is just a variety; to see what a closed set of laws is, we need another definition. We shall say that a congruence $\mathfrak{q}$ on an Ω-algebra A is *fully invariant*, if it admits every endomorphism of A.

Theorem 1.2

Let $W_0 = W_\Omega(X_0)$ be the Ω-word algebra on the standard alphabet. Then the Galois connexion between Ω-algebras and laws establishes a natural bijection between varieties of Ω-algebras and fully invariant congruences on W_0.

Proof:

For any class $\mathscr{C}$ of Ω-algebras, let $\mathscr{C}'$ be the set of laws holding in all $\mathscr{C}$-algebras, and for any set Σ of laws, let $\Sigma' = \mathscr{V}_\Omega(\Sigma)$ be the variety defined by Σ. We note first that $\mathscr{C}'$ is a fully invariant congruence on W_0. The congruence properties are clear: in every $\mathscr{C}$-algebra we have $w = w$, for any $w \in W_0$; if $w_1 = w_2$ holds, then so does $w_2 = w_1$, and if $w_1 = w_2$, $w_2 = w_3$ hold, then $w_1 = w_3$ holds too. Further, if $u_i = v_i$ $(i = 1,\cdots,n)$ are laws holding in A and $\omega \in \Omega(n)$, then

$$u_1 \cdots u_n \omega = v_1 \cdots v_n \omega$$

holds in A. Now let $(w_1,w_2) \in \mathscr{C}'$ and let θ be any endomorphism of W_0. If $\alpha: W_0 \to A$, where $A \in \mathscr{C}$, is any homomorphism, then so is $\theta\alpha$, whence $w_1\theta\alpha = w_2\theta\alpha$. This shows that the law $w_1\theta = w_2\theta$ holds in A, and so $(w_1\theta,w_2\theta) \in \mathscr{C}'$; thus $\mathscr{C}'$ is fully invariant.

To complete the proof, we show that

$$\mathscr{V}'' = \mathscr{V} \tag{2}$$

for any variety $\mathscr{V}$, and

$$\mathfrak{q}'' = \mathfrak{q} \tag{3}$$

for any fully invariant congruence $\mathfrak{q}$ on W_0.

By the definition of a variety, $\mathscr{V} = \Sigma'$ for some $\Sigma \subseteq W_0^2$; hence $\mathscr{V}'' = \Sigma''' = \Sigma' = \mathscr{V}$, i.e., (2). Next, let $\mathfrak{q}$ be a fully invariant congruence on W_0. We assert that

$$W_0/\mathfrak{q} \in \mathfrak{q}'. \tag{4}$$

To prove this it is enough to show that the laws corresponding to the elements of $\mathfrak{q}$ hold in $W_0/\mathfrak{q}$. Thus, let $(w_1,w_2) \in \mathfrak{q}$ and let $\alpha: W_0 \to W_0/\mathfrak{q}$ be any homomorphism. By Proposition III.5.7 there is an endomorphism $\alpha' : W_0 \to W_0$ which makes the diagram

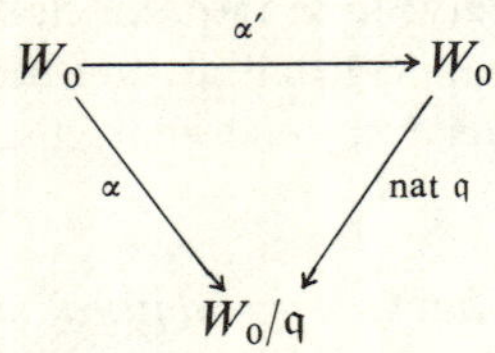

commutative. Now $\mathfrak{q}$ is fully invariant; hence $(w_1\alpha', w_2\alpha') \in \mathfrak{q}$, and so $w_1\alpha = w_1\alpha'(\text{nat } \mathfrak{q}) = w_2\alpha'(\text{nat } \mathfrak{q}) = w_2\alpha$, as we wished to show. Thus (4) holds. Turning to (3), we note that in any case $\mathfrak{q}'' \supseteq \mathfrak{q}$. Now, if $(w_1, w_2) \notin \mathfrak{q}$, then $w_1 = w_2$ is not a law in $W_0/\mathfrak{q}$, because $w_1 \not\equiv w_2 (\text{mod } \mathfrak{q})$, but $W_0/\mathfrak{q} \in \mathfrak{q}'$ by (4), and so $(w_1, w_2) \notin \mathfrak{q}''$; therefore $\mathfrak{q}'' \subseteq \mathfrak{q}$ and (3) follows. ∎

From the definitions in II.2, we see that the following classes of algebras are varieties: groupoids, semigroups (with neutral element), groups (with operators), abelian groups, rings (with unit element), R-modules (for a given ring R), and lattices. Other classes of algebras which do not appear to be varieties at first sight may sometimes be defined as varieties. This was already done for groups in II.2. Similarly, we may define quasigroups as a variety of algebras with three binary operators μ, ρ, λ by writing the multiplication as $ab\mu$, and if

(5) $$ab\mu = c,$$

putting $a = cb\rho$, $b = ca\lambda$. In a quasigroup these operators satisfy the laws

(6) $$xy\mu y\rho = x, \quad xy\mu x\lambda = y, \quad zy\rho y\mu = z, \quad xzx\lambda\mu = z.$$

Conversely, any (μ,ρ,λ)-algebra satisfying (6) is a quasigroup, as is easily verified. Of course this is not entirely equivalent to the previous definition of quasigroup, because the notion of subalgebra is different. For this reason, the algebras of the variety defined by (6) are sometimes called *equasigroups*. In the same way, loops (or more precisely, eloops) can be defined as a variety of algebras with three binary operators and one 0-ary operator.

If C is any fixed Ω-algebra, then the Ω-algebras over C form a variety; for, as we saw in II.2, an Ω-algebra over C may be defined as an algebra with operators Ω and certain 0-ary operators (corresponding to the elements of C) satisfying certain laws.

EXERCISES

1. Show that the algebra $W_0/\mathfrak{q}$ in (4) if non-trivial is the free $\mathfrak{q}'$-algebra on X_0.

2. Show that modular lattices form a variety.

3. Show that a variety of groups definable by a finite set of laws is also definable by a single law.

4. (Malcev [54].) Let Q be a nonempty quasigroup and $u \in Q$. Show that Q is a loop with neutral element u, with respect to the derived operations $xy\mu' = xuu\lambda\mu uy\lambda\rho$, $xy\rho' = xuy\lambda\mu uu\lambda\rho$, $xy\lambda' = uyuu\lambda\mu x\lambda\rho$.

5. Show that groups may be defined, in terms of right division as binary operator, as nonempty algebras satisfying $xz\rho yz\rho\rho = xy\rho$, $xx\rho yy\rho y\rho\rho = y$. Similarly, abelian groups are defined by the laws $xxy\rho\rho = y$, $xy\rho z\rho = xz\rho y\rho$.

6. (Higman & Neumann [52].) Show that groups may be defined by the single law $xxx\rho y\rho z\rho xx\rho x\rho z\rho\rho\rho = y$, and abelian groups by the single law $xyz\rho yx\rho\rho\rho = z$.

2. FREE GROUPS AND FREE RINGS

A variety containing algebras with more than one element always has free algebras. This will follow from the characterization of varieties given in the next section, but it can also be seen directly: the free $\mathscr{V}$-algebra on X is obtained by taking the Ω-algebra

$$A = \Omega\{X \mid \Phi\}, \tag{1}$$

where Φ consists of all the instances of all the laws of $\mathscr{V}$. In particular, this ensures the existence of free algebras for the categories mentioned in IV.1. For the free algebra to be of real use, however, there must be a simple normal form for its elements, i.e., the word problem should admit a simple solution. This is the case for groups and associative rings, and we shall describe this normal form here, partly as an application of the techniques of III.8–9, and partly to throw more light on the concept of a free algebra.

It is best to begin the discussion with semigroups. The normal form for the elements of a free semigroup is extremely easy to obtain and is of use in discussing groups and even general algebras (cf. IV.4).

Let Sg be the category of semigroups. To obtain the free semigroup on a set X, we consider the groupoid Φ_X whose elements are all the nonempty rows of elements of X, with juxtaposition as multiplication. It is easily seen that Φ_X is associative, i.e. a semigroup, and since it is generated by X, it must be a homomorphic image of the free groupoid Γ_X on X, i.e. the word algebra on X. But any two elements of Γ_X which are identified by this homomorphism can only differ in the placing of brackets (i.e., the order in which the multiplications are performed), and therefore they agree in every homomorphism into a semigroup. This shows that Φ_X is in fact the free semigroup on X; note that we have, in effect, used the presentation (1), which dispenses with the verification of the universal property.

If instead of Sg we consider the category Sg* of semigroups with neutral element 1, the free semigroup with 1 on X is obtained by adjoining a neutral element to Φ_X, i.e. by taking $\Phi_X^1 = \Phi_X \sqcup \{1\}$ and defining

$$a1 = 1a = a \qquad (a \in \Phi_X^1).$$

Interpreting 1 as the empty row, we may regard Φ_X^1 as the set of all rows in X (including the empty row) with juxtaposition as multiplication. Sometimes it is more convenient to take X to be indexed, say $X = (x_i)$; then the elements of Φ_X^1 are the *monomials*

$$x_I = x_{i_1} \cdots x_{i_n}$$

corresponding to the different rows $I = (i_1, \cdots, i_n)$ obtainable from the index set.

Coming now to groups, let us take as operator domain a set consisting of a binary operator (multiplication), a unary operator (inversion), and a 0-ary operator 1 (the neutral element), and write down the free group on X in the presentation (1). This presentation may still be simplified by the following almost trivial

Lemma 2.1

If a group G is generated by a set Y admitting inversion, then the semigroup with 1 *generated by Y is also G.*

We need only check that the inverse of any product of elements of Y is itself a product of elements of Y, and this follows from the formula

$$(y_1 \cdots y_n)^{-1} = y_n^{-1} \cdots y_1^{-1}, \tag{2}$$

because Y admits inversion. ∎

Now let F_X be the free group on X and write $X^{-1} = \{x^{-1} \mid x \in X\}$; then the set $Y = X \cup X^{-1}$ admits inversion and generates F_X, qua group.

On the other hand, consider the semigroup with 1 defined by the presentation

(3) $$E_X = \mathrm{Sg}^*\{X \cup X^{-1} \mid xx^{-1} = x^{-1}x = 1(x \in X)\}.$$

Since the relations of E_X all hold in F_X, it follows that F_X is a homomorphic image of E_X, and if we use (2) to define inverses generally in E_X, then E_X is actually a group and is therefore isomorphic to F_X. In this way we obtain a presentation of the free group on X as a semigroup on $Y = X \cup X^{-1}$; i.e., as a homomorphic image of Φ_Y^1. To obtain a normal form for the elements of F_X, we consider the graph on Φ_Y^1, whose segments are defined by the direct moves

$$uxx^{-1}v \to uv, \quad ux^{-1}xv \to uv \; (u,v \in \Phi_Y^1,\, x \in X)$$

and their inverses. Since a direct move decreases the length of each word (= element of Φ_Y^1) to which it is applied, the number of direct moves which can be applied to any word w is bounded by the length of w. Moreover, if w_1, w_2 are each obtained by a direct move from the same element u, then *either*: (i) these moves affect nonoverlapping regions of u, and so can be carried out independently to yield an element v obtained from each of w_1, w_2 by a direct move: $u = u_1yy^{-1}u_2zz^{-1}u_3$, $w_1 = u_1u_2zz^{-1}u_3$, $w_2 = u_1yy^{-1}u_2u_3$, and $v = u_1u_2u_3$, where $y,z \in Y$; *or* (ii) it may happen that the moves affect overlapping regions, say $u = u_1yy^{-1}yu_2$ $(y \in Y)$ and w_1, w_2 are obtained by applying the moves $yy^{-1} \to 1$, $y^{-1}y \to 1$ to the occurrence shown. In both cases, we reach u_1yu_2, and so $w_1 = w_2$. Thus, the conditions of Theorem III.9.3 are satisfied, and we therefore have established a normal form for the reduced words:

Theorem 2.2

If F_X is the free group on X, then each element of F_X can be uniquely expressed as a word of the form

(4) $$y_1y_2\cdots y_n \qquad (y_i \in X \cup X^{-1},\, y_{i-1} \neq y_i^{-1}). \quad \blacksquare$$

There are many other ways of proving this theorem; some of them, using features special to groups, are possibly more direct than the above proof, but none of them makes the assertion as trivial as it appears to be at first sight. The crux of any proof is to establish that the elements (4) as written down all represent different elements of F_X. A direct way of doing this would be to take the expressions (4) themselves as elements of a group and define a multiplication between them so as to obtain F_X, but then the associative law has to be verified (this would be an application of Theorem III.9.1). A slightly quicker way is to represent F_X as a permutation group on the elements (4); here it is only necessary to check the conditions for a representation.

As a second example we take the case of associative rings with unit element 1. Since any such ring may be regarded as a linear algebra over the ring of integers, we shall more generally consider linear K-algebras, where K is a commutative and associative ring with 1. A linear K-algebra A with 1 is then essentially a ring A with a canonical homomorphism $K \to Z(A)$, where $Z(A) = \{z \in A \mid zx = xz \text{ for all } x \in A\}$ is the centre of A. The category As_K of associative linear K-algebras with 1 is subordinate to Sg*, and we can therefore represent Sg* in As_K. The universal functor $U(S)$ for this representation is called the *semigroup algebra of* S *over* K. In terms of a presentation $S = \mathrm{Sg}^*\{X \mid \Phi\}$ of S we have

$$U(S) = \mathrm{As}_K\{X \mid \Phi\}.$$

This becomes particularly simple if we take $X = S$ and Φ the set of equations which make up the multiplication table of S. The elements of $U(S)$ are then expressed uniquely in the form $\Sigma s\alpha_s$, where $\alpha_s \in K$, $s \in S$ and $\alpha_s = 0$ for all but a finite number of s. Addition and multiplication are defined by the equations

$$\Sigma s\alpha_s + \Sigma t\beta_t = \Sigma s(\alpha_s + \beta_s), \quad (\Sigma s\alpha_s)(\Sigma t\beta_t) = \Sigma st\alpha_s\beta_t,$$

where st is the product of s and t in S. Thus $U(S)$ may be described as the free K-module on S as basis, with multiplication induced by the multiplication in S (i.e., basis elements multiply as in S and general elements by linearity). We note in passing that this description does not depend on the associative law in any way, so we may in the same way represent groupoids in (nonassociative) linear K-algebras; for any groupoid Γ, the *groupoid algebra of* Γ *over* K is the free K-module on Γ as basis, with mutliplication induced by the multiplication in Γ.

To obtain the free K-algebra on X, we form the free semigroup with 1 on X and take its semigroup algebra over K. Taking X to be indexed, for convenience, we may state the result as

Theorem 2.3

Let K be any commutative and associative ring with 1. *Then, for any set $X = (x_i)$, the free associative K-algebra with* 1 *on X is the free K-module on the monomials x_I as basis.* ∎

The free commutative associative K-algebra on X ($\bar{A}_X$, say) is obtained similarly as the semigroup algebra over K of the free commutative semigroup $\bar{\Phi}_X$ on X. Taking $X = (x_i)$ again to be indexed, with a totally ordered index set, we see that by the commutative law, the elements of $\bar{\Phi}_X$ may be uniquely expressed as *ascending* monomials

$$x_I = x_{i_1} \cdots x_{i_n}, \qquad \text{where } i_1 \leqslant \cdots \leqslant i_n.$$

Hence we obtain

Theorem 2.4

If K is as in Theorem 2.3 and $X = (x_i)$ is any set on a totally ordered index set, then the free commutative and associative K-algebra with 1 on X is a free K-module on the ascending monomials in X as basis. ∎

EXERCISES

1. (Morimoto.) If $f(x,y)$ is a derived operator in a group, such that $f(x,f(y,z)) = f(f(x,y),z)$, show that f is of one of the forms xy, yx, x, y, 1.

2. If $f(x, y)$ is a derived operator in a variety of associative K-algebras with 1 (K an integral domain) such that $f(x, f(y, z)) = f(f(x, y), z)$, show that f is of one of the forms x, y, $\alpha + \beta(x + y) + \gamma xy$ or $\alpha + \beta(x + y) + \gamma yx$, where $\alpha\gamma + \beta - \beta^2 = 0$.

3. Show that the number of elements of length k in the free semigroup on q free generators is q^k, and in the free group on q free generators the number is $2q(2q - 1)^{k-1}$. What are the corresponding numbers for the free commutative semigroup and the free abelian group?

3. THE GENERATION OF VARIETIES

From the definition of a variety it is not easy to tell whether a given class of Ω-algebras is a variety or not. We now give necessary and sufficient conditions, due to Birkhoff [35], for this to be the case.

Theorem 3.1

A full subcategory $\mathscr{K}$ of the category (Ω) of all Ω-algebras is a variety if and only if the following three conditions are satisfied:

(i) *$\mathscr{K}$ is hereditary.*
(ii) *$\mathscr{K}$ admits homomorphic images.*
(iii) *$\mathscr{K}$ admits direct products.*

Note that any category satisfying (ii) is abstract; if it satisfies (iii), it contains the trivial algebra, as empty product.

Proof:

The necessity of the conditions is clear and may be left to the reader to verify. To prove the sufficiency, we shall assume $\mathscr{K}$ to be nontrivial, since the trivial abstract category is clearly a variety.

If $\mathscr{V}$ denotes the class of all $\mathscr{K}$-algebras, then it is clear that

$$\mathscr{V}'' \supseteq \mathscr{V},$$

and we have to establish equality. Let $A \in \mathscr{V}''$ and express A as a homomorphic image of an Ω-word algebra W on a sufficiently large set X (by Theorem III.2.7):

(1) $$\theta : W \to A.$$

By Theorem III.5.3 and the remark following it, $\mathscr{K}$ has free algebras of any rank; let F be the free $\mathscr{K}$-algebra on X; then the identity mapping on X may be extended to an epimorphism $\alpha : W \to F$ with kernel $\mathfrak{q}$, say. If $(u,v) \in \mathfrak{q}$, then the Ω-words u and v have the same value in F, and hence in any $\mathscr{K}$-algebra; i.e., $u = v$ is a law of $\mathscr{K}$, and therefore this law is satisfied by A. Thus, $u\theta = v\theta$; hence $\mathfrak{q} \subseteq \ker \theta$, and dividing (1) by nat $\mathfrak{q}$, we obtain the epimorphism

$$\theta^* : F \to A.$$

Since $F \in \mathscr{V}$, it follows from (ii) that $A \in \mathscr{V}$. ▮

By examining the last part of the proof we obtain

Corollary 3.2

Let F be a free algebra of a category $\mathscr{K}$; then every relation in F is a law in $\mathscr{K}$ and conversely. ▮

We also have the following consequences of Theorem 3.1:

Corollary 3.3

Every nontrivial variety[1] *has free algebras.* ▮

[1] A variety is nontrivial if it is nontrivial as a category, i.e., if it contains algebras with more than one element.

Corollary 3.4

Every variety is local, residual, and admits free and direct composition.

This follows by Proposition II.7.4, Theorem III.6.1, and Theorem III.6.4. ∎

Let $\mathscr{K}$ be any category of Ω-algebras and $\mathscr{K}'$ the set of laws holding in every $\mathscr{K}$-algebra. Then $\mathscr{K}''$ is the variety defined by $\mathscr{K}'$ and is therefore the least variety containing every $\mathscr{K}$-algebra. We shall also say: $\mathscr{K}''$ is the variety *generated* by $\mathscr{K}$, and we write

$$\mathscr{K}'' = \mathrm{V}\mathscr{K}.$$

From its definition in terms of a Galois connexion, it follows that V is a closure operator on categories; in particular, $\mathscr{K}$ is V-closed if and only if $\mathscr{K}$ is itself a variety. An explicit expression for V, due to P. Hall, is given in

Theorem 3.5

If $\mathscr{K}$ is any category of Ω-algebras, then the variety generated by $\mathscr{K}$ consists of all homomorphic images of subdirect products of $\mathscr{K}$-algebras, i.e.

$$\mathrm{V} = \mathrm{QR}. \tag{2}$$

Proof:

The two sides of (2), applied to a trivial category, evidently produce the same result, so let $\mathscr{K}$ be a nontrivial category. By Theorem 3.1, $\mathrm{R}\mathscr{K} \subseteq \mathrm{V}\mathscr{K}$ and $\mathrm{QR}\mathscr{K} \subseteq \mathrm{V}\mathscr{K}$; so it remains to prove the opposite inclusion. By Theorem III.5.3, $\mathrm{R}\mathscr{K}$ is a category which has free algebras of all ranks exceeding some cardinal α, and every $\mathrm{R}\mathscr{K}$-algebra is a homomorphic image of one of these free algebras. Now consider $\mathrm{QR}\mathscr{K}$; this consists of all the homomorphic images of all the free $\mathrm{R}\mathscr{K}$-algebras. Thus $\mathrm{QR}\mathscr{K}$ has free algebras of rank $>\alpha$, and so $\mathrm{SQR}\mathscr{K}$ has free algebras of any rank. When the rank is positive, these algebras already lie in $\mathrm{QR}\mathscr{K}$ (by Proposition III. 5.5), so $\mathrm{QR}\mathscr{K}$ has free algebras of all positive ranks. Now let $A \in \mathrm{V}\mathscr{K}$, $A \neq \varnothing$, take any generating set $X \neq \varnothing$ and let F be the free $\mathrm{QR}\mathscr{K}$-algebra on X. Then any relation in F is a law in $\mathscr{K}$ (by Corollary 3.2) and therefore holds in A; by Dyck's theorem (Theorem III. 8.3) the identity mapping on X extends to an epimorphism $F \to A$, and so $A \in \mathrm{QR}\mathscr{K}$. In the excluded case ($A = \varnothing$) we can (formally) express A as direct product of the empty family. ∎

From this theorem we obtain the following strengthening of Birkhoff's criterion (Theorem 3.1).

Corollary 3.6

If $\mathscr{K}$ is any residual category admitting homomorphic images, then the $\mathscr{K}$-algebras form a variety.

For then $\mathrm{V}\mathscr{K} = \mathrm{QR}\mathscr{K} = \mathrm{Q}\mathscr{K} = \mathscr{K}$. ∎

We remark that neither Theorem 3.5 nor Corollary 3.6 contains an assertion about $\mathscr{K}$-homomorphisms; in fact, it can be shown that every residual category admitting homomorphic images is a full subcategory of (Ω), cf. Stanley [66].

The proof of Theorem 3.5 shows that if $\mathscr{K}$ is any category with free algebras, then the class of all homomorphic images of all free $\mathscr{K}$-algebras forms a variety, which can only be $\mathrm{V}\mathscr{K}$. It follows that the free $\mathscr{K}$-algebra on a set X is isomorphic to the free $\mathrm{V}\mathscr{K}$-algebra on X, or, in other words, since free algebras are determined only up to isomorphism and categories are abstract, every free $\mathrm{V}\mathscr{K}$-algebra belongs to $\mathscr{K}$. Conversely, if a free $\mathrm{V}\mathscr{K}$-algebra is in $\mathscr{K}$, it is also a free $\mathscr{K}$-algebra. We thus obtain the following criterion for a category to possess free algebras.

Proposition 3.7

A category $\mathscr{K}$ of Ω-algebras possesses free algebras if and only if every free $\mathrm{V}\mathscr{K}$-algebra is a $\mathscr{K}$-algebra. ∎

This proposition applies not only to subcategories of (Ω), but to any class $\mathscr{C}$ of Ω-algebras. Formally this could be brought within the above framework by considering the category $[\mathscr{C}]$ consisting of all algebras isomorphic to algebras in $\mathscr{C}$, with all homomorphisms between them. In applying the closure operators Q, R, V, we shall often omit the brackets and write $\mathrm{V}\mathscr{C}$ instead of $\mathrm{V}[\mathscr{C}]$, etc.

A case of special interest arises when $\mathscr{C}$ consists of a single algebra A. Then $\mathrm{V}\mathscr{C}$ is the least variety containing A; we also say that A is a *generic algebra* in $\mathrm{V}\mathscr{C}$, and write $\hat{A}$ instead of $\mathrm{V}\{A\}$. Expressed differently, if $\mathscr{V}$ is a variety, then a generic algebra for $\mathscr{V}$ is a $\mathscr{V}$-algebra A such that every law of A holds in $\mathscr{V}$.

Proposition 3.8

Every variety has generic algebras; in particular, if $\mathscr{V}$ is a nontrivial variety, then any free $\mathscr{V}$-algebra on an infinite alphabet is generic for $\mathscr{V}$.

Proof:

Clearly, the trivial algebra is generic for the trivial variety; so let $\mathscr{V}$ be a nontrivial variety and F a free $\mathscr{V}$-algebra on an infinite alphabet; then $\hat{F} \subseteq \mathscr{V}$. Now let $A \in \mathscr{V}$; then every finitely generated subalgebra of A is a $\mathscr{V}$-algebra, and hence is a homomorphic image of F. Therefore A is locally $\hat{F}$, but $\hat{F}$ is local (by Corollary 3.4), and so $A \in \hat{F}$. ▮

We have seen that a variety $\mathscr{V}$ may be defined either (i) by the set of all its laws in some infinite alphabet (Theorem 1.1), or (ii) by the free $\mathscr{V}$-algebra on some infinite alphabet (Proposition 3.8). When the alphabet is finite, these two descriptions need no longer be equivalent. To discuss the relation between them we first need some definitions.

Let $\mathscr{V}$ be any variety and denote the free $\mathscr{V}$-algebra of rank n by $F_n(\mathscr{V})$. For any cardinal n, we shall associate with $\mathscr{V}$ two varieties $\mathscr{V}^n$ and $\mathscr{V}_n$, the one containing $\mathscr{V}$ and the other contained in it. Namely, $\mathscr{V}^n$ is the class of all Ω-algebras satisfying all laws of $\mathscr{V}$ in at most n letters, and $\mathscr{V}_n$ is the class of all Ω-algebras satisfying all the laws in $F_n(\mathscr{V})$. It is clear from this definition that $\mathscr{V}^n$ and $\mathscr{V}_n$ are varieties, and the above remarks show that $\mathscr{V}^n = \mathscr{V}_n = \mathscr{V}$ for all infinite n. Further, it is easily verified that

$$\mathscr{V}_1 \subseteq \mathscr{V}_2 \subseteq \cdots \subseteq \mathscr{V} \subseteq \cdots \subseteq \mathscr{V}^2 \subseteq \mathscr{V}^1; \quad \bigcup \mathscr{V}_n = \bigcap \mathscr{V}^n = \mathscr{V}.$$

The least value of n such that $\mathscr{V}^n = \mathscr{V}$ is called the *axiom rank* of $\mathscr{V}$ and is denoted by $r_a(\mathscr{V})$; the least n such that $\mathscr{V}_n = \mathscr{V}$ is called the *base rank* of $\mathscr{V}$ and is denoted by $r_b(\mathscr{V})$. From the definition we see that r_a is the least integer such that $\mathscr{V}$ may be defined by laws in r_a letters (or $\aleph_0$, if no such integer exists), while r_b is the least integer such that every $\mathscr{V}$-algebra satisfies all the laws holding in the free $\mathscr{V}$-algebra of rank r_b, i.e., such that $F_{r_b}(\mathscr{V})$ is generic for $\mathscr{V}$ (or $\aleph_0$, if no such integer exists). In general neither r_a nor r_b need be finite; we refer to Higman [59] for an example of a variety of groups for which r_b is infinite and to Lyndon [54] for an example of a variety of algebras for which r_a is infinite. It is not known whether r_a is finite for every variety of groups, but for many well-known varieties of algebras both r_a and r_b are finite. To facilitate the calculation of $r_a(\mathscr{V})$ we may use

Proposition 3.9

Let $\mathscr{V}$ be any variety of Ω-algebras; then an algebra A belongs to $\mathscr{V}^n$ if and only if all its n-generator subalgebras belong to $\mathscr{V}$.

Proof:

Put $F_n = F_n(\mathscr{V})$ and let X_n be a free generating set of F_n. By Corollary 3.2 any relation in F_n is a law of $\mathscr{V}$, and it clearly involves at most n letters. Now assume that $A \in \mathscr{V}^n$ and let

(3) $$\theta : X_n \to A$$

be any mapping; then any relation in F_n is a law of $\mathscr{V}$ in at most n letters, and therefore holds in A; hence θ extends to a homomorphism. This holds for any mapping θ; therefore, all n-generator subalgebras of A are in $\mathscr{V}$. Conversely, when this condition is satisfied, any mapping (3) extends to a homomorphism, hence any law in at most n letters holds in A because it holds in F_n, and so $A \in \mathscr{V}^n$. ∎

As an application we determine the axiom and base ranks for the variety Gp of all groups. Since groups can be defined by laws in 3 letters, we have $r_a(\mathrm{Gp}) \leqslant 3$. We assert that equality holds; to establish this, it is enough to show that $\mathrm{Gp}^2 \supset \mathrm{Gp}$, and this will follow if we exhibit a loop whose 2-generator subloops are all groups. Such a loop is obtained by taking the elements $\pm e_i$, where e_i $(i = 1, \cdots, 8)$ runs over a basis of the Cayley-Dickson algebra, in the usual form (cf. e.g. Kuroš [63], ch. V). Next consider the base rank: since the free group of rank 2 contains free groups of infinite rank as subgroups, we have $r_b(\mathrm{Gp}) \leqslant 2$, and here equality holds because the free group of rank 1 is abelian and so is certainly not generic for Gp. Thus we have $r_a(\mathrm{Gp}) = 3$, $r_b(\mathrm{Gp}) = 2$.

Consider now a set Σ of laws in the n letters $x_1, \cdots, x_n$. We may regard Σ as a subset of W_n^2, where W_n is the Ω-word algebra on $x_1, \cdots, x_n$; thus we may replace Σ by the fully invariant congruence on W_n which it generates. If the congruence is denoted by $\mathfrak{q}$, then $F_n = W_n/\mathfrak{q}$ is the free algebra of rank n for the variety Σ'. With these notations, we have (since $\mathfrak{q}$ can be any fully invariant congruence on W_n)

Proposition 3.10

Let $\mathfrak{q}$ be any fully invariant congruence on W_n and put $F_n = W_n/\mathfrak{q}$. Given any variety $\mathscr{V}$, let $\mathfrak{r}_n$ be the set of all its laws in $x_1, \cdots, x_n$. Then

(i) $\mathfrak{r}_n \subseteq \mathfrak{q}$ *if and only if* $F_n \in \mathscr{V}^n$.
(ii) $\mathfrak{r}_n \supseteq \mathfrak{q}$ *if and only if* $\mathscr{V}_n \subseteq \mathfrak{q}'$.

In particular, $\mathfrak{r}_n = \mathfrak{q}$ *if and only if* $\mathscr{V}^n = \mathfrak{q}'$, $\mathscr{V}_n = \hat{F}_n$, *and this holds whenever*

(4) $$\hat{F}_n \subseteq \mathscr{V} \subseteq \mathfrak{q}'.$$

Proof:

By definition, $\mathscr{V}^n = \mathfrak{r}'_n$, from which (i) follows by Theorem 1.2, because $\mathfrak{q}$ and $\mathfrak{r}_n$ are both fully invariant. Next put $G_n = W_n/\mathfrak{r}_n$; then G_n is the free $\mathscr{V}$-algebra on $x_1,\cdots,x_n$, and so $\mathscr{V}_n = \hat{G}_n$, from which (ii) follows. Now the rest is clear if we observe that (4) implies $\mathscr{V}_n = \hat{F}_n$, $\mathscr{V}^n = \mathfrak{q}'$. ∎

For a more detailed study of $\mathscr{V}^n$ and $\mathscr{V}_n$ when $\mathscr{V}$ is a variety of groups, we refer to H. Neumann [56].

The free algebras of the variety generated by a class $\mathscr{C}$ of algebras may be characterized as follows:

Theorem 3.11

Let $\mathscr{C}$ be a class of Ω-algebras and $\mathscr{V}$ the variety generated by $\mathscr{C}$. Then a $\mathscr{V}$-algebra F is free on a subset X if and only if every mapping

$$\phi: X \to A \qquad (A \in \mathscr{C})$$

of X into an algebra A of $\mathscr{C}$ may be extended to a homomorphism $\phi^*: F \to A$.

Proof:

The necessity of the condition is clear. Conversely, suppose that it is satisfied and let $B \in \mathscr{V}$. Then every mapping $\theta: X \to B$ may be extended to a homomorphism

(5) $$\bar{\theta}: W \to B,$$

where $W = W_\Omega(X)$ is the Ω-word algebra on X. Further, the inclusion mapping $X \to F$ extends to an epimorphism $W \to F$, so that we may assume $F = W/\mathfrak{q}$, making an obvious identification. To complete the proof we show that $\bar{\theta}$ can be factored by nat $\mathfrak{q}$, giving a mapping $\theta^*: F \to B$ which extends θ. This will follow if we can show that

(6) $$\mathfrak{q} \subseteq \ker \bar{\theta}.$$

Let $(u,v) \in \mathfrak{q}$; to establish (6) we must show that $u\bar{\theta} = v\bar{\theta}$. This will certainly follow if we can show that

(7) $$u = v$$

is a law in $\mathscr{C}$, for then it must also be a law in B. Now let $\phi: X \to A$ be any mapping, where $A \in \mathscr{C}$, and extend this to a homomorphism $\phi^*: F \to A$; then $u^\mathfrak{q}\phi^* = v^\mathfrak{q}\phi^*$, because $u^\mathfrak{q} = v^\mathfrak{q}$. This shows (7) to be a law in A, as asserted; therefore F is $\mathscr{V}$-free on X. ∎

This result enables us to give a simple characterization of Ω-algebras which are free for *some* variety $\mathscr{V}$:

Corollary 3.12

Let A be an Ω-algebra on the generating set X. Then A is free on X (for some class of Ω-algebras) if and only if every mapping $X \to A$ extends to an endomorphism of A.

For this condition is clearly necessary, and the sufficiency follows from Theorem 3.11. ∎

An Ω-algebra satisfying the condition of this corollary is often called *relatively free*; thus A is relatively free on X if and only if A is $\hat{A}$-free on X. As examples of relatively free algebras, we have the free algebras of a given variety; an example of an algebra which is not relatively free on any subset is given in Exercise 2.

When A is a generic algebra of a variety $\mathscr{V}$, then by Theorem 3.5 the $\mathscr{V}$-algebras can all be obtained as homomorphic images of subdirect powers of A. It is of interest to have a more explicit description of the free $\mathscr{V}$-algebras in this case:

Theorem 3.13

Let A be any Ω-algebra; then the free $\hat{A}$-algebra of rank α is obtained as follows: Let I be a set of cardinal α and for each $i \in I$ define a mapping $\delta_i : A^I \to A$ by

(8) $$((a_j)_{j \in I})\delta_i = a_i.$$

Then the subalgebra of A^{A^I} generated by the elements δ_i $(i \in I)$ is the free $\hat{A}$-algebra on these elements.

Proof:

Let F be the subalgebra generated by the δ_i $(i \in I)$; given any $a \in A^I$, say $a = (a_i)$, denote by ε_a the projection of A^{A^I} on the corresponding factor. If we restrict ε_a to F, this defines a homomorphism

$$\varepsilon'_a : F \to A,$$

and by (8) this maps δ_i to a_i. Thus every mapping $(\delta_i) \to A$ extends to a homomorphism $F \to A$, and by Theorem 3.11, this shows that F is $\hat{A}$-free on the δ_i. ∎

When α is finite, say $\alpha = n$, then the δ_i are just the unit operators in the clone $\mathcal{O}(A)$ (cf. III.3); thus, the free $\hat{A}$-algebra of rank n is the subalgebra of A^{A^n} generated by the n-ary unit operators. This is necessarily finite when n and A are finite, and we thus obtain the following result, due to B. H. Neumann [37], in the case of groups:

Corollary 3.14

If A is a finite Ω-algebra, then every free $\hat{A}$-algebra of finite rank is finite, and hence every finitely generated $\hat{A}$-algebra is finite. ▮

The varieties of Ω-algebras form a complete lattice, with greatest element (Ω) and least element the trivial variety. In the case of groups, a multiplication can be defined on the set of subvarieties of Gp, and it turns out, rather surprisingly, that the resulting groupoid is a free semigroup (B. H., H., and P. M. Neumann [62]). For general Ω, there is no multiplication, but we still have the lattice structure. The atoms of this lattice (i.e. the minimal nontrivial varieties) will be referred to as *minimal varieties.* They are also called *equationally complete* classes, since they are defined by maximal proper fully invariant congruences on the Ω-word algebra (corresponding to a maximal consistent set of laws). If W is the Ω-word algebra on a set X with more than one element, then a fully invariant congruence $\mathfrak{q}$ on W is proper if and only if it does not contain (x,y), where x and y are any distinct elements of W. It follows that any proper fully invariant congruence can be embedded in a maximal proper fully invariant congruence on W. This proves

Proposition 3.15

Every nontrivial variety contains minimal varieties. ▮

In a minimal variety, any algebra with more than one element is generic; we shall call the generic algebras of minimal varieties *elementary*; thus A is elementary if and only if $\hat{A}$ is minimal.

Let $\mathcal{V}$ be any nontrivial variety and $\mathcal{M}$ a minimal variety contained in $\mathcal{V}$. Clearly $F_2(\mathcal{M})$ is a homomorphic image of $F_2(\mathcal{V})$, and since $F_2(\mathcal{M})$ is generic for $\mathcal{M}$ it follows that the minimal subvarieties of $\mathcal{V}$ are entirely determined by the relatively free homomorphic images of $F_2(\mathcal{V})$. In particular, if $F_2(\mathcal{V})$ is finite, then there are only finitely many homomorphic images, and we obtain the following result, due to D. Scott [56].

Theorem 3.16

Let $\mathscr{V}$ be any nontrivial variety such that $F_2(\mathscr{V})$ is finite. Then $\mathscr{V}$ contains only finitely many minimal subvarieties. In particular, if A is a finite Ω-algebra with more than one element, then $\hat{A}$ contains only finitely many minimal subvarieties.

The last part follows because $F_2(\hat{A})$ is finite, by Corollary 3.14. ∎

Let us call A *F_n-simple* if A is nontrivial, and for every nontrivial homomorphic image B of $F_n(\hat{A})$, we have

$$F_n(\hat{A}) \cong F_n(\hat{B}).$$

The determination of elementary algebras is facilitated by the following criterion, also due to Scott [56].

Theorem 3.17

An Ω-algebra A is elementary if and only if, for some $n > 1$, A is F_n-simple and belongs to $(\hat{A})_n$.

Proof:

The conditions on A state that (i) any n-generator $\hat{A}$-algebra B satisfies $(\hat{B})_n \supseteq (\hat{A})_n$, and (ii) $(\hat{A})_n \supseteq \hat{A}$. Since in any case, $(\hat{B})_n \subseteq (\hat{A})_n \subseteq \hat{A}$, the conditions imply that $\hat{B} \supseteq. \hat{A}$, whence

$$\hat{B} = \hat{A}. \tag{9}$$

Taking B to be any nontrivial algebra of some minimal subvariety of $\hat{A}$, we see from (9) that $\hat{A}$ must be minimal, and hence A is elementary. Conversely, when A is elementary, then (9) holds for every nontrivial $\hat{A}$-algebra, and hence (i) and (ii) hold for all $n > 1$. ∎

When A is finite, these conditions enable one to test in a finite number of steps whether A is elementary. As an illustration, we determine the elementary groups. If G is an elementary group, then G is nontrivial, and so contains nontrivial cyclic subgroups; taking a suitable factor of G, we obtain a cyclic group Z_p of prime order p in $\hat{G}$. Since Z_p is generic for $\hat{G}$, every $\hat{G}$-group, and in particular G itself, is abelian and satisfies $x^p = 1$, i.e., G is elementary abelian. Conversely, an elementary abelian group is easily seen to be elementary, using Theorem 3.17.

EXERCISES

1. Show that the conditions of Theorem 3.1 are independent, by verifying that the class under (n) below satisfies all the conditions except (n) (n = i,⋯,iv):

(i) $\emptyset$; (ii) the class of divisible abelian groups (G is *divisible*, if $x^n = a$ has a solution x in G for every a in G and every positive integer n); (iii) the class of torsion-free abelian groups; (iv) the class of finite abelian groups.

2. Show that a finite abelian p-group is relatively free if and only if it is elementary abelian.

3. Let $\mathscr{K}$ be the category of all groups in which every maximal subgroup is of infinite index, with all homomorphisms. Show that $\mathscr{K}$ is not residual. (Use Theorem 3.5.)

4. Let $A = W/\mathfrak{q}$ be any Ω-algebra, in a given presentation, where W is the Ω-word algebra on the standard alphabet, and denote by $\mathfrak{r}$ the greatest fully invariant congruence contained in $\mathfrak{q}$. Then $W/\mathfrak{r}$ is the free $\hat{A}$-algebra of countable rank.

5. (G. T. Haigh.) Let A be a finitely generated Ω-algebra, where Ω is finite. If $\mathfrak{q}$ is a congruence such that $A/\mathfrak{q}$ is finite, then $\mathfrak{q}$ is finitely generated.

6. If $\mathscr{V}$ is any variety of algebras, show that the free algebras of rank m of the varieties $\mathscr{V}^{n+1}$, $\mathscr{V}$, $\mathscr{V}_n$ are all isomorphic, provided that $n \geqslant m$. Give an example where all three varieties are distinct. (Take $n = 1$, $\mathscr{V}$ = associative rings.)

7. Let A be relatively free on X. Then if X is infinite, $\hat{A}$ is the only variety for which A is the free algebra on X. If X is finite with n elements, show that for any variety $\mathscr{V}$, the free $\mathscr{V}$-algebra of rank n is isomorphic to A if and only if

$$\hat{A} \subseteq \mathscr{V} \subseteq (\hat{A})^n.$$

8. For any variety $\mathscr{V}$, if $\mathscr{V}^n \neq \mathscr{V}$ and $r_a(\mathscr{V}^n) \leqslant n$, show that $r_b(\mathscr{V}^n) > n$.

9. Show that for the variety of abelian groups, $r_a = 3$, $r_b = 1$. (Use the loop associated with the Pappus configuration, Exercise II.2.12.)

10. Show that for the variety of associative rings, $r_a = 3$, $r_b = 2$.

11. If As denotes the variety of associative rings, then As^2 is the variety of alternative rings and As^1 is the variety of power-associative rings; show that for As^2, $r_a = 2$, $r_b \geqslant 3$, and for As^1, $r_a = 1$, $r_b \geqslant 2$.

12. An Ω-algebra A is said to be *n-primal*, if the clone of action of Ω on A includes all n-ary operations on A (cf. III.3). Show that A is n-primal if and only if A^{A^n} is generated, qua Ω-algebra, by the unit operators.

13. (A. L. Foster.) If Ω is countable, show that any Ω-algebra which is n-primal for some n must be finite. (Observe that an infinite set has uncountably many n-ary operations.)

14. (A. L. Foster.) Let A be a nontrivial Ω-algebra which is n-primal for all n, and assume that Ω is at most countable. Show that the extensions of A in the variety $\hat{A}$ are the subdirect powers of A containing the constant functions.

15. (A. Tarski.) Show that an associative ring R is elementary if and only if it is commutative and for some prime p, $px = 0$ for all $x \in R$ and either $xy = 0$ for all $x,y \in R$ or $x^p = x$ for all $x \in R$.

16. (J. Kalicki & D. Scott.) Determine all elementary semigroups with two elements.

17. (M. P. Schützenberger.) Show that a lattice is elementary if and only if it is distributive. (Use Exercise II.7.9.)

4. REPRESENTATIONS IN VARIETIES OF ALGEBRAS

We now apply the representation theory of III.4 to varieties. Since every variety is residual, a residual representation of a category $\mathscr{L}$ in a variety $\mathscr{V}$ of Ω-algebras has a universal functor, by Theorem III.4.2. Now just as a variety is a special type of residual category, so there is a special type of residual representation, which may be defined by identities in case $\mathscr{L}$ is a category of algebras (not necessarily with the same operator domain as $\mathscr{V}$).

Thus, we now have two operator domains Ω, Θ, a variety $\mathscr{V}$ of Ω-algebras, and a category $\mathscr{L}$ of Θ-algebras with free algebras. Consider any mapping

$$\rho : B \to A \tag{1}$$

from an $\mathscr{L}$-algebra B to a $\mathscr{V}$-algebra A; given a set $w_1,\cdots,w_s$ of Θ-words in $x_1,\cdots,x_r$ and two Ω-words f,g in $y_1,\cdots,y_s$, we say that the mapping (1) satisfies the identity

$$f(w_1\rho,\cdots,w_s\rho) = g(w_1\rho,\cdots,w_s\rho), \tag{2}$$

if the two sides of (2) are equal whenever $x_1,\cdots,x_r$ are replaced by arbitrary elements of B. It is clear from the definition that for any set Σ of identities of the form (2), the set of all mappings from $\mathscr{L}$-algebras to $\mathscr{V}$-algebras satisfying the identities Σ defines a representation of $\mathscr{L}$ in $\mathscr{V}$. These representations defined by identities may be characterized as follows:

Theorem 4.1

Let $\mathscr{V}$ be a variety of algebras and $\mathscr{L}$ a category of algebras (not necessarily with the same operator domain as $\mathscr{V}$) with free algebras. Then a given *representation of $\mathscr{L}$ in $\mathscr{V}$ may be defined by identities if and only if it is residual and satisfies the following condition:*

If $\rho: B \to A$ is a mapping from an $\mathscr{L}$-algebra B to a $\mathscr{V}$-algebra A such that for some $\mathscr{L}$-epimorphism $\beta: B_1 \to B$ the mapping $\beta\rho: B_1 \to A$ is admissible, then ρ is itself admissible.

A representation of $\mathscr{L}$ in $\mathscr{V}$ satisfying this last condition is said to *admit homomorphisms.*

Proof:

The necessity of the conditions is clear, and their verification may be left to the reader. Now assume that they are satisfied; then, by Theorem III.4.2, there is a universal functor (U,u) for the representation of $\mathscr{L}$ in $\mathscr{V}$, and the mapping

(3) $$u: B \to U(B)$$

is admissible. Moreover, $U(B)$ is generated by im u, by Proposition III.4.1. Let G be the free $\mathscr{L}$-algebra on a set X, and consider any relation

(4) $$f(x_i u) = g(x_i u) \qquad (x_i \in X)$$

in $U(G)$. We assert that (4) is identically satisfied by the representation. If $\beta: G \to B$ is any $\mathscr{L}$-homomorphism and $\beta': U(G) \to U(B)$ the induced $\mathscr{V}$-homomorphism, then putting $x_i\beta = b_i$, we have

(5) $$f(b_i u) = f(x_i \beta u) = f(x_i u)\beta' = g(x_i u)\beta' = g(x_i \beta u) = g(b_i u).$$

Now given any admissible mapping $\rho: B \to A (A \in \mathscr{V})$, we have $\rho = u\rho^*$, where $\rho^*: U(B) \to A$, and by (4) and (5),

$$f(b_i \rho) = f(b_i u)\rho^* = g(b_i u)\rho^* = g(b_i \rho).$$

This shows that (4) is identically satisfied. Now let Σ be the set of all identities satisfied by the representation, and consider a mapping

(6) $$\rho: B \to A \qquad (B \in \text{Ob } \mathscr{L},\ A \in \mathscr{V}),$$

which satisfies all the identities of Σ. Take a free $\mathscr{L}$-algebra G on a set X, with an $\mathscr{L}$-epimorphism $\beta: G \to B$; then $U(G)$ is generated by im u and any relation in $U(G)$ between the elements xu $(x \in X)$ is an identity of Σ and therefore also holds between the elements $x\beta\rho$, by the definition of ρ.

By Lemma III.8.1 there exists a $\mathscr{V}$-homomorphism $\beta^*: U(G) \to A$ which makes the diagram

$$\begin{array}{ccc} G & \xrightarrow{u} & U(G) \\ \beta\downarrow & & \downarrow\beta^* \\ B & \xrightarrow[\rho]{} & A \end{array}$$

commutative. Thus $\beta\rho$ is admissible, and since β is an $\mathscr{L}$-epimorphism, ρ is admissible. ▮

From the necessity of the conditions we obtain

Corollary 4.2

If $\mathscr{V}$ and $\mathscr{W}$ are any varieties of algebras, then the universal functor exists for any representation of $\mathscr{W}$ in $\mathscr{V}$ which is defined by identities. ▮

An important particular case is obtained if we take $\mathscr{V} \prec \mathscr{L}$ and consider the natural representation of $\mathscr{L}$ in $\mathscr{V}$. The admissible mappings are by definition the mappings $\rho: B \to A$ from an $\mathscr{L}$-algebra B to a $\mathscr{V}$-algebra A, such that for any n-ary operator ω acting in $\mathscr{L}$, and any $b_1, \cdots, b_n \in B$, we have

(7) $$(b_1\rho \cdots b_n\rho)\omega = (b_1 \cdots b_n\omega)\rho.$$

Thus the natural representation of $\mathscr{L}$ in $\mathscr{V}$, i.e. the representation of $\mathscr{L}$ in $\mathscr{V}$ by $\mathscr{L}$-homomorphisms, is defined by the identities (7) and we have proved

Corollary 4.3

Let $\mathscr{V}$ be a variety of algebras which is subordinate to a category $\mathscr{L}$ of algebras (not necessarily with the same operator domain as $\mathscr{V}$) with free algebras. Then the natural representation is defined by identities and has a universal functor. ▮

Given any variety $\mathscr{V}$ of Ω-algebras, let $\mathscr{V}'$ be the derived category obtained by restricting Ω to Ω'. Then the $\mathscr{V}'$-algebras are (up to isomorphism) the Ω'-subalgebras of $\mathscr{V}$-algebras; hence $\mathscr{V}'$ is hereditary. Moreover, if $(B_\lambda)_{\lambda \in \Lambda}$ is any family of $\mathscr{V}'$-algebras, we may take B_λ to be an Ω'-subalgebra of a $\mathscr{V}$-algebra A_λ. Hence the direct product ΠB_λ is an Ω'-subalgebra of ΠA_λ, and this shows that $\mathscr{V}'$ admits direct products. By Corollary III.5.2, $\mathscr{V}'$ is a category with free algebras, and this proves the first part of

Theorem 4.4

Let $\mathscr{V}$ be a nontrivial variety of Ω-algebras and $\mathscr{V}'$ the derived category of Ω'-algebras obtained by restricting Ω to Ω'; then $\mathscr{V}'$ admits subalgebras and direct products. Moreover, if F is the free $\mathscr{V}$-algebra on a set X, then the Ω'-subalgebra of F generated by X is the free $\mathscr{V}'$-algebra on X.

To prove the last assertion, let F' be the Ω'-subalgebra of F generated by X; then, by definition, $F' \in \mathscr{V}'$. Now any $\mathscr{V}'$-algebra B is isomorphic to an Ω'-subalgebra of some $\mathscr{V}$-algebra A, and we may take B to be embedded in A, without loss of generality. Any mapping $\phi: X \to B$ induces a mapping $\phi i: X \to A$, where $i: B \to A$ is the inclusion mapping, and ϕ may therefore be extended to an Ω-homomorphism $\bar{\phi}: F \to A$. The restriction $\phi' = \bar{\phi} \mid F'$ is an Ω'-homomorphism extending ϕ, and since it is determined by its values on X, is unique. This shows F' to be the free $\mathscr{V}'$-algebra on X. ∎

We note that $\mathscr{V}'$ need not be a variety. For example, if $\mathscr{V}$ is the variety of associative algebras and $\mathscr{V}'$ the derived category consisting of special Jordan algebras (cf. VII.7), then $\mathscr{V}'$ does not admit homomorphic images, and so is not a variety, although by Theorem 4.4 we may speak of the 'free special Jordan algebra' on any set.

In Corollary 4.3 we saw that the natural representation of a derived category always has a universal functor. There is a simple criterion for this functor to be injective, which is often useful.

Theorem 4.5

Let $\mathscr{V}$ be a variety of Ω-algebras and $\mathscr{V}'$ the derived category obtained by restricting the operator domain from Ω to Ω'. If B is a homomorphic image of a free $\mathscr{V}'$-algebra, say

$$B = F'/\mathfrak{q},$$

where F' is the $\mathscr{V}'$-subalgebra generated by X in F, the free $\mathscr{V}$-algebra on X, and if $\bar{\mathfrak{q}}$ is the Ω-congruence on F generated by $\mathfrak{q}$, then the universal $\mathscr{V}$-algebra for B is

$$U(B) \cong F/\bar{\mathfrak{q}}, \tag{8}$$

and the canonical mapping $u: B \to U(B)$ is injective if and only if

$$\bar{\mathfrak{q}} \cap F'^2 = \mathfrak{q}. \tag{9}$$

Proof:

The canonical homomorphism $\theta : F' \to B$, when restricted to X and combined with the mapping $u : B \to U(B)$, defines a mapping from X to $U(B)$ which extends to an Ω-homomorphism $\bar{\theta} : F \to U(B)$. Since $\ker \theta = \mathfrak{q}$, it follows that $\ker \bar{\theta} \supseteq \mathfrak{q}$, and therefore $\ker \bar{\theta} \supseteq \bar{\mathfrak{q}}$, because $\bar{\mathfrak{q}}$ is the least Ω-congruence containing $\mathfrak{q}$. Dividing by nat $\bar{\mathfrak{q}}$, we find

$$(10) \qquad \theta^* : F/\bar{\mathfrak{q}} \to U(B).$$

Now in any case, $\mathfrak{q} \subseteq \bar{\mathfrak{q}} \cap F'^2$, hence there is an epimorphism

$$F'/\mathfrak{q} \to F'/\bar{\mathfrak{q}} \cap F'^2 \cong F'^{\bar{\mathfrak{q}}}/\bar{\mathfrak{q}},$$

(by the second isomorphism theorem), while

$$F'^{\bar{\mathfrak{q}}}/\bar{\mathfrak{q}} \subseteq F/\bar{\mathfrak{q}},$$

so altogether we obtain a homomorphism

$$(11) \qquad \phi : B = F'/\mathfrak{q} \to F/\bar{\mathfrak{q}};$$

from the definition of θ^* it follows that

$$(12) \qquad u = \phi\theta^*.$$

Now any admissible mapping $\alpha : B \to A$ ($A \in \mathscr{V}$) can be factored by u, so that $\alpha = u\alpha'$; hence, by (12), $\alpha = \phi\theta^*\alpha'$. If we also had $\alpha = \phi\beta$, then β and $\theta^*\alpha'$ would agree on im ϕ, and hence on $F/\bar{\mathfrak{q}}$, which shows that $\theta^*\alpha'$ is unique. Thus $F/\bar{\mathfrak{q}}$ has the universal mapping property and may therefore be identified with $U(B)$, with universal mapping (11). Now $\ker \phi = (\bar{\mathfrak{q}} \cap F'^2)/\mathfrak{q}$, and this is the diagonal if and only if (9) holds. ∎

These results will be applied to linear algebras in Chapter VII; another application, which is described below, provides a way of embedding arbitrary Ω-algebras in semigroups. The method is based on the construction of Ω-word algebras in III.2, and is due in essence to Malcev [52], who uses it to represent linear algebras in associative algebras.

Let Ω be any operator domain and denote by Sg(Ω) the class of all semigroups over Ω, i.e. semigroups with the elements of Ω as constant operators. Clearly, this is a variety; we shall denote the semigroup multiplication by '$\circ$' to avoid confusion. Any semigroup in Sg(Ω) may be regarded as an Ω-algebra by defining

$$(13) \qquad a_1 a_2 \cdots a_n \omega = a_1 \circ a_2 \circ \cdots \circ a_n \circ \omega \qquad (\omega \in \Omega(n)).$$

Thus, $\mathrm{Sg}(\Omega) \prec (\Omega)$, and so (Ω) may be represented in $\mathrm{Sg}(\Omega)$ by the natural representation. This has a universal functor, by Corollary 4.3; thus, to each Ω-algebra A there corresponds a semigroup $U(A)$ over Ω, together with a mapping

$$u: A \to U(A) \tag{14}$$

which is an Ω-homomorphism (regarding $U(A)$ as Ω-algebra by (13)), and which is universal for homomorphisms of A into semigroups over Ω. We assert that (14) is injective; to prove this, we need only show that A can be faithfully represented in some semigroup over Ω. Such a semigroup may be constructed as follows. Let Σ be the set of all finite rows of elements of the disjoint sum $A \sqcup \Omega$ (including the empty row), such that no consecutive block of n elements of A is followed by an element of $\Omega(n)$. We represent A by mappings of Σ into itself by associating with the element $a \in A$ the operation ρ_a defined by

$$(x_1,\cdots,x_r)\rho_a = (x_1,\cdots,x_r,a) \qquad (x_i \in A \sqcup \Omega).$$

It is clear that $(x_1,\cdots,x_r,a)$ again belongs to Σ. Now Ω is represented by associating with $\omega \in \Omega(n)$ the operation σ_ω defined by

$$(x_1,\cdots,x_r)\sigma_\omega = \begin{cases} (x_1,\cdots,x_m,x_{m+1}\cdots x_r\omega) & \text{if } r = m+n \geqslant n \\ & \text{and } x_i \in A \text{ for } i > m, \\ (x_1,\cdots,x_r,\omega) & \text{otherwise.} \end{cases}$$

Now it is easily verified that

$$\rho_{a_1}\rho_{a_2}\cdots\rho_{a_n}\sigma_\omega = \rho_{a_1a_2\ldots a_n\omega}.$$

Hence we have an Ω-homomorphism of A into the semigroup of unary operations on Σ. Moreover, if $(\ )$ is the empty row, then

$$(\ )\rho_a = (a) \qquad (a \in A);$$

therefore $\rho_a = \rho_b$ holds if and only if $a = b$. The result thus established may be summed up as

Proposition 4.6

Any Ω-algebra A may be embedded in a semigroup over Ω, and there is a universal semigroup over Ω for all such embeddings. ▌

In the special case $A = W_\Omega(X)$ this is, of course, merely the embedding in the semigroup of all Ω-rows in X.

In a variety, the free product need not exist (unlike the free composition), and the problem of its existence is essentially one of showing that a certain universal functor is injective. An alternative form of the definition of free products in varieties is given in

Theorem 4.7

Let $\mathscr{V}$ be a variety; then a $\mathscr{V}$-algebra A is the free product of a family $(A_\lambda)_{\lambda \in \Lambda}$ of subalgebras, provided there exists for each $\lambda \in \Lambda$ a presentation

$$A_\lambda = \mathscr{V}\{X_\lambda \mid \Phi_\lambda\}, \tag{15}$$

such that $\mathscr{V}\{X \mid \Phi\}$ is a presentation for A, where $X = \sqcup X_\lambda$, $\Phi = \sqcup \Phi_\lambda$ and for any pair $\lambda \neq \mu$, $A_\lambda \cap A_\mu$ is the minimal subalgebra of A.

Proof:

We first show that A is the free composition of the A_λ. Since A_λ is only determined up to isomorphism by (15), we may take the X_λ to be pairwise disjoint, and then put $X = \bigcup X_\lambda$. Let $\phi_\lambda : A_\lambda \to B$ be any family of homomorphisms and consider the mapping $\phi : X \to B$ such that $\phi \mid X_\lambda = \phi_\lambda \mid X_\lambda$. Any relation between the elements of X in A is a consequence of the relations Φ_λ, and therefore also holds between their images in B. Hence (by Lemma III.8.1) ϕ can be extended to a homomorphism $\bar{\phi} : A \to B$ which agrees with ϕ_λ on X_λ and therefore also agrees with ϕ_λ on A_λ. Moreover, $\bar{\phi}$ is unique because its effect on the generating set X of A is given. This shows A to be the free composition, with the inclusions $A_\lambda \to A$ as canonical homomorphisms. By hypothesis, different factors intersect in the minimal subalgebra of A, and this shows A to be the free product. ∎

From Corollary III.6.3 we obtain

Corollary 4.8

Let $\mathscr{V}$ be a variety in which every minimal subalgebra is trivial; then the free product of any family of $\mathscr{V}$-algebras exists. ∎

This corollary shows that the free product of groups, semigroups with 1, rings, etc., exists. Of course it is necessary to specify the variety in which we operate; thus, the free product of abelian groups is different according to whether we are in the variety of all groups or that of all abelian groups.

EXERCISES

1. Show that the free product of a family of $\mathscr{V}$-algebras exists if and only if the free product exists for every finite subfamily; deduce that the property of admitting free products is local. (Use Theorem 4.7.)

2. Let $\mathscr{V}$ be a variety of algebras and $\mathscr{L}$ a category of algebras with free algebras which has two representations R_1, R_2 in $\mathscr{V}$ such that every R_2-admissible mapping is also R_1-admissible. If for each representation R_i the universal functor (U_i,u_i) exists, show that there is a canonical epimorphism

$$v: U_1(B) \to U_2(B) \qquad (B \in \text{Ob } \mathscr{L}). \tag{16}$$

If R_2 admits homomorphic images and if for some $\mathscr{L}$-algebra B, (16) is an isomorphism, show that v is an isomorphism for every homomorphic image of B.

3. Let $\mathscr{V}$ be any variety of algebras and $\mathscr{L} = \mathscr{F}(\mathscr{V})$ the category of families of $\mathscr{V}$-algebras (with totally unordered index set). Show that the universal functor exists for the representation of $\mathscr{F}(\mathscr{V})$ in $\mathscr{V}$ by functions which are homomorphic in each argument. (When $\mathscr{V}$ is the variety of abelian groups, this functor is the tensor product.)

4. For any positive integer n, denote by Z_n the ring of integers mod n. Obtain necessary and sufficient conditions for the free product of Z_m and Z_n to exist in the category of: (i) commutative (and associative) rings; (ii) commutative rings without zero divisors; (iii) commutative rings with 1. (In each case the conditions must be such as to ensure that Z_n,Z_m belong to the category.)

5. If H is a fixed group, show that the category of extensions of H admits free products (this is also called the free product *amalgamating* H).

6. In any category a morphism μ is called left (right) regular, if $\mu\alpha = \mu\beta$ ($\alpha\mu = \beta\mu$) implies $\alpha = \beta$. Show that in the category Gp a left regular morphism is an epimorphism and a right regular morphism is a monomorphism. (Use Exercise 5.)

7. Show that any ordered set A may be embedded in a free lattice L in such a way that the ordering in A is induced by that of L.

8. Given any variety $\mathscr{V}$ of Ω-algebras, let F_X be the free $\mathscr{V}$-algebra on a set X, and for any $A \in \mathscr{V}$ denote by $A(X)$ the free composition (in $\mathscr{V}$) of A and F_X. Show that the canonical mapping $A \to A(X)$ is injective (the algebra $A(X)$ is called the *extension of A obtained by adjoining the indeterminates* $x \in X$).

Chapter V

Relational Structures and Models

The concept of an algebra developed in Chapters II–IV includes most of the algebraic structures encountered in practice, but there are some important exceptions. Thus, although groups, rings, and vector spaces are included, neither ordered groups nor fields satisfy the definition. To see what modifications are necessary, we note that an ordered group is a structure with certain operations, and besides, a relation. A field is a structure with 'operations' which are not everywhere defined. As we saw in II.2, it would be formally possible to define both ordered groups and fields as algebras, but only at the cost of some artificiality. For a natural development one needs to have relations as well as operations, or at least relations alone, since it turns out that operations may be obtained as a special case. Furthermore, in writing down the definition of particular structures it may be necessary to use inequalities as well as equations. This is made possible by introducing logical connectives, and it leads to the consideration of classes of algebras other than varieties.

1. RELATIONAL STRUCTURES OVER A PREDICATE DOMAIN

As in Chapter II we begin by taking a set Ω of symbols such that with each $\omega \in \Omega$ a nonnegative integer $a = a(\omega)$ is associated. However, instead of using ω to define an n-ary operation, we now wish to define more generally an $(n+1)$-ary relation. For this reason, we shall refer to the elements of Ω in the present context as *predicates* rather than operators, and Ω is called a *predicate domain*. If $a(\omega) = m - 1$, then ω is said to be an m-ary predicate, and the *arity* of ω, as predicate, is m. As before, we put $\Omega(n) = \{\omega \in \Omega \mid a(\omega) = n\}$.

Definition

A *relational structure* over the predicate domain Ω, or more briefly, an Ω-*structure*, is a nonempty set M with a rule which assigns to each n-ary predicate $\omega \in \Omega$ an n-ary relation in M. As in algebras, the set M is called the *carrier* of the Ω-structure, and we shall not use a special notation to distinguish between a structure and its carrier, as the intended meaning will always be clear from the context.

Since an n-ary operation is a special case of an $(n+1)$-ary relation, we see that algebras may be regarded as a special case of relational structures. Accordingly, an Ω-structure M is said to be an algebra with respect to the subdomain Ω' of Ω if for each $\omega \in \Omega'$ the relation in M defined by ω is an operation. In the general case, if ω_M is the relation in M defined by $\omega \in \Omega(m-1)$ and $a \in M^m$, we often write

$$M \vDash \omega(a)$$

and say that $\omega(a)$ *holds* in M, to indicate that $a \in \omega_M$. If $\omega(a)$ holds for all $a \in M^m$, we say that ω is *valid* in M, and write

$$M \vdash \omega.$$

To indicate that $\omega(a)$ does not hold in M, we write $M \sim \vDash \omega(a)$, and if ω is not valid in M, we write $M \sim \vdash \omega$; here it must be borne in mind that $M \sim \vdash \omega$ does not mean that $M \sim \vDash \omega(a)$ for all $a \in M^m$ (but merely that this holds for some a). An Ω-structure M in which every predicate is valid is said to be *full*; clearly, such a structure is completely determined by its carrier. By the *trivial* Ω-structure we understand the full Ω-structure consisting of a single element.

Consider two Ω-structures M and N. For any $\omega \in \Omega$, we denote the relations defined by ω in M and N by ω_M and ω_N respectively; further, we write $m = a(\omega) + 1$. We say that N is a *substructure* of M if the carrier of N is a subset of that of M and

$$\omega_N = \omega_M \cap N^m \qquad (\omega \in \Omega(m-1);\ m = 1, 2, \cdots). \tag{1}$$

Given M, if we take any nonempty subset N of M and define ω_N by (1), we clearly obtain an Ω-structure on N; thus, every subset N $(\neq \emptyset)$ of an

Ω-structure M may be defined as a substructure of M in just one way, and it is always this Ω-structure on N which we have in mind, unless the contrary is stated.

A mapping ϕ from one Ω-structure M to another, N, is called a *homomorphism* if for each $\omega \in \Omega(m-1)$ and $a \in M^m$,

$$\text{if } M \models \omega(a), \text{ then } N \models \omega(a\phi),$$

where $a\phi = (a_1\phi,\cdots,a_m\phi)$ if $a = (a_1,\cdots,a_m)$. If the mapping ϕ has an inverse which is also a homomorphism, we call ϕ an *isomorphism* from M to N; we also say in this case that M and N are *isomorphic*, and write $M \cong N$. If M is isomorphic to a substructure of N, we say that M is *embeddable* in N. The terms 'monomorphism', 'epimorphism', 'endomorphism', and 'automorphism' are defined analogously to the case of algebras. It should be noticed, however, that a bijective homomorphism need not be an isomorphism, i.e., the analogue of Lemma II.3.6 is not true.

Given two Ω-structures M and N, we say that N is a *quotient* of M if there is an epimorphism $\theta : M \to N$ such that for any $\omega \in \Omega(m-1)$ and $b \in N^m$,

$$N \models \omega(b) \text{ if and only if there exists } a \in M^m \text{ such that } a\theta = b \text{ and } M \models \omega(a).$$

These conditions on θ, together with the Ω-structure on M, serve to determine the Ω-structure on N uniquely. Thus if M is any Ω-structure and $\mathfrak{q}$ an equivalence on M, then the quotient set $M/\mathfrak{q}$ together with the mapping nat $\mathfrak{q}$ determines a unique quotient structure on $M/\mathfrak{q}$, and in speaking of $M/\mathfrak{q}$ as an Ω-structure it is always this quotient structure we have in mind, unless the contrary is stated.

We remark that any product of homomorphisms, when defined, is again a homomorphism; any mapping to a full Ω-structure is a homomorphism; and a mapping from a full Ω-structure is a homomorphism if and only if the image is a full Ω-structure.

Proposition 1.1

(i) *Let $\phi : M \to N$ be a homomorphism of Ω-structures. Then N is a quotient of M, provided that ϕ is surjective and the Ω-structure on N is the least for which ϕ is a homomorphism.* (ii) *A quotient of a quotient of M is again a quotient of M* (*up to isomorphism*).

The proof of (i) is clear from the definitions. To prove (ii), let N be a quotient of M and P a quotient of N, with epimorphisms $\alpha : M \to N$, $\beta : N \to P$. Denote by P' the least structure with the same carrier as P which makes $\alpha\beta$ a homomorphism. Since α, β are homomorphisms, the structure P' is contained in the structure P (qua set of relations), i.e. the identity mapping of the carrier induces a homomorphism $P' \to P$. Now let $\omega \in \Omega(m-1)$ and $c \in P^m$ be such that $P \vDash \omega(c)$. Then by definition, there exists $b \in N^m$ such that $b\beta = c$ and $N \vDash \omega(b)$, and hence there exists $a \in M^m$ such that $a\alpha = b$ and $M \vDash \omega(a)$. Now $a\alpha\beta = c$; therefore $P' \vDash \omega(c)$, and this shows that the homomorphism $P' \to P$ induced by the identity is an isomorphism, i.e., P' and P are equal. ∎

Let $(M_\lambda)_{\lambda \in \Lambda}$ be a family of Ω-structures; then the Cartesian product $M = \Pi M_\lambda$, with the projections $\varepsilon_\lambda : M \to M_\lambda$, may be given an Ω-structure by the rule:

(2) $$M \models \omega(a) \text{ if and only if } M_\lambda \models \omega(a\varepsilon_\lambda) \text{ for all } \lambda \in \Lambda,$$

where $a \in M^m$, $\omega \in \Omega(m-1)$ $(m = 1,2,\cdots)$. In other words, M carries the greatest Ω-structure for which all the projections are homomorphisms.

2. BOOLEAN ALGEBRAS

For the further study of Ω-structures it would be useful to have an analogue of the Ω-word algebra. Now the Ω-word algebra on a set X may be thought of as consisting of all derived operators applied to X. Similarly, in the case of Ω-structures one introduces *derived predicates*; however, the set of these predicates is not an Ω-structure, but an algebra, with operators which are quite independent of Ω. We shall therefore first study these algebras abstractly; they are named after G. Boole, who first used them in a systematic study of the propositional calculus.

We begin by recalling that a lattice L is distributive if and only if relative complements in L (when they exist) are unique (Proposition II.4.5). In particular, if L has 0 and 1 and every element x has a complement x', we may regard $x \to x'$ as a unary operator. Such a lattice is complemented in the sense of the following

Definition

A lattice L is *complemented* if it has a least element 0 and a unary operator $x \to x'$ such that

(1) $$x'' = x,$$

(2) $$(x \vee y)' = x' \wedge y',$$

(3) $$x \wedge x' = 0.$$

Clearly, complemented lattices form a variety with operators $(\vee, \wedge, ', 0)$ and laws (1)–(3), together with the lattice laws and the law

(4) $$0 \vee x = x,$$

characterizing 0 as the least element of L. Moreover, by (1) and (2), one of $\vee, \wedge$ can be expressed in terms of the other so that complemented lattices may be defined in terms of $(\vee, ', 0)$ alone. We shall not carry this out, but merely note the following consequences of (1)–(4):

By (4), $0 \leqslant x$ for all $x \in L$, i.e.

(5) $$0 \wedge x = 0.$$

If we put $1 = 0'$ and substitute from (5) and (4) into (2), we find

(6) $$1 \wedge x = x, \quad 1 \vee x = 1.$$

Therefore 1 is the greatest element of L. Further, from (3) and (1),

(7) $$x \vee x' = 1.$$

We now make the following

Definition

A *Boolean algebra* is a complemented distributive lattice.

From the definition it is clear that Boolean algebras form a variety. Moreover, any order-isomorphism between Boolean algebras is a lattice-isomorphism, and since this takes complementary pairs into complementary pairs, we have

Proposition 2.1

Any order-isomorphism between Boolean algebras is an isomorphism of Boolean algebras. ∎

Examples of Boolean Algebras

(i) The set $\mathscr{B}(A)$ of all subsets of a given set A is a Boolean algebra if we take $0 = \emptyset$ and for $X \subseteq A$, $X' = A \backslash X$. More generally, any system of subsets of A including $\emptyset$ and closed under complements and finite unions is a Boolean algebra; this is merely a subalgebra of $\mathscr{B}(A)$ as just defined (such a subalgebra is called a *field of sets* on A).

(ii) In any interval $I = [a,b]$ of a distributive lattice L, the set of elements of I which have a complement in I forms a Boolean algebra.

(iii) The two-element lattice is a Boolean algebra. If the elements of this lattice are a and b, and $a \vee b = b$ say, then $a \leqslant b$, and this becomes a

Boolean algebra if we put $a' = b$, $b' = a$. This lattice will be denoted by 2.

(iv) Propositional logic analyses the form of statements which can be made in a given context. If 'P' and 'Q' are any propositions, denoted by the symbols p and q respectively, then one can form the propositions 'P and Q', 'P or Q', 'not P'; these may be denoted by $p \wedge q$, $p \vee q$, and p' respectively. With these definitions the set of propositions becomes a Boolean algebra, if we regard two propositions as equal if they have the same truth-value in all circumstances. Here the least element 0 is taken to be 'nobody is reading this book just now' or some other proposition known to be false.

An important example of a Boolean algebra is the following: Let R be a ring in which every element is idempotent:

$$x^2 = x \quad \text{for all } x \in R. \tag{8}$$

Then $x + y = (x + y)^2 = x^2 + xy + yx + y^2 = x + xy + yx + y$, hence $xy + yx = 0$. Putting $y = x$, we find $2x = 0$, i.e. $x = -x$, and so

$$xy = yx.$$

In other words, every ring satisfying (8) is commutative and of characteristic 2. Assume now that in addition R has a unit element 1 and is associative; if we define in R the operations

$$x \wedge y = xy, \quad x' = 1 - x,$$

then R is a Boolean algebra with respect to these operations and the zero of R as least element. The verification may be left to the reader. Conversely, in any Boolean algebra we may define ring operations by putting

$$xy = x \wedge y, \quad x + y = (x \wedge y') \vee (x' \wedge y);$$

the resulting algebra is then an associative ring with 1 satisfying the idempotent law (8). Such a ring is said to be *Boolean*; with these definitions we see that Boolean rings are entirely equivalent to Boolean algebras. Although we shall mostly use the lattice operations, it is useful to bear the ring operations in mind. For example, in a ring generated by a set X, the distributive law enables us to express every element as a linear combination of products of elements of X. Similarly, in a Boolean algebra generated by X, every element can be expressed as a sup of infs of elements x and x' $(x \in X)$; by duality, the same holds with $\vee$ and $\wedge$ interchanged. We state this as

Proposition 2.2 (conjunctive normal form)

Let B be a Boolean algebra generated by $x_1, \cdots, x_n$; *then every element of B is of the form* $w_1 \wedge \cdots \wedge w_r$, *where*

$$w_k = y_{k1} \vee \cdots \vee y_{kh_k}, \quad y_{kl} = x_i \text{ or } x_i'.$$

A Boolean algebra which is complete, as a lattice, is called a *complete* Boolean algebra. For instance, $\mathscr{B}(A)$ is complete, for any set A; on the other hand, the subalgebra of $\mathscr{B}(A)$ generated by the singletons is not complete unless A is finite. In any Boolean algebra, complete or not, we have the following laws, known as *De Morgan's laws*:

Proposition 2.3

Let (a_i) *be any family of elements of a Boolean algebra B. If in either of the equations*

$$(\bigvee a_i)' = \bigwedge a_i', \tag{9}$$

$$(\bigwedge a_i)' = \bigvee a_i' \tag{10}$$

one side exists, then so does the other, and the two are equal.

Proof:

Suppose that $a = \bigvee a_i$; then $a_i \leqslant a$ for all $i \in I$, and hence

$$a' \leqslant a_i' \qquad \text{for all } i \in I.$$

If also $b \leqslant a_i'$ for all $i \in I$, then $b' \geqslant a_i$, whence $b' \geqslant a$, and so $b \leqslant a'$. This shows that $a' = \bigwedge a_i'$, i.e., (9). Thus, (9) holds whenever the left-hand side exists; by duality, (10) holds whenever the left-hand side exists. Now assume that the right-hand side of (9) exists; then, by what has been proved about (10), $\bigvee a_i'' = \bigvee a_i$ exists and (9) again holds; (10) now follows similarly.

Any Boolean algebra satisfies the following infinite distributive law:

$$\bigvee (a_i \wedge b) = (\bigvee a_i) \wedge b, \tag{11}$$

whenever both sides exist. There is, however, a more general distributive law which need not hold. We shall say that a Boolean algebra B is *completely distributive* if, for any family of elements $(a_{ij})_{i \in I, j \in J}$ of A, whenever one side of the equation

$$\bigwedge_I \left(\bigvee_J a_{ij} \right) = \bigvee_{\alpha \in J^I} \left(\bigwedge_I a_{i\alpha(i)} \right) \tag{12}$$

exists, then so does the other and both are equal. For example, the algebra $\mathscr{B}(A)$ is complete and completely distributive, for any set A. It turns out that this is essentially the only case, by the following result of Tarski:

Theorem 2.4

A Boolean algebra which is complete and completely distributive is isomorphic to $\mathscr{B}(A)$, for some set A.

Proof:

Let B be the given algebra and write

$$1 = \bigwedge_{a \in B} (a \vee a').$$

Expanding the right-hand side by the distributive law, we have

$$1 = \bigvee_{C \subseteq B} f(C), \tag{13}$$

where for any $C \subseteq B$,

$$f(C) = \Big(\bigwedge_{a \in C} a\Big) \wedge \Big(\bigwedge_{b \notin C} b'\Big).$$

We assert that every $f(C)$ is either 0 or an *atom*, i.e., a minimal element different from 0. For, if $f(C) \neq 0$ and $0 \neq c \leqslant f(C)$, then either $c \in C$, in which case $f(C) \leqslant c$ and so $f(C) = c$, or $c \notin C$, and then $f(C) \leqslant c'$, whence $c \leqslant c'$, which means that $c = c \wedge c' = 0$. This shows that in any case $c = f(C)$, so that $f(C)$ is indeed an atom.

Let A be the set of all $f(C) \neq 0$; we shall show that $B \cong \mathscr{B}(A)$. To every subset of A there corresponds an element of B, namely, its sup in B (which exists because B is complete); conversely, if $c \in B$, then $c = c \wedge (\bigvee f(C))$, by (13), and hence

$$c = \bigvee \{f(C) \mid f(C) \leqslant c\}.$$

Thus we have a bijection between B and $\mathscr{B}(A)$; clearly, this is order-preserving and therefore is an isomorphism. ∎

We remark that A is the set of atoms of B. In particular, if B is finite, it is clearly complete and completely distributive and we obtain

Corollary 2.5

Any finite Boolean algebra is isomorphic to $\mathscr{B}(A)$ for some finite set A. ∎

We now consider homomorphisms of Boolean algebras. Since the operations of a Boolean algebra may be expressed in terms of the operations of a Boolean ring, and vice versa, it follows that a homomorphism of Boolean algebras is the same thing as a homomorphism of Boolean rings. In particular, the kernel of such a homomorphism

$$\phi : A \to B$$

is completely determined by the inverse image, N say, of 0. We shall therefore call N an *ideal* (as in ring theory) even when we are considering A and B as Boolean algebras. In terms of the lattice operations, an ideal is given by the properties:

(i) if $a,b \in N$, then $a \vee b \in N$,
(ii) if $a \in N$ and $b \leqslant a$, then $b \in N$.

The ideal is *proper*, if

(iii) $1 \notin N$.

This last condition amounts to saying that $N \neq A$. The inverse image of 1 in a homomorphism is called a *dual ideal*; if it is proper, i.e. associated with a nontrivial homomorphism, it is called a *filter* of A. Thus, a filter is determined by the properties dual to (i)–(iii) above:

(i′) if $a,b \in F$, then $a \wedge b \in F$,
(ii′) if $a \in F$ and $b \geqslant a$, then $b \in F$,
(iii′) $0 \notin F$.

Let R be a ring (not necessarily Boolean); then a *maximal ideal* of R is understood to be an ideal of R which is maximal in the set of all proper ideals of R. When R has a unit element 1, the proper ideals are just those ideals which do not contain 1. Applying Corollary II.6.4, we thus obtain

Theorem 2.6 (Krull)

Every proper ideal of a ring with 1 *is contained in a maximal ideal.* ▮

In particular, this result may be applied to Boolean algebras. A maximal filter in a Boolean algebra is usually called an *ultrafilter*; taking duals in Theorem 2.6 we thus obtain

Theorem 2.7

Every filter of a Boolean algebra is contained in an ultrafilter. ▮

There is another way of characterizing maximal ideals in a Boolean algebra which is of great importance. It rests on the fact that simple Boolean algebras are very easy to determine explicitly.

Proposition 2.8

A Boolean algebra is simple if and only if it is isomorphic to 2.

For clearly 2 is simple; conversely, if A is simple, then $A \neq 0$ and A has no proper homomorphic images; i.e., no ideals apart from 0 and A. Let $a \in A$, $a \neq 0$; then the ideal generated by a must be nonzero and therefore equals A, whence $ab = 1$ for some $b \in A$. Hence $a = a(ab) = a^2b = ab = 1$. This shows that $A = \{0,1\} = 2$, as asserted. ∎

We thus obtain the important

Corollary 2.9 (Tarski)

Let I be an ideal of a Boolean algebra A; then I is maximal if and only if $A/I \cong 2$. ∎

Expressed in terms of A itself, this means that an ideal I is maximal in A if and only if for every $a \in A$, either $a \in I$ or $a' \in I$. By going over to duals, we see that ultrafilters of A are characterized by the same property.

Later we shall mainly be concerned with filters of algebras which have the form $\mathscr{B}(A)$. If I is any set, we shall understand by a filter $\mathscr{F}$ *on* I a system of subsets of I which is a filter of $\mathscr{B}(I)$. The members of $\mathscr{F}$ are also referred to as $\mathscr{F}$-sets. A subsystem $\mathscr{S}$ of $\mathscr{F}$ which generates $\mathscr{F}$ (as filter) is called a *base* of $\mathscr{F}$. To obtain the dual ideal generated by $\mathscr{S}$ we form the system $\mathscr{S}^*$ consisting of all intersections of finite subsystems of $\mathscr{S}$ and take the system of all sets containing an $\mathscr{S}^*$-set. The resulting dual ideal will be a filter if and only if $\emptyset \notin \mathscr{S}^*$. When this conditions holds, $\mathscr{S}$ is said to have the *finite intersection property*. Thus a system $\mathscr{S}$ is a filter base if and only if it has the finite intersection property.

Let I be an infinite set of cardinal α, and denote by Φ the system of all subsets of I whose complement in I has cardinal less than α. Clearly, Φ is a filter on I, called the *minimal Fréchet filter*. By a *Fréchet filter* on I we understand any filter containing Φ. We shall sometimes want to know when a given filter base is contained in a Fréchet filter. This is answered by

Proposition 2.10

A system $\mathscr{S}$ of subsets of I is contained in a Fréchet filter if and only if every finite intersection of sets of $\mathscr{S}$ has cardinal equal to $|I|$.

For, clearly, $\mathscr{S}$ is contained in a Fréchet filter if and only if $\mathscr{S} \cup \Phi$ is a filter base; i.e., writing again $\mathscr{S}^*$ for the system of finite intersections of sets of $\mathscr{S}$, if and only if every $\mathscr{S}^*$-set meets every Φ-set. Now a subset J of

I meets every Φ-set if and only if J is not contained in the complement of any Φ-set, i.e., if and only if $|J| = |I|$. ∎

A filter $\mathscr{F}$ on I is said to be *principal* if it is generated by a single element J, say. Clearly, such a filter is an ultrafilter if and only if J consists of a single element, and on a finite set I, these are the only ultrafilters. An infinite set always has nonprincipal ultrafilters, since e.g. the minimal Fréchet filter is contained in an ultrafilter which is nonprincipal. However, this proof of the existence of nonprincipal ultrafilters depends essentially on Zorn's lemma, and no explicit construction is known of a nonprincipal ultrafilter.

This brief outline of Boolean algebras will suffice for the applications we have in mind; of course it is not intended to be complete on any aspect, or even a representative selection. The reader who wishes to read a more comprehensive introduction is referred to Dwinger [61], or for a more detailed account to Sikorski [60]; the papers of Halmos [62] indicate a development specially suited to the applications to logic, including an alternative treatment of the subject matter of this chapter.

EXERCISES

1. Give a set of laws for Boolean algebras in terms of $(\vee, ', 0)$.

2. Let R be a commutative ring with 1, and denote by I the set of idempotents of R. Show that I is a Boolean algebra relative to the operations: $x \wedge y = xy$, $x' = 1 - x$. Describe the operations of I as a Boolean ring, in terms of the ring operations on R. Under what conditions is I, as a Boolean ring, a subring of R?

3. Prove (11) in a complete Boolean algebra; more generally, establish the law $(\bigwedge a_i) \vee (\bigwedge b_j) = \bigwedge (a_i \vee b_j)$.

4. Determine all complete subalgebras of $\mathscr{B}(I)$, for any set I.

5. If I is an infinite set and $\mathscr{F}$ is the system of finite subsets of I, verify that $\mathscr{F}$ is an ideal in $\mathscr{B}(I)$. Verify that the algebra $\mathscr{B}(I)/\mathscr{F}$ has no atoms: show that $\mathscr{B}(I) / \mathscr{F}$ is not complete.

Show that the Boolean algebra of Lebesgue measurable sets in the unit interval is complete but not completely distributive.

6. (M. H. Stone.) Show that the intersection of all maximal ideals of a Boolean algebra is 0. Deduce that any Boolean algebra B is isomorphic to a subalgebra of $\mathscr{B}(M)$, where M may be taken to be the set of all maximal ideals of B. (If $a \neq 0$, observe that any maximal ideal containing a' does not contain a; now apply Proposition II.7.1.)

7. Show that the base rank for Boolean algebras is zero, i.e., that 2 is generic in the variety of Boolean algebras. (Use Exercise 6.)

8. Show that every finitely generated Boolean algebra is finite. (Use Corollary IV.3.14.)

9. Show that any two finite Boolean algebras with the same number of elements are isomorphic. (Use Corollary 2.5 and the remark preceding it.)

10. Show that any two countably infinite Boolean algebras without atoms are isomorphic.

11. Show that the free Boolean algebra on k free generators has 2^{2^k} elements. (Take a set I of k elements and show that any mapping $\mathscr{B}(I) \to 2$ can be built up by Boolean operations from the characteristic functions on the singletons; then apply Theorem IV.3.13.)

12. (A. L. Foster.) Show that finite Boolean algebras are n-primal for every n. (Use Exercise 11 and Theorem IV.3.13.)

13. Give an example of a complemented lattice which is not distributive and show that Proposition 2.1 need not hold for such lattices. (Try a finite modular lattice of length two.)

14. Show that with one exception, a finite complemented lattice has an even number of elements.

15. In a Boolean algebra B, the mapping $x \to x'$ is an isomorphism of B with its dual. Restate this fact in terms of Boolean rings.

16. Show that every nonprincipal ultrafilter on an infinite set includes all subsets with a finite complement. Deduce that on a countable set every nonprincipal ultrafilter is a Fréchet filter.

17. If A is a directed set, show that the right segments of A have the finite intersection property.

18. If G is a group with trivial centre, show that the direct factors of G form a Boolean algebra.

19. If C is any subalgebra of a Boolean algebra B, show that any homomorphism $\phi: C \to 2$ extends to a homomorphism $\bar{\phi}: B \to 2$. (Embed $\ker \phi$ in a maximal ideal of B and use Corollary 2.9.)

20. (Dwinger & Yaqub [63].) Show that the free product of any family of Boolean algebra extensions of a fixed Boolean algebra C exists. (If $C = 2$, use Exercise 19 to prove that any Boolean algebra is retractable and apply Proposition III.6.2; for the general case, use the first part of Exercise 6.)

3. DERIVED PREDICATES

The Ω-structures have considerably less 'structure' than do the Ω-algebras; this can already be seen from the fact that every subset and every quotient of an Ω-structure again carries an Ω-structure. It is therefore necessary to restrict the notion of Ω-structure by imposing further conditions. The simplest form that such a condition can take is one which asserts that a given predicate is valid in a given structure. More generally, we can derive other predicates from the given ones, and our conditions may then take the form of asserting certain derived predicates.

The set of derived predicates will be defined as a Boolean algebra as follows. Let I be an infinite set of symbols called *variables*, and consider the expressions

(1) $$\varepsilon(i,j) \qquad (i,j \in I),$$

(2) $$\omega(i_1,\cdots,i_m) \qquad (i_k \in I,\ \omega \in \Omega(m-1)).$$

Now a derived Ω-predicate is essentially an element of the free Boolean algebra on the expressions (1), (2) as free generators. This statement requires some modification, to take account of quantifiers. Before introducing these, we note some definitions which, although not indispensable, are helpful because they suggest the concepts to be defined later.

A derived predicate is also called a *formula*; a formula of the form (1) or (2) is said to be *atomic*. If P, Q are any formulae, then $P \vee Q$ is called the *disjunction* of P and Q, $P \wedge Q$ is called their *conjunction*, and P' is the *negation* of P; this is also denoted by $\sim P$. Any formula built up from atomic formulae by using $\vee$ and $\wedge$ alone (without $\sim$) is called *positive*. Further, $P' \vee Q$ is also denoted by $P \Rightarrow Q$ and is called an *implication*, and $(P \Rightarrow Q) \wedge (Q \Rightarrow P)$ is denoted by $P \Leftrightarrow Q$ and is called an *equivalence*. Instead of 1 and 0, we write t and f respectively (for 'truth' and 'falsity').

Before we can describe the relations corresponding to these predicates, we have to determine the arity of each predicate. With each formula P, we shall associate a finite set I_P of variables, called the *free variables* of P; the arity of P is then defined as the number of free variables of P. The set I_P of free variables of P is defined by induction on the length of P (as a word in the Boolean algebra of formulae). For the atomic formula (1), the free variables are i, j, and for (2) they are $i_1,\cdots,i_m$. Thus, in particular, the arity of (2) is m if and only if the variables occuring in ω are distinct; this shows that there is no ambiguity in the use of this term. Generally, if P and Q are any formulae and I_P, I_Q their sets of free variables, then $P \vee Q$ and $P \wedge Q$ have $I_P \cup I_Q$ and $\sim P$ has I_P as set of free variables.

We now introduce two further operators on the set of formulae; they are to be unary operators $\bigvee_i$, $\bigwedge_i$ called *existential quantification* and *universal quantification* with respect to i. Thus, for each $i \in I$, there is one existential quantifier $\bigvee_i$ and one universal quantifier $\bigwedge_i$; they are related by the equation

(3) $$\bigwedge_i P = \sim\left(\bigvee_i \sim P\right),$$

which shows each of them to be determined by the other. We also impose the laws

(4) $$\bigvee_i \bigvee_i P = \bigvee_i P, \quad \bigvee_i \bigvee_j P = \bigvee_j \bigvee_i P, \quad \left(\bigvee_i P\right) \vee \left(\bigvee_i Q\right) = \bigvee_i (P \vee Q),$$

and dually, the laws obtained by applying (3) to (4). In this way our set of derived predicates is enlarged to include formulae with quantifiers. We shall use the term 'formula' to mean any derived predicate in the new sense, while a predicate in the former sense, i.e., one without quantifiers, will now be called an *open formula*. If I_P is the set of free variables of P, then that of $\bigvee_i P$ $\left(\text{or of} \bigwedge_i P\right)$ is $I_P \backslash \{i\}$. Any variable occurring in a formula as other than a free variable is said to be *bound*; e.g., i is bound in $\bigvee_i P$. Of course a variable may occur both free and bound: if i is free in P and Q, then i is bound in $\bigvee_i P$ and hence both free and bound in $\left(\bigvee_i P\right) \wedge Q$. In such cases we can always rename the bound occurrence of i so as to avoid confusion; i.e., instead of

(5) $$\left(\bigvee_i P(i)\right) \wedge Q(i)$$

we may consider

(6) $$\left(\bigvee_j P(j)\right) \wedge Q(i),$$

where $j \in I$ does not occur elsewhere in P or Q. When we come to define the relations corresponding to (5) and (6) we shall see that there is no difference between them (just as laws in different variables may be equivalent, cf. IV.1). Finally, a formula without free variables is called an *elementary sentence*, or simply a *sentence*. Examples of sentences are

t, f, and the formulae obtained by binding the free variables, i.e. applying any quantifiers to all the free variables in some formula.

Let P be a formula with the free variables $I_P = \{i_1, \cdots, i_n\}$; we write this as $P(i_1, \cdots, i_n)$ or $P(I_P)$ and define an n-ary relation in each Ω-structure M, corresponding to P. If P is atomic, then P has the form (1) or (2). In the first case, for any $a,b \in M$, we put

$$M \vDash \varepsilon(a,b) \qquad \text{if and only if } a = b.$$

For this reason we shall usually write '$i = j$' instead of '$\varepsilon(i,j)$'. When P has the form (2), the meaning of $P(a_1, \cdots, a_n)$ $(a_k \in M)$ is given by the definition of M as Ω-structure. Now let P, Q be any formulae with I_P, I_Q as sets of free variables, for which relations in M have been defined, and let $\theta : I_P \cup I_Q \to M$ be any mapping; then we put

$$M \vDash P(I_P\theta) \vee Q(I_Q\theta) \qquad \text{if and only if } M \vDash P(I_P\theta) \text{ or } M \vDash Q(I_Q\theta),$$

and

$$M \vDash \sim P(I_P\theta) \qquad \text{if and only if } M \sim \vDash P(I_P\theta).$$

In particular, $M \vDash \sim \varepsilon(a,b)$ if and only if $a \neq b$; for this reason we usually write '$i \neq j$' instead of '$\sim \varepsilon(i,j)$'. Finally, we put

$$M \vDash \bigvee_i P(I_P\theta) \text{ if and only if } M \vDash P(I_P\theta') \text{ for some mapping}$$
$$\theta' : I_P \to M \text{ such that } j\theta = j\theta' \text{ for } j \neq i.$$

Thus if $I_P = \{i_1, \cdots, i_n\}$ and $i = i_1$, then $M \vDash \bigvee_i P(i, a_2, \cdots, a_n)$ if and only if $M \vDash P(b, a_2, \cdots, a_n)$ for some $b \in M$. We also put

$$M \vDash t.$$

By induction on the length of words this defines, for each formula with n distinct free variables, an n-ary relation in M. In particular, the effects of $\wedge$, $\Rightarrow$, $\Leftrightarrow$ and $\bigwedge$ are thus determined. It is often convenient to take θ to be a mapping from the whole of I to M and to write $P\theta$ instead of $P(I_P\theta)$; with this notation it is easily verified that:

$$M \vDash P\theta \wedge Q\theta \qquad \text{if and only if } M \vDash P\theta \text{ and } M \vDash Q\theta,$$
$$M \vDash P\theta \Rightarrow Q\theta \qquad \text{if and only if } M \vDash Q\theta \text{ whenever } M \vDash P\theta,$$
$$M \vDash P\theta \Leftrightarrow Q\theta \qquad \text{if and only if either both } P\theta \text{ and } Q\theta \text{ hold or neither holds, in } M,$$
$$M \vDash \bigwedge_i P\theta \qquad \text{if and only if } M \vDash P\theta' \text{ for every mapping } \theta' : I \to M \text{ such that } j\theta' = j\theta \text{ for all } j \neq i.$$

We also note that f never holds in M, and that a sentence P is valid in M if and only if $M \models P \Leftrightarrow t$. Further, if a sentence holds in M, then it is valid in M. Thus, for any sentence P,

$$\text{either } M \vdash P \text{ or } M \vdash \sim P.$$

From any formula we obtain a sentence by applying universal quantifiers to all the free variables, which we indicate by prefixing $\bigwedge$. Then for any formula P,

$$M \vdash P \text{ if and only if } M \models \bigwedge P.$$

Two sentences P, Q are equivalent in M, in symbols $M \vdash P \Leftrightarrow Q$, precisely if

$$M \vdash P \text{ if and only if } M \vdash Q.$$

This means that as far as M is concerned, any sentence P may be replaced by an equivalent sentence Q. For example, the sentences formed by applying universal quantifiers to (5) and (6) are equivalent in any Ω-structure. Frequently, certain pairs of sentences which are equivalent in any Ω-structure are identified. This amounts to considering, not the Boolean algebra of derived predicates, but a certain homomorphic image. The defining relations are usually in the form of laws, so that we are in effect dealing with a relatively free Boolean algebra. The defining laws of this variety are called *tautologies*; it is not hard to see that every tautology is a consequence of tautologies of the form

$$P = t,$$

and any sentence P occurring in a law of this form is also called a tautology. The detailed study of sets of defining laws for tautologies belongs to the predicate calculus (cf. Kleene [52], Church [56]). We shall not enter into this subject, but merely note that each of the laws of a Boolean algebra stated above corresponds in fact to a tautology which has to be verified, a task which may be left to the reader. In what follows, we shall confine ourselves to the Boolean algebra of derived predicates and its effect on Ω-structures. In other words, we shall only be concerned with the semantic side of the predicate calculus, namely, the particular interpretation of the formulae given above, and not with the syntactic side, which makes deductions about the validity of a formula from its internal structure.

We note in passing that from the definition of a homomorphism as a mapping preserving the Ω-predicates, it follows that a homomorphism need not preserve the derived predicates. This accounts for the fact that homomorphisms play a subordinate role in relational structures; we shall return to this point in VI.5.

A sentence of the form

$$(i_1)\cdots(i_n)\,P(i_1,\cdots,i_n),$$

where $P(i_1,\cdots,i_n)$ is an open formula with the free variables $i_1,\cdots,i_n$ and (i_r) is a quantifier, standing for $\bigwedge$ or $\bigvee$, is said to be a *prenex sentence*, or a sentence in *prenex normal form*. It may be shown that every sentence is equivalent to one in prenex normal form. This is not very difficult, but we shall confine ourselves to giving an example. Thus the sentence

$$\bigwedge_i \bigwedge_j [R(i,j) \vee R(j,i)] \wedge \left[\bigvee_i \sim R(i,i)\right] \tag{7}$$

is not in prenex normal form as it stands, but it is equivalent to the prenex sentence

$$\bigwedge_i \bigwedge_j \bigvee_k [R(i,j) \vee R(j,i)] \wedge [\sim R(k,k)]. \tag{8}$$

From the definitions we saw that any existential quantifier can be expressed in terms of universal quantifiers, and vice versa, but in the process of doing so the property of being prenex is usually lost. Therefore the type of quantifier which occurs in a prenex sentence cannot generally be varied at will. A prenex sentence is said to be *universal* if all its quantifiers are universal, and *existential* if all its quantifiers are existential. In particular, the prenex sentence obtained by applying $\bigwedge$ ($\bigvee$) to an open formula is called its *universal* (*existential*) *closure*.

EXERCISES

1. Show that the relation $P \Rightarrow Q$ between derived predicates is a preordering and $P \Leftrightarrow Q$ is the associated equivalence relation.

2. Verify that the defining laws of Boolean algebras and (4) are tautologies, i.e. that the two sides of the equations define the same relation in each case. Show also that

$$\left(\bigvee_i P\right) \wedge \left(\bigvee_i Q\right) = \bigvee_i (P \wedge Q)$$

is not generally a tautology.

3. Show that the sentences form a subalgebra S of the algebra of all derived predicates and that for any Ω-structure M, the sentences which do not hold in M form a maximal ideal $\mathfrak{q}$ in S. (Verify that $S/\mathfrak{q} \cong 2$.)

4. CLOSED SENTENCE CLASSES AND AXIOMATIC MODEL CLASSES

Henceforth we take Ω to be a fixed predicate domain and all Ω-structures occurring are understood to have their carrier in the universe U, fixed once and for all. If M is any Ω-structure and P a formula in the standard alphabet I over Ω, then we have defined the relation

(1) $$M \vdash P.$$

In this way we obtain a Galois connexion between the class $[\Omega]$ of all Ω-structures and $\overline{\Omega}_I$, the set of all derived predicates (in I). By this connexion,

(i) to any class $\mathscr{C}$ of Ω-structures, there corresponds the set $\mathscr{C}^*$ of all formulae which are valid in each $M \in \mathscr{C}$, and

(ii) to any set Σ of formulae there corresponds the class Σ^* of all those Ω-structures in which all the formulae of Σ are valid.

Any Ω-structure in Σ^* is called a *model* for Σ, or a Σ-*model*; any sentence in $\mathscr{C}^*$ is called a *theorem* in $\mathscr{C}$, or a $\mathscr{C}$-*theorem*. We note that not all formulae in $\mathscr{C}^*$ are $\mathscr{C}$-theorems, but only those which have no free variables. On the other hand, $\mathscr{C}^*$ is completely determined by its theorems, since a formula P belongs to $\mathscr{C}^*$ if and only if its universal closure is a $\mathscr{C}$-theorem.

A class of Ω-structures which is of the form Σ^*, for some set Σ of sentences, is said to be an *axiomatic class*, and Σ a set of *axioms*. If Σ is finite, the class Σ^* is said to be *elementary*; replacing the finite set Σ by the conjunction of its elements, we see that an elementary class can always be defined by a single axiom. A set of formulae of the form $\mathscr{C}^*$, where $\mathscr{C}$ is any class of Ω-structures, is said to be *model-closed*; if $\mathscr{C} \neq \emptyset$, the set $\mathscr{C}^*$ is called a *theory*. The Galois connexion defined by (1) may now be expressed as

Theorem 4.1

The model-closed sets of formulae over Ω form a closure system which is equipotent with the axiomatic classes of Ω-structures by means of the natural bijection

$$\Sigma \to \Sigma^*,$$
$$\mathscr{C} \to \mathscr{C}^*. \quad \blacksquare$$

We remark that a model-closed set is a theory if and only if it is a proper subset of $\overline{\Omega}_I$. For a theory Σ is valid for some model, and hence $f \notin \Sigma$; conversely, a model-closed set which is not a theory is of the form $\emptyset^* = \overline{\Omega}_I$. Equivalently, we see that Σ is a theory if and only if $f \notin \Sigma$.

It is of fundamental importance for the study of relational structures that the system of model-closed sets is an algebraic closure system. In formal logic this is proved by defining explicitly a set of finitary operators. These have the form of rules of deduction (enabling one to derive new valid formulae from given ones):

(i) If $M \vdash (P \Rightarrow Q)$ and $M \vdash P$, then $M \vdash Q$ (modus ponens).
(ii) If $M \vdash P$, then $M \vdash \bigwedge_i P$ (generalization).

The connexion with models is established by Gödel's completeness theorem, which states that a set of formulae which admits the rules of deduction, and is a proper subset of $\overline{\Omega}_I$, necessarily has a model. A complete discussion of these matters, including a proof of Gödel's theorem, may be found in Church [56]. We shall not prove Gödel's theorem here, since an independent proof that the closure system of model-closed sets is algebraic will be given in V.5. For the present we shall only indicate briefly how Gödel's theorem implies that any set of formulae admitting the rules of deduction is model-closed.

Let Σ be a set of formulae admitting all the rules of deduction; we have to show that $\Sigma = \Sigma^{**}$. Since $\Sigma \subseteq \Sigma^{**}$ clearly holds, we need only prove that $\Sigma^{**} \subseteq \Sigma$; thus, given any formula $P \notin \Sigma$, we have to find a model for Σ in which P is not valid. Here P may be taken to be a sentence, without loss of generality; we then have to find a model for $\Sigma \cup \{\sim P\}$. By Gödel's theorem, it is enough to show that not every sentence can be derived from $\Sigma \cup \{\sim P\}$. Suppose that f can be derived from Σ and $\sim P$. By the finitary character of the rules of deduction, it follows that f can be derived from $S_1, \cdots, S_n, \sim P$ ($S_i \in \Sigma$). Applying the deduction theorem (cf. Church [56], p. 196; this is essentially the converse of modus ponens), we find that $(\sim P) \Rightarrow f$ can be derived from $S_1, \cdots, S_n$, and since Σ admits the rules of deduction, $[(\sim P) \Rightarrow f] \in \Sigma$, but

$$[(\sim P) \Rightarrow f] = [(\sim \sim P) \vee f] = P \vee f = P;$$

thus $P \in \Sigma$, which contradicts the choice of P. ∎

Since a closed set of formulae admits all the rules of deduction, it follows that the model-closed sets are precisely the sets admitting the rules of deduction. This state of affairs (like Gödel's theorem itself) prevails only in the predicate calculus of the first order. In more general situations, such as the predicate calculus of the second order, where quantifiers may be applied to predicate variables, the closure system of model-closed sets of formulae is no longer algebraic, and therefore cannot be described by finitary rules of deduction. Here it is understood that in any model the variable n-ary predicates range over *all* n-tuples of elements (giving rise to what is called a *standard model*). If more general models are allowed, where the predicates may range over certain specified subsets of n-tuples, the analogue of Gödel's completeness theorem again holds, and it follows as above that the closure system of model-closed sets is algebraic (cf. Henkin [50]). Now the celebrated incompleteness theorem of Gödel (cf. e.g. Kleene [52]) states that for any consistent set of formulae which includes a model of the integers, there are sentences P which are undecidable in the sense that neither P nor $\sim P$ can be derived from the formulae of the system. Thus for any axiomatic system which includes the integers there are certain (nonelementary) sentences which hold in all standard models but are not provable, and so do not hold in all general models.

The Galois connexion described in Theorem 4.1 can be used in two ways, either to study the structure of the formulae over Ω in terms of their models, or to study the Ω-structures by means of their theorems. However, this method has certain limitations; thus it will not enable us to distinguish between two formulae which have the same models, or between two Ω-structures which have the same theorems. We therefore define:

(i) Two formulae P and Q are said to be *congruent*, $P \approx Q$, if for every Ω-structure M,

$$M \models P\theta \text{ for all } \theta: I \to M \text{ if and only if } M \models Q\theta \text{ for all } \theta: I \to M.$$

(ii) Two Ω-structures M and N are said to be *elementarily equivalent* or *indiscernible*, $M \equiv N$, whenever

$$M \vdash P \text{ if and only if } N \vdash P \text{ for all } P \in \overline{\Omega}_I.$$

Clearly it is enough to demand that this hold for all sentences P. Since for any sentence P, either $M \vdash P$ or $M \vdash \sim P$, it follows that $M \equiv N$ provided that $N \vdash P$ whenever $M \vdash P$.

It is easily verified that the relation $\approx$, when restricted to the subalgebra of all sentences, is an equivalence, and in fact a congruence, i.e. if $P_1 \approx Q_1$ and $P_2 \approx Q_2$, then $P_1 \wedge P_2 \approx Q_1 \wedge Q_2$ and $\sim P_1 \approx \sim Q_1$. The quotient algebra by this congruence is denoted by $\mathscr{L}(\Omega)$ and is called the *Lindenbaum algebra* over Ω. Its elements will generally be denoted by small

Greek letters; we shall also write (P) for the class of the sentence P, and put $M \vdash \alpha$ to mean $M \vdash P$ for some (and hence all) $P \in \alpha$. The properties of $\mathscr{L}(\Omega)$ are summed up in

Theorem 4.2

The Lindenbaum algebra $\mathscr{L}(\Omega)$ is a Boolean algebra which is a homomorphic image of the algebra of all sentences over Ω. Each element of $\mathscr{L}(\Omega)$ defines an elementary model class, and distinct elements of $\mathscr{L}(\Omega)$ lead to distinct model classes. Further, if $\alpha,\beta \in \mathscr{L}(\Omega)$, then

$$\alpha \leqslant \beta \text{ if and only if } M \vdash \beta \text{ for any structure } M \text{ such that } M \vdash \alpha.$$

Only the last assertion requires proof. The statement that $M \vdash \beta$ whenever $M \vdash \alpha$ means that $M \vdash \alpha$ if and only if $M \vdash \alpha \wedge \beta$, i.e., $\alpha = \alpha \wedge \beta$, and this just means that $\alpha \leqslant \beta$. ∎

Instead of $\mathscr{L}(\Omega)$ one may wish to consider $\bar{\Omega}_I$, the Boolean algebra of all derived predicates, but then it is necessary to take account of the additional structure provided by the presence of quantifiers. The resulting structures are called *polyadic algebras* and have been studied by Halmos [62] and others. If one wants to build the notion of equality: '=' into the structure as well, one obtains essentially the *cylindrical algebras* of Tarski (cf. Henkin & Tarski [61]); for a concise summary of the definitions and the connexion between the two notions, see Galler [57].

Turning now to (ii) above, we see that '$\equiv$' is an equivalence on the class $[\Omega]$ of all Ω-structures. The equivalence classes are just the minimal axiomatic classes of Ω-structures; these are called the *axiomatic types.* The set of all axiomatic types is called the *model space* over Ω and will be denoted by $\mathscr{T}[\Omega]$. The nature of this equivalence relation will concern us rather more closely in the sequel, but it is considerably more complex than the congruence relation between sentences, and not much is known about it. For the moment we merely note

Proposition 4.3

If M and N are any Ω-structures which are indiscernible and if M is finite, then N is finite and $|M| = |N|$.

To prove this assertion, we need only write down a derived predicate which expresses the fact that M has at least n elements. For any integer n, write

$$E(n) = \bigvee (i_1 \neq i_2) \wedge (i_1 \neq i_3) \wedge \cdots \wedge (i_1 \neq i_n) \wedge (i_2 \neq i_3) \wedge \cdots \wedge (i_{n-1} \neq i_n);$$

then $E(n)$ holds in M if and only if M has at least n elements. Therefore, if M is finite, with m elements, then $E(n)$ holds in M precisely for $n = 1, \cdots, m$. The same must be true for N and so N too has m elements. ▮

From the above definitions it is clear that the Galois connexion of Theorem 4.1 may equally well be considered as a Galois connexion between the Lindenbaum algebra and the model space. Our next aim is to identify the corresponding closure systems. We shall find that the theories correspond to filters in the Lindenbaum algebra (Theorem 5.4), while the closure system of axiomatic classes defines a certain topology on the model space, which will be the subject of V.6.

EXERCISES

1. Show that any two distinct atomic formulae are incongruent.

2. Show that any formula is congruent to its universal closure; deduce that '$\approx$' is not a congruence on the whole of $\bar{\Omega}_I$.

3. Verify that the congruence classes of sentences of a given theory form a filter in the Lindenbaum algebra.

5. ULTRAPRODUCTS AND THE COMPACTNESS THEOREM

Consider again the product of a family $(M_\lambda)_{\lambda \in \Lambda}$ of Ω-structures: $M = \Pi M_\lambda$, with projections $\varepsilon_\lambda : M \to M_\lambda$. We recall that the Ω-structure on M was defined by the rule

(1) $$M \models \omega\theta \text{ if and only if } M_\lambda \models \omega\theta\varepsilon_\lambda \text{ for all } \lambda \in \Lambda,$$

and the definition of equality in M may be stated in the form

(2) $$M \models \varepsilon\theta \text{ if and only if } M_\lambda \models \varepsilon\theta\varepsilon_\lambda \text{ for all } \lambda \in \Lambda.$$

Here $\theta : I \to M$ is any mapping. We now ask for what derived predicates it is true that

(3) $$M \models P\theta \text{ if and only if } M_\lambda \models P\theta\varepsilon_\lambda \text{ for all } \lambda \in \Lambda.$$

This certainly holds for all positive formulae; we need only show that

$$M \models P\theta \text{ or } M \models Q\theta \text{ if and only if } M \models P\theta \vee Q\theta,$$
$$M \models P\theta \text{ and } M \models Q\theta \text{ if and only if } M \models P\theta \wedge Q\theta,$$

and this is clear from the definitions. Similarly, it may be shown that if (3) holds for P, then it holds for $\bigvee_i P$ and $\bigwedge_i P$; but for formulae involving negations (3) is no longer true. Thus for any formula P,

$$\text{`}M \models {\sim}P\theta\text{' means that `}M_\lambda \models {\sim}P\theta\varepsilon_\lambda \text{ for all } \lambda \in \Lambda\text{'},$$

while

$$\text{`}M \sim\models P\theta\text{' means that `}M_\lambda \models {\sim}P\theta\varepsilon_\lambda \text{ for some } \lambda \in \Lambda\text{'}.$$

The two statements on the right are not equivalent, except in trivial cases; our object is to remedy this defect, i.e., to modify the definition of product structure in some way such that '$M \models {\sim}P\theta$' comes to mean the same thing as '$M \sim\models P\theta$'. This is achieved by introducing the notion of a reduced product structure:

Let $(M_\lambda)_{\lambda \in \Lambda}$ be a family of Ω-structures; then, for any filter $\mathscr{D}$ on the index-set Λ, the *reduced product* $M_{\mathscr{D}}$ of the M_λ (modulo $\mathscr{D}$) is the quotient of the Cartesian product $M = \Pi M_\lambda$ defined by the rule

(4) $$a\mathscr{D} = b\mathscr{D} \text{ if and only if } \{\lambda \in \Lambda \mid a\varepsilon_\lambda = b\varepsilon_\lambda\} \in \mathscr{D};$$

together with the Ω-structure defined by

(5) $$M_{\mathscr{D}} \models \omega\theta \text{ if and only if } \{\lambda \in \Lambda \mid M_\lambda \models \omega\theta\varepsilon_\lambda\} \in \mathscr{D}.$$

The natural mapping $M \to M_{\mathscr{D}}$ will be denoted by $a \to a\mathscr{D}$ or nat $\mathscr{D}$, and we shall also write $M/\mathscr{D}$ for $M_{\mathscr{D}}$ and $a \equiv b \pmod{\mathscr{D}}$ instead of $a\mathscr{D} = b\mathscr{D}$. It is easily seen that a reduced product taken modulo a principal filter is isomorphic to a direct product, taken over some of the factors; such a product is said to be *trivial*. A reduced product taken modulo an ultrafilter is called an *ultraproduct*, or in case all the factors are equal, an *ultrapower*. The usefulness of ultraproducts derives from the following fundamental result (Frayne, Morel, & Scott [62], Kochen [61], Łos [55′]):

Theorem 5.1 (ultraproduct theorem)

Let $(M_\lambda)_{\lambda \in \Lambda}$ be a family of Ω-structures, $\mathscr{D}$ an ultrafilter on Λ, and write $M = \Pi M_\lambda$, $M_{\mathscr{D}} = \Pi M_\lambda/\mathscr{D}$. Then for any formula P and any $\theta: I \to M$,

(6) $$M_{\mathscr{D}} \models P\theta(\text{nat } \mathscr{D}) \text{ if and only if } \{\lambda \in \Lambda \mid M_\lambda \models P\theta\varepsilon_\lambda\} \in \mathscr{D}.$$

Proof:

Let Σ be the set of all formulae for which (6) holds; then Σ contains t and all atomic formulae, by the definition of ultraproduct (cf. (4) and (5)), so in order to show that $\Sigma = \bar{\Omega}_I$, we need only verify that Σ admits $\wedge$, $\sim$, and $\bigvee_i$.

Given any $P,Q \in \Sigma$ and any $\theta: I \to M$, let us write

(7) $$X = \{\lambda \in \Lambda \mid M_\lambda \models P\theta\varepsilon_\lambda\}, \quad Y = \{\lambda \in \Lambda \mid M_\lambda \models Q\theta\varepsilon_\lambda\},$$
$$Z = \{\lambda \in \Lambda \mid M_\lambda \models P\theta\varepsilon_\lambda \wedge Q\theta\varepsilon_\lambda\}.$$

It is clear that $Z = X \cap Y$; now $M_{\mathscr{D}} \models P\theta(\text{nat } \mathscr{D}) \wedge Q\theta(\text{nat } \mathscr{D})$ if and only if $M_{\mathscr{D}} \models P\theta(\text{nat } \mathscr{D})$ and $M_{\mathscr{D}} \models Q\theta(\text{nat } \mathscr{D})$, i.e. if and only if $X \in \mathscr{D}$ and $Y \in \mathscr{D}$ (because P, Q satisfy (6)). But this is true precisely if $Z = X \cap Y \in \mathscr{D}$, and so (6) holds for $P \wedge Q$.

Secondly, if $P \in \Sigma$, then $M_{\mathscr{D}} \models {\sim}P\theta(\text{nat } \mathscr{D})$ if and only if $M_{\mathscr{D}} \sim\models P\theta(\text{nat } \mathscr{D})$, i.e. $X \notin \mathscr{D}$, where X is given by (7); and since $\mathscr{D}$ is an ultrafilter, this holds if and only if $\Lambda \backslash X \in \mathscr{D}$, i.e. $\{\lambda \in \Lambda \mid M_\lambda \models {\sim}P\theta(\text{nat } \mathscr{D})\} \in \mathscr{D}$. Therefore ${\sim}P \in \Sigma$.

Finally, let $P \in \Sigma$ and suppose that $M_{\mathscr{D}} \models \bigvee_i P\theta(\text{nat } \mathscr{D})$; then there exists $\theta': I \to M$ such that $j\theta' = j\theta$ for $j \neq i$, and $M_{\mathscr{D}} \models P\theta'(\text{nat } \mathscr{D})$; hence $\{\lambda \in \Lambda \mid M_\lambda \models P\theta'\varepsilon_\lambda\} \in \mathscr{D}$, i.e.

(8) $$X_0 = \{\lambda \in \Lambda \mid M_\lambda \models \bigvee_i P\theta\varepsilon_\lambda\} \in \mathscr{D}.$$

Conversely, if (8) holds, choose $\theta': I \to M$ such that $j\theta' = j\theta$ for $j \neq i$ and $M_\lambda \models P\theta'\varepsilon_\lambda$ for $\lambda \in X_0$. Such θ' exists by (8), because no M_λ is empty. Then, by the choice of θ', $\{\lambda \in \Lambda \mid M_\lambda \models P\theta'\varepsilon_\lambda\} = X_0$, and so $M_{\mathscr{D}} \models P\theta'(\text{nat } \mathscr{D})$, i.e. $M_{\mathscr{D}} \models \bigvee_i P\theta(\text{nat } \mathscr{D})$. ∎

If in Theorem 5.1 we take the formula P to be closed and put $\{P\}^* = \mathscr{K}$, we obtain

Corollary 5.2

If $\mathscr{K}$ is an elementary class, then $\Pi M_\lambda/\mathscr{D} \in \mathscr{K}$ if and only if $\{\lambda \in \Lambda \mid M_\lambda \in \mathscr{K}\} \in \mathscr{D}$. ∎

In particular, if each $M_\lambda \in \mathscr{K}$, then $M_{\mathscr{D}} \in \mathscr{K}$. More generally, if $\mathscr{K}$ is an axiomatic class, say $\mathscr{K} = \Sigma^*$, then $\mathscr{K} = \bigcap_{\alpha \in \Sigma} \{\alpha\}^*$, i.e. $\mathscr{K}$ is an intersection of elementary classes and we find

Corollary 5.3

Every axiomatic class admits ultraproducts. ∎

It may be shown by examples that in general Corollary 5.2 no longer holds for axiomatic classes (cf. Exercise 4).

Using ultraproducts we can now show that the closure system of model-closed sets of formulae is algebraic. It will be sufficient to prove this in the Lindenbaum algebra $\mathscr{L}(\Omega)$ rather than the algebra of formulae $\bar{\Omega}_I$ itself. We shall in fact prove a more precise result:

Theorem 5.4

Let Σ be any subset of the Lindenbaum algebra $\mathscr{L}(\Omega)$; then the dual ideal generated by Σ is just the model-closure Σ^{**}.

Proof:

Denote by $\bar{\Sigma}$ the dual ideal generated by Σ; thus $\bar{\Sigma}$ consists of all $\alpha \in \mathscr{L}(\Omega)$ which satisfy

$$\alpha \geqslant \beta_1 \wedge \cdots \wedge \beta_n \qquad (\beta_r \in \Sigma).$$

Clearly any Σ-model is also a $\bar{\Sigma}$-model, so that

$$\bar{\Sigma} \subseteq \Sigma^{**}; \tag{9}$$

it remains to establish the reverse inclusion. Let Φ be the family of sets in which the principal ideals generated by the elements of $\bar{\Sigma}$ meet $\bar{\Sigma}$; then Φ has the finite intersection property (because the infimum of any finite subset of Σ belongs to $\bar{\Sigma}$); therefore Φ is contained in an ultrafilter $\mathscr{D}$ on $\bar{\Sigma}$ (qua set). Now to prove equality in (9), take any $\lambda \notin \bar{\Sigma}$; then for each $\alpha \in \bar{\Sigma}$ we have $\alpha \nleqslant \lambda$, and hence (by Theorem 4.2), there exists an Ω-structure M_α in which α is valid, but not λ. Put

$$M_{\mathscr{D}} = \prod_{\alpha \in \bar{\Sigma}} M_\alpha / \mathscr{D};$$

then, given $\alpha \in \bar{\Sigma}$, we have $M_\beta \vdash \alpha$ for all $\beta \in \bar{\Sigma}$ such that $\beta \leqslant \alpha$, and so $M_{\mathscr{D}} \vdash \alpha$, by the construction of Φ. This holds for all $\alpha \in \Sigma$, hence $M_{\mathscr{D}} \in \Sigma^*$. On the other hand, $M_\alpha \sim\vdash \lambda$ for all $\alpha \in \bar{\Sigma}$; therefore $M_{\mathscr{D}} \sim\vdash \lambda$. This proves that $\lambda \notin \Sigma^{**}$ and the equality in (9) is established. ∎

This theorem shows that the model-closed sets are merely the dual ideals of $\mathscr{L}(\Omega)$; moreover the filters in $\mathscr{L}(\Omega)$ are the theories. A theory is said to be *complete* if it is maximal in the set of theories. Combining Theorems 5.4, II.5.2, and V.2.7 we now have

Theorem 5.5

The closure system of model-closed sets of formulae is algebraic. Moreover, every theory can be extended to a complete theory. ∎

A set of formulae is said to be *consistent* if it has a model. Thus Σ is consistent if and only if $f \notin \Sigma^{**}$, or, equivalently, if Σ can be extended to a theory. By Theorem 5.5 and the definition of algebraic closure systems we now have

Corollary 5.6

A set of formulae is consistent if and only if every finite subset is consistent. ▮

This result is sometimes called the *compactness theorem*, for reasons which will become clear in V.6. It was first obtained as a theorem in logic by Gödel (via his completeness theorem) and independently by Skolem and Malcev (cf. Church [56]).

EXERCISES

1. Show that under the Galois connexion of Theorem 5.1, complete theories correspond to axiomatic types.

2. Verify that an ultraproduct modulo a principal filter is isomorphic to one of the factors (this explains why ultraproducts are only of interest in the case of infinitely many factors).

3. If $\mathscr{K}$ is an axiomatic class of Ω-structures, show that the infinite $\mathscr{K}$-structures again form an axiomatic class. (Use the sentences $E(n)$ defined in Proposition 4.3.)

4. Let $\mathscr{K}$ be any axiomatic class of Ω-structures; if $\mathscr{K}$ contains finite structures with arbitrarily large numbers of elements, show that Corollary 5.2 does not hold for $\mathscr{K}$.

5. Show that any ultrapower of a finite structure of n elements again has n elements.

6. Show that any ultraproduct of infinite structures is again infinite.

7. Show that a theory is complete if and only if for every sentence P, either P or $\sim P$ belongs to the theory. Deduce that for any Ω-structure M, the set of sentences valid in M is a complete theory.

8. If Θ is any theory and P is a sentence such that $\Theta \cup \{\sim P\}$ is inconsistent, show that $P \in \Theta$.

6. THE MODEL SPACE

We now consider the closure system of axiomatic model classes in more detail. Since the formulae do not allow us to distinguish between

Ω-structures of a given type, we may as well take as objects of the discussion the axiomatic types rather than the individual Ω-structures. Thus we have the natural mapping $M \to (M)$ from $[\Omega]$ to $\mathscr{T}[\Omega]$, which associates with each Ω-structure its type. The inverse images of subsets of $\mathscr{T}[\Omega]$ under this mapping are those classes of Ω-structures which are unions of types, called *type classes* for short. Thus a type class is a class $\mathscr{K}$ of Ω-structures such that

$$\text{if } M \in \mathscr{K} \text{ and } M \equiv N, \text{ then } N \in \mathscr{K}.$$

The axiomatic model classes may be used to define a topology of $\mathscr{T}[\Omega]$, which is described in

Theorem 6.1

The axiomatic model classes form a topological closure system on the model space $\mathscr{T}[\Omega]$. Under the corresponding topology $\mathscr{T}[\Omega]$ is a totally disconnected compact Hausdorff space.

A topological space with the properties named is sometimes called a *Boolean space*.

Proof:

Let $\mathscr{C}_1$, $\mathscr{C}_2$ be axiomatic model classes, say $\mathscr{C}_r = \Sigma_r^*$ $(r = 1,2)$; we assert that $\mathscr{C} = \mathscr{C}_1 \cup \mathscr{C}_2$ is axiomatic. This will follow if we show that $\mathscr{C}^{**} \subseteq \mathscr{C}$. Let M be an Ω-structure which is not in $\mathscr{C}$; then $M \notin \mathscr{C}_r (r = 1,2)$, so there is a $\mathscr{C}_r$-theorem P_r which is not valid in M. Hence $P_1 \vee P_2$ is a $\mathscr{C}$-theorem which is not valid in M, i.e. $M \notin \mathscr{C}^{**}$. Thus $\mathscr{C}^{**} = \mathscr{C}$, which shows that the axiomatic classes form a topological closure system. Moreover, the empty class is axiomatic, since it may be defined by f, say, and we therefore have a topology. The closed sets are intersections of sets of the form $\{P\}^*$, where P is any sentence. Since the complement of $\{P\}^*$ is of the same form, namely $\{\sim P\}^*$, it follows that there is a base consisting of closed and open sets ('clopen sets'), so that the topology is totally disconnected. Further, if (M) and (N) are distinct types, then there is a sentence P which is valid in M but not in N. Hence $\sim P$ is valid in N but not in M, and $\{P\}^*$, $\{\sim P\}^*$ are nonintersecting neighbourhoods of (M), (N) respectively, which shows that we have a Hausdorff space.

To prove compactness, let $(\mathscr{C}_\lambda)_{\lambda \in \Lambda}$ be a family of closed sets with the finite intersection property. If $\mathscr{C}_\lambda = \Sigma_\lambda^*$, then for any finite subset Λ_0 of Λ,

$$(\bigcup_{\Lambda_0} \Sigma_\lambda)^* = \bigcap_{\Lambda_0} \mathscr{C}_\lambda \neq \emptyset;$$

in particular, every finite subset of $\Sigma = \bigcup \Sigma_\lambda$ is consistent. Therefore so is Σ itself, by Corollary 5.6, and hence

$$\bigcap \mathscr{C}_\lambda = \Sigma^* \neq \emptyset.$$

This establishes the compactness. ∎

Corollary 6.2

The elementary classes correspond to those subsets of $\mathscr{T}[\Omega]$ which are both open and closed. In other words, an axiomatic model class is elementary if and only if its complement is also axiomatic.

For clearly every elementary class defines a subset of $\mathscr{T}[\Omega]$ which is both open and closed. Conversely, suppose that $\mathscr{C}$ is both open and closed; being open, $\mathscr{C}$ can be expressed as a union of elementary classes, $\mathscr{C} = \bigcup \mathscr{C}_\lambda$ say. Since $\mathscr{C}$ is also closed, it is compact and is therefore covered by a finite subfamily of the $\mathscr{C}_\lambda$, say $\mathscr{C} = \mathscr{C}_1 \cup \cdots \cup \mathscr{C}_n$; hence $\mathscr{C}$ is itself elementary. ∎

In connexion with this corollary we note that the elementary classes correspond to the elements of the Lindenbaum-algebra $\mathscr{L}(\Omega)$, and in fact this correspondence is bijective.

The closure operation in the model space may be described in terms of ultraproducts as follows (Taimanov [62]):

Proposition 6.3

Let S be any subset of $\mathscr{T}[\Omega]$ and N any Ω-structure; then the type (N) belongs to the closure of S if and only if there is a family $(M_\lambda)_{\lambda \in \Lambda}$ of S-structures and an ultrafilter $\mathscr{D}$ on Λ such that

(1) $$N \equiv \Pi M_\lambda / \mathscr{D}.$$

Proof:

By definition, $(N) \in \bar{S}$ if and only if every S-theorem is valid in N. Therefore if (1) holds, it follows by the ultraproduct theorem (Theorem 5.1) that $(N) \in \bar{S}$. Conversely, let $(N) \in \bar{S}$ and take a model M_λ in each type λ belonging to S. Since every S-theorem is valid in N, any sentence P valid in N has a model in S (because otherwise $\sim P$ would be an S-theorem). For each $\alpha \in \mathscr{L}(\Omega)$ such that $N \vdash \alpha$, define a subset E_α of Λ by

$$E_\alpha = \{\lambda \in \Lambda \mid M_\lambda \vdash \alpha\};$$

then $E_\alpha \neq \emptyset$ and the E_α admit finite intersections, for if $P \in \alpha$, $Q \in \beta$, then $E_\alpha \cap E_\beta = E_\gamma$, where $\gamma = (P \wedge Q)$. Hence, there is an ultrafilter $\mathscr{D}$ on Λ

containing all the E_α. We assert that (1) holds for this choice of $M_{\mathscr{D}} = \Pi M_\lambda/\mathscr{D}$. For if $N \vdash \alpha$, then $E_\alpha \in \mathscr{D}$, and hence $M_{\mathscr{D}} \vdash \alpha$; thus $N \equiv M_{\mathscr{D}}$, i.e., (1). ▮

From this proposition we obtain the following generalization of Corollary 5.3.

Corollary 6.4

A type class is axiomatic if and only if it admits ultraproducts. ▮

For a more precise result we refer to VI.6; here we content ourselves with pointing out another consequence of Proposition 6.3. An axiomatic type is just a point of the space $\mathscr{T}[\Omega]$ and is therefore closed. Since ultrapowers are a special case of ultraproducts, taken within a single type, we obtain from Corollary 6.4

Corollary 6.5

Any type class admits ultrapowers. ▮

Any axiomatic class $\mathscr{K}$ is a closed subspace of $\mathscr{T}[\Omega]$, and under the induced topology this is again a totally disconnected compact Hausdorff space. The intersections of arbitrary axiomatic (or elementary) classes with $\mathscr{K}$ will be called *relatively axiomatic* (or *elementary*) classes in $\mathscr{K}$. The most important use made of relative classes is in the case of Ω-structures which are algebras with respect to a subdomain Ω^* of Ω. We have to show that the class of Ω-structures which are algebras with respect to Ω^* is axiomatic. This is done by constructing sentences which express the fact that the predicates in Ω^* define operations (not merely relations). Let $\omega \in \Omega$ be an $(n + 1)$-ary predicate. Then the sentence

(2) $$\bigwedge \bigvee_j \omega(i_1,\cdots,i_n,j)$$

is called the *determinateness condition* for ω. It is valid in an Ω-structure M if and only if, for every n-tuple $(a_1,\cdots,a_n)$ in M, there is at least one $b \in M$ such that $\omega(a_1,\cdots,a_n,b)$ holds; in other words, if ω defines a—possibly many-valued—operation in M. To obtain an operation in our sense we need a second sentence, the *uniqueness condition* for ω:

(3) $$\bigwedge [(\omega(i_1,\cdots,i_n,j) \wedge \omega(i_1,\cdots,i_n,k)) \Rightarrow (j = k)].$$

This expresses the fact that to each n-tuple $(a_1,\cdots,a_n)$ in M there is at most one $b \in M$ such that $M \models \omega(a_1,\cdots,a_n,b)$. Thus (2) and (3) together just state that ω defines an n-ary operation in M.

Now let Ω^* be a subdomain of Ω and Σ the set consisting of the determinateness and uniqueness conditions for all the elements of Ω^*. Then the axiomatic class defined by Σ is just the class of all Ω-structures which are Ω^*-algebras. Applying Theorem 6.1, we thus obtain

Theorem 6.6

The types of all Ω-structures which are Ω^*-algebras form a closed (and hence compact) subspace of the model space $\mathscr{T}[\Omega]$. Any axiomatic algebra class is relatively elementary if and only if its complement is also relatively elementary. ▮

In practice we shall write operators in the operational form of Chapter II rather than the relational form used here. This is done merely for convenience; the reader may, if he wishes, transcribe all sentences in purely relational form. For example, if ω is a ternary relation which in a certain structure M defines a binary operation, then the commutative law for this operation, written out in full, reads

$$\bigwedge [(\omega(i_1,i_2,i_3) \wedge \omega(i_2,i_1,i_4)) \Rightarrow (i_3 = i_4)],$$

while the associative law is given by

$$\bigwedge [(\omega(i_1,i_2,i_4) \wedge \omega(i_4,i_3,i_5) \wedge \omega(i_2,i_3,i_6) \wedge \omega(i_1,i_6,i_7)) \Rightarrow (i_5 = i_7)].$$

It may be noted that both these laws are described by universal sentences, although the determinateness condition needed to define ω as an operation was not universal.

We now give some examples of relatively axiomatic classes.

(i) *Ordered sets*. Let Ω consist of a single binary relation ρ; then the elementary class defined by

$$(4) \qquad \bigwedge [\sim \rho(i,i)]$$

$$(5) \qquad \bigwedge [(\rho(i,j) \wedge \rho(j,k)) \Rightarrow \rho(i,k)]$$

is essentially the class of ordered sets. Lattices may be singled out by adding the axioms

$$(6)_r \qquad \bigwedge \bigvee_k \bigwedge_l [(\rho_r(i,k) \wedge \rho_r(j,k)) \wedge (\rho_r(i,l) \wedge \rho_r(j,l) \Rightarrow \rho_r(k,l))] \quad (r = 1,2);$$

Here we have put '$\rho_1(i,j)$' for '$\rho(i,j) \vee (i = j)$' and '$\rho_2(i,j)$' for '$\rho_1(j,i)$', for brevity. It is important here to bear in mind the difference between substructures and subalgebras; thus a substructure of a lattice as defined above is not in general a sublattice.

(ii) *Ordered groups.* We now take Ω to consist of the group operators and a binary predicate ρ. Then ordered groups may be defined by the group axioms, together with (4) and (5) above and

(7) $$\bigwedge [\rho(i,j) \Rightarrow (\rho(ik,jk) \wedge \rho(ki,kj))].$$

(iii) *Fields.* A field may be defined as a commutative (and associative) ring satisfying the condition

(8) $$\bigwedge \bigvee_k [(i \neq 0) \Rightarrow (ik = j)].$$

As a rule the ring consisting of 0 alone is excluded, e.g. by adding the sentence

(9) $$\bigvee (i \neq 0).$$

As a further example we have the fields of a given finite characteristic. For every natural number n we have in any field (as in any additive group) the derived operator $x \rightarrow nx$, where

$$nx = x + x + \cdots + x \qquad (n \text{ summands}).$$

Now a field of characteristic n may be defined by the sentences

(10) $$\bigwedge (ni = 0),$$

(11) $$\bigvee [(i \neq 0) \wedge (2i \neq 0) \wedge \cdots \wedge ((n-1)i \neq 0)].$$

As is well known (cf. e.g. v.d.Waerden [37]), this class is nonempty if and only if n is prime. The fields of characteristic zero may be defined by taking all the sentences (11), for $n = 1,2,\cdots$. Hence they also form an axiomatic class, but as we shall see in a moment, this class is no longer elementary.

We first note the following consequence of Theorem 6.6 (cf. Robinson [63]).

Theorem 6.7

Let Σ be any set of sentences about fields. If there are fields of arbitrarily high characteristic satisfying Σ, then there are fields of characteristic zero satisfying Σ.

Proof:

Let $\mathscr{K}_n$ be the class of fields defined by Σ and (11); thus $\mathscr{K}_n$ consists of all Σ-models of characteristic at least n. The fields of characteristic zero which satisfy Σ form the intersection $\bigcap \mathscr{K}_n$. Now by hypothesis,

$$\mathscr{K}_1 \supseteq \mathscr{K}_2 \supseteq \cdots$$

is a chain of nonempty closed subsets of a compact space, hence their intersection is also nonempty, as we wished to show. ∎

Let P be any sentence which is valid in every field of characteristic zero. If we apply the theorem with $\Sigma = \{\sim P\}$, we obtain

Corollary 6.8

An elementary sentence which holds in every field of characteristic zero also holds in every field of sufficiently high characteristic. ▮

Now if the class of all fields of characteristic zero were elementary, let P be a sentence defining this class. By Corollary 6.8, P would then also hold for all fields whose characteristic exceeds a certain number n_0; in other words, there would be no fields of finite characteristic greater than n_0. This is a contradiction and it proves

Corollary 6.9

The class of fields of characteristic zero is not relatively elementary. ▮

EXERCISES

1. Show that the model space $\mathscr{T}[\Omega]$ has a countable base if and only if Ω is at most countable.

2. Let A be any Boolean algebra and denote by A^* the subset of 2^A consisting of all homomorphisms $A \to 2$. Show that A^* is a Boolean space with respect to the topology induced by the product topology in 2^A. Use this fact to give another proof of Theorem 6.1.

3. If X is any Boolean space, then the set X^* of clopen (= closed and open) subsets forms a Boolean algebra. In particular, if $X = A^*$, for some Boolean algebra A (cf. Exercise 2), then $A^{**} \cong A$; moreover, X^{**} is homeomorphic to X.

4. Show that the model classes admitting ultraproducts form a topological closure system and that the resulting topology is compact.

5. Show that a type class $\mathscr{K}$ is elementary if and only if both $\mathscr{K}$ and its complement in $\mathscr{T}[\Omega]$ admit ultraproducts.

6. Show that the class of fields of finite characteristic is not axiomatic.

7. Show that the class of all finite groups is not axiomatic.

8. Show that the class of all finite ordered sets satisfying the axiom

$$\bigwedge \bigvee_{k} [(\rho(i,k) \wedge \rho(k,j)) \vee (\rho(j,k) \wedge \rho(k,i)) \vee (i = j)]$$

is elementary.

Chapter VI

Axiomatic Model Classes

As we saw in Chapter V, some properties of Ω-structures can be expressed by means of elementary sentences, while others cannot. We now take the analysis one step further: we consider various properties of axiomatic model classes and enquire to what extent these may be characterized by the form of the defining sentences. The chapter concludes with an elegant characterization, due to Keisler [61], of axiomatic model classes in terms of ultraproducts.

1. REDUCTS AND ENLARGEMENTS

Let Ω be any predicate domain and $\mathscr{L}$ an axiomatic model class, which is taken to be fixed in what follows. The sentences defining $\mathscr{L}$ will be called the *basic axioms*; $\mathscr{L}$-structures will simply be called *models*, and the terms *submodel*, *quotient model*, etc. will all be understood to refer to $\mathscr{L}$, when nothing else is said. All classes of Ω-structures are henceforth considered relative to $\mathscr{L}$; thus if $\mathscr{C}$ is any class of Ω-structures, by a $\mathscr{C}$-model is meant a member of $\mathscr{C} \cap \mathscr{L}$. Here $\mathscr{L}$ may be the class of all Ω-algebras, or more generally, the class of Ω-structures which are algebras with respect to $\Omega^* \subseteq \Omega$, or also the class of all Ω-structures.

If Ω' is any subdomain of Ω, then to every model M there corresponds a unique Ω'-structure $M \mid \Omega'$, called the Ω'-*reduct* of M, which is obtained from M by ignoring the predicates which do not belong to Ω', and any basic axioms containing such predicates. In particular, a *finite reduct* of M is a reduct $M \mid \Omega_f$, where Ω_f is a finite subset of Ω. Conversely, given an Ω'-structure N, if there exists a model M whose Ω'-reduct is N, we call M an Ω-*enlargement* of N. Such an enlargement is said to be *finite* if $\Omega \backslash \Omega'$ is finite, *unary* if $\Omega \backslash \Omega'$ contains only unary predicates; Ω itself is also referred to as an enlargement of Ω'.

For classes of models we can define closure operations as for Ω-algebras. A model class will be called *abstract*, if it contains with any model all its isomorphic copies; for any abstract class $\mathscr{C}$ of models we write:

S$\mathscr{C}$ for the class of submodels of $\mathscr{C}$-models.
H$\mathscr{C}$ for the class of homomorphic images of $\mathscr{C}$-models.
P$\mathscr{C}$ for the class of products of $\mathscr{C}$-models.
L$\mathscr{C}$ for the class of locally $\mathscr{C}$ models: $M \in \mathrm{L}\mathscr{C}$ whenever M has a local system of S$\mathscr{C}$-models.
$\bar{\mathrm{L}}\mathscr{C}$ for the class of sublocally $\mathscr{C}$ models: $M \in \bar{\mathrm{L}}\mathscr{C}$ whenever every finite substructure of M can be embedded in a $\mathscr{C}$-model.

An S-closed class is said to be *hereditary*; it will usually be clear from the context whether this term refers to subalgebras or substructures. An L-closed class is said to be *locally defined* and an $\bar{\mathrm{L}}$-closed class is said to be *sublocally defined.*

It is clear that

$$\mathrm{L}\mathscr{C} \subseteq \bar{\mathrm{L}}\mathscr{C},$$

but equality need not hold. E.g., if $\mathscr{L}$ is the class of abelian groups and $\mathscr{C}$ the class of periodic groups (in $\mathscr{L}$), then $\mathrm{L}\mathscr{C} = \mathscr{C}$ while $\bar{\mathrm{L}}\mathscr{C} = \mathscr{L}$. In general, every locally $\mathscr{C}$ model is sublocally $\mathscr{C}$, and every sublocally defined class is locally defined.

If $\mathscr{C}$ is any class of models and $\Omega' \subseteq \Omega$, we denote by $\mathscr{C} \mid \Omega'$ the class of all Ω'-reducts of $\mathscr{C}$-models. It is easily verified that for any axiomatic class $\mathscr{C}$, M is a $\mathscr{C}$-model if and only if, for every finite subdomain Ω_f of Ω, $M \mid \Omega_f$ is a $(\mathscr{C} \mid \Omega_f)$-model. From this remark it follows that an axiomatic class $\mathscr{C}$ is closed under any of the operations S, H, P, L, $\bar{\mathrm{L}}$ if and only if $\mathscr{C} \mid \Omega_f$ is closed (within $\mathscr{L} \mid \Omega_f$) under the operation, for all finite subdomains Ω_f of Ω. This allows one in most cases to make a reduction to the case where Ω is finite. Of course in most algebraic systems Ω is finite in any case; an exception is the class of vector spaces over an infinite field.

The following criterion for hereditary classes to be locally defined is often useful. We recall that a class is *inductive* if it admits suprema, i.e., unions of chains.

Proposition 1.1

Let $\mathscr{C}$ be a hereditary class of models; then $\mathscr{C}$ is locally defined if and only if $\mathscr{C}$ is inductive.

Proof:

Assume that $\mathscr{C}$ is locally defined and let $(M_\lambda)_{\lambda \in \Lambda}$ be any chain of $\mathscr{C}$-models. Such a chain, to be admissible, must belong to our universe U, qua collection of sets; this is the case if $M_\lambda \in U$ for each $\lambda \in \Lambda$ and $\Lambda \in U$ (cf. I.10), and it ensures that $M = \bigcup M_\lambda \in U$. Now (M_λ) is a local system of $\mathscr{C}$-models for M; therefore $M \in \mathrm{L}\mathscr{C} = \mathscr{C}$. Conversely, suppose that $\mathscr{C}$ is inductive and let M be a model with a local system of $\mathscr{C}$-models ($= \mathrm{S}\mathscr{C}$-models). By Proposition I.5.9, $\mathscr{C}$ admits unions of directed sets, and so $M \in \mathscr{C}$. ∎

A characterization of sentences defining inductive model classes has been given by Chang [59].

EXERCISES

1. Let A be any closure operator on classes of models, and define an operator A_f by the rule: $M \in \mathrm{A}_f\mathscr{C}$ if and only if $M \mid \Omega_f \in \mathrm{A}(\mathscr{C} \mid \Omega_f)$ for all finite subdomains Ω_f of Ω. Determine which of the operators defined in VI.1 satisfy $\mathrm{A}_f = \mathrm{A}$.

2. A class $\mathscr{C}$ of Ω-structures is said to be *reductive*, if $M \in \mathscr{C}$ holds if and only if $M \mid \Omega_f \in \mathscr{C} \mid \Omega_f$ for all finite subdomains Ω_f of Ω. Show that for any reductive class $\mathscr{C}$ and any operator A, $\mathrm{A}_f\mathscr{C} = \mathrm{A}\mathscr{C}$.

2. THE LOCAL DETERMINATION OF CLASSES

Axiomatic classes have the important property that they are determined locally in the sense that $\mathrm{S}\mathscr{C}$ is sublocally defined for any axiomatic class $\mathscr{C}$. Naturally this is of particular interest for hereditary classes; moreover there is a simple criterion for these classes to be axiomatic. These two results are not directly related, but both depend on the possibility of describing the notion of inclusion by elementary sentences. We therefore begin with this description.

Let M be any model and Λ a set containing the carrier of M. Our object is to construct a set of sentences Δ over the predicate domain

$\Omega \sqcup \Lambda$, where the elements of Λ are distinct unary predicates, such that the models of Δ are precisely the extensions of a model with carrier Λ in which M is embedded as a submodel. Thus Δ consists of the determinateness and uniqueness conditions for Λ,

(1) $\bigvee \lambda(i) \qquad (\lambda \in \Lambda)$,

(2) $\bigwedge \{(\lambda(i) \wedge \lambda(j)) \Rightarrow (i = j)\} \qquad (\lambda \in \Lambda)$,

together with sentences expressing that different elements of Λ define different predicates:

(3) $\bigwedge \{(\lambda(i) \wedge \mu(j)) \Rightarrow (i \neq j)\}$ for all $\lambda, \mu \in \Lambda$ such that $\lambda \neq \mu$.

In particular, to each element a of M there corresponds a unary predicate $a(i)$. Now for each $a = (a_1,\cdots,a_n) \in M^n$ and each $\omega \in \Omega(n-1)$, we include in Δ the sentences

(4) $\bigwedge \{a_1(i_1) \wedge \cdots \wedge a_n(i_n) \Rightarrow \omega(i_1,\cdots,i_n)\} \qquad$ if $M \models \omega(a_1,\cdots,a_n)$,

(5) $\bigwedge \{a_1(i_1) \wedge \cdots \wedge a_n(i_n) \Rightarrow \sim \omega(i_1,\cdots,i_n)\} \qquad$ if $M \sim\models \omega(a_1,\cdots,a_n)$.

If we identify the unique element for which λ is valid with λ, we see that each Δ-model N contains Λ as a subset. Moreover, (4) and (5) state that the restriction to M of the inclusion mapping $\Lambda \rightarrow N$ is actually an embedding. The set of sentences (1)–(5) is called the *diagram of M with constants* Λ. It is clear that (4) and (5) are valid with every open formula in place of ω. The set $\bar{\Delta}$ obtained by taking (1)–(3) and (4), (5) for *every* formula of $\Omega \sqcup \Lambda$ (not necessarily open) is called the *complete diagram* of M with constants Λ. A model of the complete diagram of M is an instance of an elementary extension of M, to be discussed in VI.3.

Using the notion of a diagram we can establish the following *principle of localization* (Henkin [53], Robinson [63]).

Theorem 2.1

If $\mathscr{C}$ is any axiomatic model class, then $\mathrm{S}\mathscr{C}$ *is sublocally defined.*

Proof:

Let $\mathscr{C} = \Sigma^*$ and consider any model M. We must show that $M \in \mathrm{S}\mathscr{C}$ whenever every finite submodel of M is embeddable in a $\mathscr{C}$-model. Let Δ be the diagram of M with constants M; then $M \in \mathrm{S}\mathscr{C}$ provided that $\Sigma \cup \Delta$ is consistent, and by compactness (Cor. V.5.6), this is true whenever every finite subset of $\Sigma \cup \Delta$ is consistent. Let $\Sigma_f \cup \Delta_f$ be such a subset, where $\Sigma_f \subseteq \Sigma$ and $\Delta_f \subseteq \Delta$, and let N be the (finite) set of elements of M occurring as constants in Δ_f; then any extension of N is a Δ_f-model. By assumption on M, $N \in s\mathscr{C}$; thus there is a $\mathscr{C}$-model containing N. This $\mathscr{C}$-model satisfies Σ and Δ_f; hence $\Sigma_f \cup \Delta_f$ is consistent. It follows that $\Sigma \cup \Delta$ is consistent and so there is a Σ-model extending M. ∎

This result may be used in two ways. First we obtain localization theorems for classes of the form $s\mathscr{C}$, where $\mathscr{C}$ is axiomatic, and in particular for the hereditary axiomatic classes. Second, if $\mathscr{C}$ is a class such that $s\mathscr{C}$ is not sublocally defined, then Theorem 2.1 shows that $\mathscr{C}$ is not axiomatic. However, we remark that the necessary condition of Theorem 2.1 for $\mathscr{C}$ to be axiomatic is not sufficient (cf. Eršov [62]).

We may apply Theorem 2.1 to algebras by singling out a subdomain Ω^* of Ω and considering those Ω-structures which are Ω^*-algebras.

Corollary 2.2

Let $\mathscr{K}$ be an axiomatic class of Ω-structures which are Ω^*-algebras; then an Ω-model M can be embedded in a $\mathscr{K}$-algebra if and only if every finite submodel of M can be embedded in a $\mathscr{K}$-algebra. ∎

Usually the structure to be embedded is itself an algebra (possibly with a different operator domain), and in that case the following consequence of Corollary 2.2 is often sufficient.

Corollary 2.3

Let $\mathscr{K}$ be an axiomatic class of Ω-structures which are Ω^*-algebras, and let $\mathscr{L}$ be a class of algebras represented in $\mathscr{K}$. Then an $\mathscr{L}$-algebra B can be embedded in a $\mathscr{K}$-algebra provided that for every finite subset X of B, there exists a $\mathscr{K}$-algebra A_X and an admissible mapping $\theta: X \to A_X$ such that the restriction $\theta \mid X$ is injective. ∎

We turn now to the problem of describing hereditary classes. This will involve the diagram of a finite reduct of a finite structure, and it is of importance here to express this diagram (without constants) by a single elementary sentence. This is accomplished by

Lemma 2.4

Let M be a finite Ω-structure, where Ω is a finite predicate domain. Then there is an existential sentence P in a number of variables not exceeding $|M|$ such that the class of P-models consists precisely of those models which contain a submodel isomorphic to M.

Proof:

Let $M = \{a_1,\cdots,a_k\}$ and consider the following set of formulae in the variables $1,2,\cdots,k$:

(6) $\qquad i \neq j \qquad$ for any pair (i,j) of distinct variables,

and for each $\omega \in \Omega(n-1)$ and each n-tuple $(i_1,\cdots,i_n)$ $(1 \leqslant i_r \leqslant k)$,

(7) $\qquad \omega(i_1,\cdots,i_n) \qquad$ if $M \models \omega(a_{i_1},\cdots,a_{i_n})$,

(8) $\qquad \sim\omega(i_1,\cdots,i_n) \qquad$ if $M \sim\models \omega(a_{i_1},\cdots,a_{i_n})$.

Since Ω and M are finite, the total number of formulae (6)–(8) is finite. If we form their conjunction and bind all the variables by existential quantifiers we obtain a sentence which is clearly of the required form. ∎

Proposition 2.5

A model class $\mathscr{C}$ is the elementary class defined by a universal sentence if and only if $\mathscr{C}$ is abstract and hereditary, and there exists a finite subdomain Ω_f of Ω and a positive integer n such that a model M lies in $\mathscr{C}$ whenever $N \mid \Omega_f$ can be embedded in a $(\mathscr{C} \mid \Omega_f)$-model, for every submodel N of M with at most n elements.

Proof:

Let P be a universal sentence which defines $\mathscr{C}$. If $M \vdash P$, then P holds in any model isomorphic to M and in any submodel of M; the third condition is also satisfied if we take for Ω_f the set of predicates occurring in P and for n the number of variables in P.

Now let $\mathscr{C}$ be a class satisfying the conditions of the proposition, with given Ω_f and n. The number of open formulae in at most n variables over Ω_f is finite; we can therefore form the conjunction P_0 of all such formulae which hold in each $(\mathscr{C} \mid \Omega_f)$-model. Let $\mathscr{D}$ be the model class defined by P_0; then $\mathscr{D}$ may also be defined by the universal sentence, P say, corresponding to P_0 (obtained from P_0 by applying universal quantifiers); hence, by the first part of the proof, $\mathscr{D}$ satisfies the conditions of the proposition. We complete the proof by showing that

(9) $$\mathscr{C} = \mathscr{D}.$$

If $M \in \mathscr{C}$, then $M \mid \Omega_f$ belongs to $\mathscr{C} \mid \Omega_f$ and is therefore a P-model. Since P only involves predicates from Ω_f, it follows that M is itself a P-model, i.e., $M \in \mathscr{D}$. This proves one half of (9). Next, let $M \in \mathscr{D}$, i.e.,

$M \vdash P$, and take any submodel N of M with at most n elements. Since P is universal, we have

(10) $$N \vdash P.$$

Applying the lemma to $N \mid \Omega_f$, we obtain an existential sentence Q in at most n variables which holds in just those Ω_f-models which have a submodel isomorphic to $N \mid \Omega_f$. The negation $\sim Q$ of Q is equivalent to a universal sentence Q' in at most n variables, as is easily seen. By (10), Q' does not follow from P, because Q' does not hold in N. From the definition of P this means that Q' does not hold in $\mathscr{C} \mid \Omega_f$; thus there is a $(\mathscr{C} \mid \Omega_f)$-model, N' say, in which Q' is not valid. Hence, Q is valid in N', but Q was the sentence describing $N \mid \Omega_f$, so we have a $(\mathscr{C} \mid \Omega_f)$-model N' in which $N \mid \Omega_f$ can be embedded. By hypothesis this means that $M \in \mathscr{C}$, i.e. (9) holds. ▮

We note two simple consequences of this result.

Corollary 2.6

An elementary sentence defines a hereditary class if and only if it is congruent to a universal sentence.

For a universal sentence clearly defines a hereditary class, and hence so does any sentence congruent to it. Conversely, if P defines a hereditary class, then $\{P\}^*$ satisfies all the conditions of Proposition 2.5 and may therefore be defined by a universal sentence Q. Now it is clear that P is congruent to Q. ▮

A property is said to be *persistent* if it holds in a model M whenever it holds in some submodel of M. By taking negations we obtain from Corollary 2.6:

Corollary 2.7

An elementary sentence expresses a persistent property if and only if it is congruent to an existential sentence. ▮

Hereditary axiomatic classes may now be characterized as follows (Łos [55], Robinson [63]):

Theorem 2.8

For any model class $\mathscr{C}$ the following three conditions are equivalent:

(i) *$\mathscr{C}$ is hereditary and axiomatic,*

(ii) *$\mathscr{C}$ is hereditary and $\mathscr{C} \mid \Omega_f$ is sublocally defined for every finite subdomain Ω_f of Ω,*

(iii) *$\mathscr{C}$ is defined by a set of universal sentences.*

A class satisfying (iii) (and hence (i) and (ii)) is called a *universal class*, or also an *open sentence class*, since it may be defined by open formulae.

Proof:

The implication (i) $\Rightarrow$ (ii) follows by Theorem 2.1. To prove that (ii) $\Rightarrow$ (iii), we take a class $\mathscr{C}$ satisfying (ii), and for every finite subset Ω_f of Ω and every positive integer n we denote by $\mathscr{C}(\Omega_f,n)$ the class of all models M such that the Ω_f-reduct of any submodel with at most n elements can be embedded in a $(\mathscr{C} \mid \Omega_f)$-model. The class $\mathscr{C}(\Omega_f,n)$ then satisfies the conditions of Proposition 2.5 and can therefore be defined by a universal sentence. By (ii), $M \in \mathscr{C}$ if and only if $M \in \mathscr{C}(\Omega_f,n)$ for all Ω_f and all n, i.e.

$$\mathscr{C} = \bigcap \mathscr{C}(\Omega_f,n),$$

where the intersection is taken over all the pairs (Ω_f,n). Hence $\mathscr{C}$ may be defined by a set of universal sentences. Finally, (iii) $\Rightarrow$ (i) holds trivially. ▮

If we specialize to algebras by taking the predicates of a subdomain Ω^* to be operators, we obtain a criterion for an axiomatic class of algebras to admit subalgebras. We leave the formulation of this result to the reader and mention only the

Corollary 2.9

If $\mathscr{C}$ is a universal class of algebras, then $A \in \mathscr{C}$ if and only if every finitely generated subalgebra of A is in $\mathscr{C}$. ▮

This is much weaker than the theorem, even for finite Ω^*. For example, the class of periodic groups is locally defined, but not sublocally defined; thus it is not a universal class (it is not even an axiomatic class) although the conditions of the corollary are satisfied.

EXERCISES

1. Show that a universal class is elementary if and only if it can be defined by a single universal sentence.

2. Show that the class of periodic groups is not axiomatic. (Apply Corollary 2.2, taking M to be an infinite cyclic group; note that Corollary 2.3 cannot be used here.)

3. If $\mathscr{C}$ is an axiomatic class, show that $\mathrm{S}\mathscr{C}$ is a universal class.

4. If $\mathscr{C}$ is an elementary class, so that $\mathrm{S}\mathscr{C}$ is universal, then $\mathrm{S}\mathscr{C}$ need not be itself elementary. For example the class of semigroups $\mathscr{C}$ such that

$$\bigvee_{h}\bigwedge_{i,j}\bigvee_{k,l}[(h=h)\wedge(ik=j)\wedge(li=j)]$$

consists of all groups, regarded as semigroups admitting division. If $\mathscr{C}$ is regarded as class of algebras, verify that $\mathrm{S}\mathscr{C}$ consists of all semigroups embeddable in groups. (A set of universal sentences defining $\mathrm{S}\mathscr{C}$ has been determined by Malcev [39,] cf. VII.3, who has also shown that $\mathrm{S}\mathscr{C}$ is not elementary (Malcev [40])).

5. (Malcev.) Show that the class of free groups is not axiomatic.

6. (Kargapolov.) Show that the class of locally free groups is not axiomatic.

7. A group G is said to satisfy the *normalizer condition* if every proper subgroup is distinct from its normalizer. Verify that the class of groups satisfying the normalizer condition is hereditary (cf. Kuroš [56]; this class is not locally defined and therefore is not axiomatic).

8. If $\mathscr{L}$ is the class of all Ω-algebras and $\mathscr{C}$ is any class of Ω-algebras, show that an algebra A is sublocally $\mathscr{C}$, provided that for every finite subset X of A, the subalgebra $J(X)$ generated by X has a congruence $\mathfrak{q}$ separating X such that $J(X)/\mathfrak{q}\in\mathscr{C}$.

3. ELEMENTARY EXTENSIONS

We have seen that the notion of homomorphism is not invariant under passage to derived predicates. This defect may be remedied by introducing elementary mappings. A mapping between Ω-structures

$$\phi: M\to N$$

is said to be *elementary*, if for any formula P and any $\theta: I\to M$,

(1) $$\text{if } M\models P\theta,\text{ then } N\models P\theta\phi.$$

Clearly an elementary mapping is a homomorphism, but in fact we can say much more:

Proposition 3.1

If $\phi: M\to N$ is an elementary mapping, then ϕ is injective and for any formula P and $\theta: I\to M$,

(2) $$M\models P\theta \quad \textit{if and only if} \quad N\models P\theta\phi.$$

To see that ϕ is injective, we need only apply (1) with '$i \neq j$' for P. Now one half of (2) is just a restatement of (1), while the other half follows by replacing P by $\sim P$ in (1). ∎

Let N be any Ω-structure and M a substructure of N. Then N is said to be an *elementary extension* of M if the inclusion mapping $\iota: M \to N$ is elementary. We shall express this relation between M and N by writing $M \prec N$. If $M \prec N$ and $N \prec M$, this means that M and N have the same carrier, and the identity mapping between them is an isomorphism. As in the case of algebras, we shall regard M and N as the same structure in this case, so that $\prec$ defines an ordering on $[\Omega]$. Under this ordering $[\Omega]$ is inductive, for if $(M_\lambda)_{\lambda \in \Lambda}$ is a chain, let M be its union and define an Ω-structure on M by the rule that the inclusion mappings $M_\lambda \to M$ be elementary. Then M becomes an Ω-structure which is an elementary extension of all the M_λ.

If $\phi: M \to N$ is any elementary mapping between Ω-structures, then M and N are indiscernible, as follows by taking P in Proposition 3.1 to be any sentence. It may happen, however, that an extension N/M of Ω-structures is not elementary, even though M and N are indiscernible. For example, the positive integers are indiscernible from the positive even integers, qua ordered set, and they form an extension, but not an elementary extension, since 2 is a successor in one model but not the other.

We shall now give some criteria for elementary extensions (for these and others cf. Tarski & Vaught [57]). If M is an Ω-structure and X a subset of M, then we shall denote by $\langle M,X \rangle$ the unary enlargement of M obtained by regarding the elements of X as constant operators (defined by unary predicates corresponding to the elements of X). Thus e.g. for $X = M$, $\langle M,M \rangle$ is a model of the diagram of M with constants M.

Theorem 3.2

For any extension N/M of Ω-structures the following three conditions are equivalent:

(i) $M \prec N$,

(ii) *For every formula P and every $\theta: I \to M$, if $N \models (\bigvee_i P)\theta$, then there exists $\theta': I \to M$ such that $j\theta' = j\theta$ for $j \neq i$ and $N \models P\theta'$.*

(iii) *For every finite subset X of M, $\langle M,X \rangle \equiv \langle N,X \rangle$.*

Proof:

(ii) $\Rightarrow$ (i). Let Φ be the set of all formulae P such that

(3) for every $\theta: I \to M$, if $N \models P\theta$, then $M \models P\theta$.

We have to show that Φ includes all formulae. Clearly, every open formula belongs to Φ; further, if $P \in \Phi$, then $\bigwedge_i P \in \Phi$, and by (ii), also $\bigvee_i P \in \Phi$. Thus (3) holds for all formulae P, whence $M \prec N$.

(i) $\Rightarrow$ (iii). Let $\langle N,X \rangle \vdash P$, where P is of the form

$$P = P(x_1,\ldots,x_n) \qquad (x_\nu \in X).$$

Take elements $i_1,\ldots,i_n \in I$ which do not occur in P; then for any $\theta: I \to M$ such that $a_{i_1} = x_1,\ldots,a_{i_n} = x_n$, if $Q = P(i_1,\ldots,i_n)$,

$$N \models (\bigvee Q)\theta, \text{ therefore } M \models (\bigvee Q)\theta,$$

i.e., $\langle M,X \rangle \vdash P$. Repeating the argument with P replaced by $\sim P$, we obtain the converse: if $\langle M,X \rangle \vdash P$, then $\langle N,X \rangle \vdash P$.

(iii) $\Rightarrow$ (ii). Assume that (iii) holds and let P, $i \in I$ and $\theta: I \to M$ be given such that $N \models (\bigvee_i P)\theta$. Let J be the subset of I occurring in P and denote by X the image of $J\backslash\{i\}$ under θ; then $\langle N,X \rangle \models (\bigvee_i P)\theta$, where on the right all the $j\theta$ $(j \neq i)$ are regarded as constant. Hence $\langle M,X \rangle \models (\bigvee_i P)\theta$, i.e., for some $\theta': I \to M$ such that $j\theta' = j\theta$ for $j \neq i$ $(j \in J)$, we have $M \models P\theta'$; and this still holds if we take $j\theta' = j\theta$ for all $j \neq i$. Therefore (ii) is satisfied. ∎

Corollary 3.3

Let B/A be an extension of Ω-structures which are Ω^*-algebras. If for every finitely generated subalgebra A_f of A and every $b \in B$, there is an automorphism of B which leaves A_f elementwise fixed and maps b into A, then $A \prec B$.

This follows by verifying that condition (ii) of Theorem 3.2 is satisfied. ∎

The notion of elementary extension may be used to construct from a given Ω-structure M, others indiscernible from M, of a prescribed cardinal λ. There are essentially two results (both due to Tarski & Vaught [57]), according as λ is greater or less than $|M|$. To avoid trivialities one has of course to assume that all the cardinals involved are infinite (cf. Proposition V.4.3).

Theorem 3.4

Let M be an infinite Ω-structure, X any subset of M, and λ any infinite cardinal such that

$$\max(|\Omega|, |X|) \leqslant \lambda \leqslant |M|.$$

Then there exists an Ω-structure L of cardinal λ such that

$$X \subseteq L \prec M.$$

Proof:

From the hypotheses it follows that M contains subsets Y of cardinal λ which contain X. Let Y be such a set, and for each formula P involving the finite subset J of I, each $i \in I$ and each $\theta : J \to Y$ such that $M \models \left(\bigvee_i P\right)\theta$, choose a definite $x \in M$ such that $M \models P\theta'$, where $\theta' : J \to M$ is given by

$$j\theta' = \begin{cases} x & \text{if } j = i, \\ j\theta & \text{if } j \neq i. \end{cases}$$

Let Y' be the set of all such $x \in M$; then $|Y'| = \lambda$, because the number of formulae is $\max\{|\Omega|, \aleph_0\} \leqslant \lambda$. Moreover, taking P to be '$i = j$', we see that $Y \subseteq Y'$. If we apply the same construction to Y', we obtain a set Y'', and continuing in this way, we obtain an ascending chain

$$Y \subseteq Y' \subseteq Y'' \subseteq \cdots$$

Its union L is again of cardinal λ and contains X. Moreover $L \prec M$, by condition (ii) of Theorem 3.2. ∎

Taking $X = \emptyset$, we obtain

Corollary 3.5

Every infinite Ω-structure, over a finite or countable domain Ω, is an elementary extension of a countable Ω-structure. ∎

In particular this implies the well-known

Löwenheim-Skolem theorem

Every consistent theory over a countable predicate domain has a countable model. ∎

More generally, it follows that any consistent theory has a model of cardinal not exceeding $\max(|\Omega|, \aleph_0)$. This shows in particular that the model space introduced in V.6 does not depend on the particular universe, but only on Ω.

To obtain Ω-structures with greater cardinal, we use

Theorem 3.6

Let M be an infinite Ω-structure and λ a cardinal such that $\lambda \geqslant \max(|M|, |\Omega|)$; then M has an elementary extension of cardinal λ.

Proof:

Let Λ be any set of cardinal λ which contains M, and denote by Δ the complete diagram of M with constants Λ (cf. VI.2). Any finite subset of Δ has a model, namely M; we need only pick suitable elements of M to represent the finite number of elements of Λ involved. Therefore Δ has a model N, by compactness. By construction, N contains Λ and is an elementary extension of M. Hence its cardinal is at least λ, and by Theorem 3.4 there is an extension L of cardinal λ which contains M. Since $M \prec N$, $L \prec N$, $M \subseteq L$, it follows that $M \prec L$. ∎

Corollary 3.7

If M is any infinite Ω-structure and $\lambda \geqslant max\{|\Omega|, \aleph_0\}$, then there is an Ω-structure of cardinal λ which is indiscernible from M. ∎

As an application, we shall give a simple test for a theory to be complete, due to Vaught [54]. We recall that a complete theory is a maximal consistent set of sentences. In other words, a theory is complete if and only if all its models are indiscernible. Thus to obtain a complete theory we need only pick an Ω-structure M and take the set of all sentences valid in M. However, this does not provide an easy test which can be applied to a given set of axioms. In order to discuss Vaught's test we need the notion of categoricity. In general, a theory is said to be *categorical*, if all its models are isomorphic. Clearly, any categorical theory is complete, but the elementary theories we are discussing can never be categorical (by Corollary 3.7) except in the trivial case of a complete theory with a finite model (Proposition V.4.3). We therefore define a theory Θ to be *α-categorical*, where α is a given cardinal, if any two Θ-models of cardinal α are isomorphic. With this definition we have

Theorem 3.8 (Vaught* [54]*)

Let Θ be any elementary theory without finite models, which is α-categorical for some cardinal α, where

$$\alpha \geqslant \max\{|\Omega|, \aleph_0\}.$$

Then Θ is complete.

Proof:

If Θ is not complete, then for some sentence P, neither P nor $\sim P$ belongs to Θ, and so both $\Theta \cup \{P\}$ and $\Theta \cup \{\sim P\}$ are consistent (cf. V.5). Let M_1 be a model for $\Theta \cup \{P\}$ and M_2 a model for $\Theta \cup \{\sim P\}$. These models are both Θ-models and are therefore infinite; by Corollary 3.7, there exist Ω-structures N_1, N_2 of cardinal α and indiscernible from M_1, M_2 respectively. But then N_1 and N_2 are Θ-models, and hence are isomorphic, which contradicts the fact that $N_1 \vdash P$, $N_2 \vdash \sim P$. ∎

EXERCISES

1. If M is any Ω-structure, show that any ultrapower of M is an elementary extension of M.

2. Show that a theory without finite models is complete provided that for some $\alpha \geqslant \max\{|\Omega|, \aleph_0\}$ all its models of cardinal α are indiscernible.

3. Apply Vaught's test to show that the following theories are complete:

(i) Densely ordered sets (with or without end-points). (Cf. Exercise I.5.9.)
(ii) Infinite-dimensional vector spaces over a given field.
(iii) Algebraically closed fields of given characteristic. (Cf. VII.2.)
(iv) Nontrivial Boolean algebras without atoms. (Cf. Exercise V.2.10.)

4. Given an Ω-structure A, there exists an elementary extension of A which is an ultraproduct of the finite substructures of A.

4. P-CLOSED CLASSES AND QUASIVARIETIES

We now consider more closely the form which universal sentences can take. Any universal sentence over Ω is of the form

$$\bigwedge P, \tag{1}$$

where P is an open formula in a finite number of free variables. Thus P is built up from atomic formulae by $\sim$, $\vee$, and $\wedge$. Using the conjunctive normal form, we can express P as a conjunction of formulae of the type

$$Q = E_1 \vee \cdots \vee E_r, \tag{2}$$

where each E_k is an atomic formula or the negation of an atomic formula. Accordingly we call E_k a *positive* or *negative* constituent of Q. Now P is valid in a given structure M if and only if all the terms Q composing it are valid in M; we may therefore limit ourselves to the consideration of formulae of the form (2). For any Q given by (2), we write

$$Q^{(k)} = E_1 \vee \cdots \vee E_{k-1} \vee E_{k+1} \vee \cdots \vee E_r \qquad (k = 1, \cdots, r).$$

Let Σ be any set of universal sentences; as long as we are only interested in the model class defined by Σ, we may replace the sentences in Σ by the corresponding open formulae (obtained by omitting all quantifiers), and these may be replaced by their conjunctive components (2). Thus we may assume Σ to consist of formulae of the form (2). A formula Q in Σ is said to be *reducible* in Σ, if Q has more than one component and $Q^{(k)} \in \Sigma^{**}$ for some $k = 1,\cdots,r$; otherwise it is called *irreducible in* Σ. Thus to say that Q is irreducible in Σ means that Q has one component only, or that for each $k = 1,\cdots,r$, there is a Σ-model M_k such that $M_k \sim\vdash Q^{(k)}$.

Proposition 4.1

Every universal class can be defined by a set Σ of open formulae which are irreducible in Σ.

For we have seen that a universal class $\mathscr{K}$ may be defined by a set Σ of formulae of the form (2). Now any reducible formula in Σ may be replaced by a formula with fewer components without affecting $\mathscr{K}$; repeating this process if necessary, we thus obtain a formula which is irreducible in Σ. ∎

Lemma 4.2

Let Σ be a set of irreducible open formulae such that the model class defined by Σ is P*-closed. Then no formula in Σ can have more than one positive component.*

Proof:

Suppose that Σ includes a formula $Q = E_1 \vee E_2 \vee \cdots \vee E_r$, where E_1 and E_2 are positive. Let M_k be a Σ-model in which $Q^{(k)}$ is not valid. Thus $M_1 \vdash \bigwedge (E_1 \vee E_2 \vee \cdots \vee E_r)$ and $M_1 \vdash \bigvee (\sim E_2 \wedge \sim E_3 \wedge \cdots \wedge \sim E_r)$, and therefore

$$M_1 \vdash \bigvee (E_1 \wedge \sim E_2 \wedge \cdots \wedge \sim E_r);$$

similarly,

$$M_2 \vdash \bigvee (\sim E_1 \wedge E_2 \wedge \sim E_3 \wedge \cdots \wedge \sim E_r).$$

Choose $\theta_1 : I \to M_1$ such that

$$M_1 \models (E_1 \wedge \sim E_2 \wedge \cdots \wedge \sim E_r)\theta_1$$

and $\theta_2 : I \to M_2$ such that

$$M_2 \models (\sim E_1 \wedge E_2 \wedge \sim E_3 \wedge \cdots \wedge \sim E_r)\theta_2;$$

then

$$M_1 \times M_2 \models (\sim E_1 \wedge \sim E_2 \wedge \cdots \wedge \sim E_r)(\theta_1,\theta_2),$$

where (θ_1,θ_2) stands for the mapping $i \to (i\theta_1, i\theta_2)$. But this means that $M_1 \times M_2 \sim\vdash Q$, which contradicts the fact that Σ is P-closed. ∎

Note that in the proof we have only used products of two factors; in fact any class admitting direct products of two factors clearly admits direct products of any finite number of factors, and such a class, if axiomatic, admits direct products of any number of factors, by a result of Vaught [54″].

The lemma shows that any formula in Σ has one of the forms

$$\sim A_1 \vee \sim A_2 \vee \cdots \vee \sim A_r, \tag{3}$$

$$A_1 \wedge \cdots \wedge A_s \Rightarrow A, \tag{4}$$

where the A's are atomic formulae. Now clearly, if (M_λ) is a family of Ω-structures satisfying a given formula of the form (3) or (4), then the direct product ΠM_λ also satisfies this formula; thus we obtain (McKinsey [43], Kogalovskii [59]):

Theorem 4.3

A universal model class admits direct products if and only if it has a defining system of formulae of the form (3) *or* (4). ∎

The same result holds if we replace 'model class' by 'class of algebras', as follows by relativising the proof of the theorem.

A formula of the form (4) is called a *Horn formula* (Horn [51]), and any sentence, not necessarily universal, obtained by quantifying a Horn formula is called a *Horn sentence*. Further, a *Horn class* is a model class defined by Horn sentences. A universal Horn class of algebras is also called a *quasivariety* (or quasiprimitive class). Theorem 4.3 shows that every quasivariety admits direct products; conversely, a universal class admitting direct products is a quasivariety if no formula (3) occurs in the definition. Now the trivial Ω-structure (i.e., the full one-element structure) satisfies every formula (4) but no formula (3). We thus find that a universal class is a Horn class if and only if it admits direct products; for then it contains the trivial structure as empty product. For algebras this yields

Corollary 4.4

A universal class of algebras is a quasivariety if and only if it admits direct products. ∎

As examples of quasivarieties we mention cancellation semigroups, or semigroups embeddable in groups (cf. Exercise 2.4 and VII.3), or ordered semigroups. On the other hand, totally ordered semigroups do not form a quasivariety; neither do integral domains.

Since quasivarieties admit subalgebras and direct products, we have, by Corollary III.5.2,

Proposition 4.5
Every quasivariety has free algebras. ∎

For further results on quasivarieties, including an abstract characterization (in terms of categories) the reader may be referred to Malcev [56], [58″].

The remarks preceding Corollary 4.4 give a convenient characterization of universal P-closed classes; no such simple criterion is known for general P-closed classes. It is true that any Horn sentence is preserved under direct products, and even under reduced direct products (cf. V.5). This was proved by Chang & Morel [58], who also showed that there are sentences preserved under direct products which are not equivalent to any Horn sentence (cf. Exercise 3).

EXERCISES

1. If $\mathscr{K}$ is any universal class of algebras, show that the subdirect products of $\mathscr{K}$-algebras form a quasivariety. Verify that the quasivariety generated by integral domains consists of all commutative rings without nilpotent elements (apart from 0).

2. Show that every quasivariety is generated by its subdirectly irreducible elements.

3. Obtain an elementary sentence expressing the fact that a Boolean algebra is trivial or contains an atom. Show that this sentence is preserved under direct products but not under reduced products. (Chang & Morel [58]; by the result quoted in the text, this sentence is not equivalent to a Horn sentence.)

5. CLASSES ADMITTING HOMOMORPHIC IMAGES

The '*duality*' between subsets and quotients would require a characterization of axiomatic model classes which are H-closed. Such a characterization indeed exists, but neither it nor its proof is in any way dual to the hereditary case discussed in VI.2. Let us call a sentence *positive* if it arises by quantification from a positive formula and *negative* if it can be expressed as the negation of a positive sentence. It was shown by Lyndon [59′] that an axiomatic model class is H-closed if and only if it can be defined by positive sentences. The proof requires a fairly detailed analysis of the structure of the formulae which can occur, and we shall therefore confine ourselves here to proving the result for the special case of universal classes over a finite predicate domain. For these classes the principle of localization (Theorem 2.1) is available, and a proof can be given which somewhat resembles the proof of the characterization of hereditary axiomatic classes (Theorem 2.8). At the same time it shows the limitations of the above mentioned duality. We begin with a lemma which takes the place of Lemma 2.4.

Lemma 5.1

Let Ω be a finite domain and M a finite Ω-structure. Then there is a positive universal sentence P such that the P-models are precisely those structures which do not possess a substructure with M as homomorphic image.

Proof:

Let $M = \{a_1,\ldots,a_m\}$ and let N be an Ω-structure which has a substructure with M as homomorphic image. This means that N contains m elements $b_1,\ldots,b_m$ such that the following hold in N:

(1) $\qquad b_i \neq b_j \qquad$ whenever $i \neq j$,

(2) $\qquad \sim\omega(b_{i_1},\ldots,b_{i_n}) \qquad$ whenever $M \models \sim\omega(a_{i_1},\ldots,a_{i_n})$.

For, we need only pick for b_i an element of N which maps to a_i in the homomorphism. Conversely, if all the formulae (1),(2) hold in N, then the mapping $b_i \to a_i$ is a homomorphism from a substructure of N to M. Now the total number of formulae (1),(2) is finite and each is the negation of an atomic formula. Their conjunction is therefore of the form $\sim P(b_1,\ldots,b_m)$, where P is a positive formula. By construction, N has a substructure with M as homomorphic image if and only if

$$N \vdash \bigvee \sim P(i_1,\ldots,i_m),$$

and therefore N does not possess such a substructure if and only if

$$N \vdash \bigwedge P(i_1,\ldots,i_m).$$

Since P is positive, this is a sentence of the required form. ∎

Theorem 5.2

A universal model class over a finite predicate domain admits homomorphic images if and only if it can be defined by means of universal positive sentences.

Proof:

Clearly, a positive sentence holding in a given Ω-model holds in all its homomorphic images; therefore, a universal model class defined by positive sentences admits homomorphic images (without any restriction on the size of Ω).

Now let $\mathscr{C}$ be a universal model class (over a finite domain Ω), admitting homomorphic images. If $\mathscr{D}$ is the model class defined by all universal positive sentences in $\mathscr{C}$, then clearly

(3) $$\mathscr{C} \subseteq \mathscr{D};$$

we complete the proof by showing that equality holds in (3). Thus suppose that there is an Ω-model M such that

(4) $$M \in \mathscr{D}, \quad M \notin \mathscr{C}.$$

Since $\mathscr{C}$ is sublocally defined (Theorem 2.1), there is a finite submodel of M which is not in $\mathscr{C}$. But any submodel of M again lies in $\mathscr{D}$, and we can thus take M in (4) to be finite. Let P be the positive universal sentence characterizing the Ω-models which have no submodel with M as homomorphic image. If there is a $\mathscr{C}$-model, N say, for which P does not hold, then M is a homomorphic image of a submodel of N and therefore belongs to $\mathscr{C}$. But this contradicts (4); hence every $\mathscr{C}$-model satisfies P. By definition of $\mathscr{D}$, every $\mathscr{D}$-model must satisfy P too, whereas M clearly does not satisfy P. This contradicts the fact that $M \in \mathscr{D}$, so that (4) is impossible, i.e. $\mathscr{C} = \mathscr{D}$. ▮

If we consider only Ω-models which are algebras (relative to some subdomain Ω^*), we obtain

Corollary 5.3

A universal class of Ω^*-algebras (over a finite predicate domain containing Ω^*) admits homomorphic (algebra-) images if and only if it can be defined (within the class of all Ω^*-algebras) by positive universal sentences. ▮

EXERCISE

1. Show that if P is a universal sentence, then the class of P-models admits homomorphic images if and only if P is equivalent to a universal positive sentence. (Use Corollary V.6.2 and compactness.)

6. THE CHARACTERIZATION OF AXIOMATIC MODEL CLASSES

We saw in Theorem V.6.4 that a type class, i.e. a union of types, is axiomatic if and only if it admits ultraproducts, but this result tells us nothing about the relation between Ω-structures belonging to the same type. We now show that two Ω-structures belong to the same type (i.e. are indiscernible) if and only if they have isomorphic ultrapowers. From this it is easy to derive a criterion for any model class to be axiomatic, without assuming it to be a type class. These results are due to Keisler [61], the basic theorem from which all others are derived being Theorem 6.2 below. In the proof of this theorem (and its consequences) it is however necessary to make a further assumption, which we shall now discuss.

If α is any infinite cardinal, then there are cardinals larger than α, e.g. 2^α, and hence there is a least such cardinal. We denote this least cardinal greater than α by α^+; by Theorem I.5.6, we have

(1) $$\alpha^+ \leqslant 2^\alpha.$$

In particular, if $\alpha = \aleph_0$, then $2^{\aleph_0}$ is the cardinal of the real line (the 'continuum') and Cantor's *continuum hypothesis* states that for $\alpha = \aleph_0$, equality holds in (1), i.e., writing $\aleph_1$ for $\aleph_0^+$,

(2) $$\aleph_1 = 2^{\aleph_0}.$$

The assertion

(3) $$\alpha^+ = 2^\alpha \qquad \text{for any infinite cardinal } \alpha$$

is called the *generalized continuum hypothesis*. It has never been proved, even for $\alpha = \aleph_0$, although Gödel [40] has shown it to be consistent with the axioms of set theory, assuming that these are consistent. Here it is not necessary to include the axiom of choice; its consistency is proved in the same way. Further, in place of the existence of universes, Gödel merely assumes the existence of infinite sets. If one adopts the axiomatic standpoint it is therefore reasonable to include (3) as an axiom (particularly in view of the fact that (3) has also been shown to be independent of the other axioms of set theory, by P. J. Cohen [63]). There is also a more naïve standpoint which argues that by analysing more closely the intuitive notion of a set one may be able to isolate a further assertion about sets, which intuitively is 'obvious' and which, taken as an axiom, will together with the other axioms entail the truth of (3). (However, if one is prepared to rely on intuition, one must also be prepared for the opposite conclusion, that (3) may turn out to be 'obviously' false). The reader may be referred to Gödel [47] for an illuminating discussion of the whole question; we shall not pursue this point here, but henceforth assume (3). Any results in whose proof (3) is used will be marked GCH (generalized continuum hypothesis).

Before coming to the main result we need a lemma, which generalizes Proposition I.5.8. If α is any ordinal number, then by an α-term sequence we mean a family indexed by α. Further, we shall identify any cardinal α with the first ordinal number which has cardinal α.

Lemma 6.1

Let α be an infinite cardinal, β an ordinal such that $\beta \leqslant \alpha$, and $(X_\xi)_{\xi<\beta}$ a β-term sequence of sets, each of cardinal α. Then each X_ξ contains a subset Y_ξ, again of cardinal α, such that the (Y_ξ) form a disjoint family.

We remark that if $\Phi_{\alpha\beta}$ denotes the statement of the lemma, then $\Phi_{\alpha\beta}$ implies the equation

$$\alpha\beta = \alpha \qquad \text{for} \quad \beta \leqslant \alpha \text{ (and } \alpha \geqslant \aleph_0), \tag{4}$$

and in particular,

$$\alpha^2 = \alpha \qquad (\alpha \geqslant \aleph_0). \tag{5}$$

For if we take any set X of cardinal α and put $X_\xi = X$ for $\xi < \beta$, then by $\Phi_{\alpha\beta}$, there is a β-term sequence of disjoint subsets of X, each of cardinal α, whence $\alpha\beta \leqslant \alpha$. Since the reverse inequality is evident, (4) follows.

To prove the lemma it is clearly enough to take the case $\beta = \alpha$; we shall use transfinite induction on α. Let γ be any ordinal such that $\gamma < \alpha$, and assume that for each pair ξ, η of ordinals less than γ we have an η-term sequence $Y_{\xi\eta}$ such that

(i) $Y_{\xi\eta}$ is a sequence of distinct elements of X_ξ.
(ii) If $\eta' < \eta$, then $Y_{\xi\eta'}$ is a left segment of $Y_{\xi\eta}$.
(iii) For $\xi' \neq \xi$, $Y_{\xi'\eta'} \cap Y_{\xi\eta} = \emptyset$.

If γ is a limit ordinal, define $Y_{\xi\gamma}$ by the formula

$$Y_{\xi\gamma} = \bigcup_{\eta<\gamma} Y_{\xi\eta};$$

then the $Y_{\xi\gamma}$ are again pairwise disjoint, and each is a γ-term sequence of distinct elements of X_ξ. If γ is not a limit ordinal, say $\gamma = \delta + 1$, we define $Y_{\xi\gamma}$ (by transfinite induction on ξ) by adjoining a single element from X_ξ to $Y_{\xi\delta}$ in such a way that (i)–(iii) remain true for $\eta = \gamma$. Since $|\bigcup_\xi Y_{\xi\gamma}| = |\gamma|^2 = |\gamma|$ by (5) (using the induction hypothesis on γ), while

$|X_\xi| = \alpha > |\gamma|$, this is always possible. Having defined $Y_{\xi\gamma}$, we can now define $Y_{\gamma\gamma}$ by taking any γ-term sequence of distinct elements of X_γ which do not belong to $\bigcup Y_{\xi\gamma}$. Again this is possible by (5), and $Y_{\gamma\eta}$ for $\eta < \gamma$ is defined as the appropriate left segment of $Y_{\gamma\gamma}$. By induction we thus obtain $Y_{\xi\eta}$, for each pair $\xi,\eta < \alpha$, to satisfy (i)–(iii). Now put

$$Y_\xi = \bigcup_{\eta<\alpha} Y_{\xi\eta};$$

then Y_ξ is an α-term sequence of distinct elements of X_ξ such that $Y_\xi \cap Y_{\xi'} = \emptyset$ for $\xi \neq \xi'$, by (iii), as required. ∎

Theorem 6.2 (GCH) (Keisler* [61]*)

Let α be a cardinal number not less than $\max\{|\Omega|, \aleph_0\}$ *and let I be any set of cardinal α. Then for any two families $(A_i)_{i\in I}$, $(B_i)_{i\in I}$ of Ω-structures of cardinal at most α^+, the following two assertions are equivalent.*

(i) *There exist Fréchet ultrafilters $\mathscr{D}$, $\mathscr{E}$ on I such that*

$$\Pi A_i/\mathscr{D} \cong \Pi B_i/\mathscr{E}.$$

(ii) *For any elementary sentence P, either*

(6) $$|\{i \in I \mid A_i \vdash P\}| = \alpha,$$

or

(7) $$|\{i \in I \mid B_i \vdash {\sim}P\}| = \alpha.$$

Proof:

(i) $\Rightarrow$ (ii). If P holds in fewer than α A's, then ${\sim}P$ holds for all A's of a $\mathscr{D}$-set, and therefore holds in the ultraproduct. By the isomorphism (i), ${\sim}P$ holds in all the B's of an $\mathscr{E}$-set, whence (7).

(ii) $\Rightarrow$ (i). Write $A = \Pi A_i$, $B = \Pi B_i$ and denote the projections by $\varepsilon_i : A \to A_i$, $\zeta_i : B \to B_i$. Further, let us well-order A and B: $A = (a_\xi)_{\xi<\alpha^+}$, $B = (b_\xi)_{\xi<\alpha^+}$; we remark that here the GCH has been used, to ensure that A and B can be indexed by the ordinals less than α^+. Now any ordinal can be uniquely expressed in the form $\lambda + n$, where λ is a limit ordinal (or zero) and n is a natural number (cf. Exercise I.5.5). Our first objective will be to construct α-term sequences (a'), (b') in A and B respectively, such that for any limit ordinal λ and any natural number n,

(a) $a'_{\lambda+2n} = a_{\lambda+n}$.

(b) $b'_{\lambda+2n+1} = b_{\lambda+n}$.

(c) For any elementary sentence P over Ω and an α-term sequence of unary predicates, denote by $\langle A_i, a'\varepsilon_i\rangle$ the unary enlargement of

A_i in which the unary predicates are constant operators with values $a'_\xi\varepsilon_i$, and analogously for $\langle B_i,b'\zeta_i\rangle$; then either

$$|\{i \in I \mid \langle A_i,a'\varepsilon_i\rangle \vdash P\}| = \alpha, \tag{8}$$

or

$$|\{i \in I \mid \langle B_i,b'\zeta_i\rangle \vdash \sim P\}| = \alpha. \tag{9}$$

The construction is by transfinite induction: if (a'_ξ), (b'_ξ) have been constructed for $\xi < \gamma$ to satisfy (a)–(c), then we define a'_γ and b'_γ as follows:

(a) $\gamma = \lambda + 2n$, *where* λ *is a limit ordinal or zero and* n *is a natural number*. We put $a'_{\lambda+2n} = a_{\lambda+n}$ and write Γ for the set of formulae $P(k)$ over Ω and a γ-term sequence $(\rho_\xi)_{\xi<\gamma}$ of unary predicates such that for fewer than α suffixes $i \in I$,

$$\langle A_i,a'\varepsilon_i\rangle \sim\vdash \bigvee_k (\rho_\gamma(k) \wedge P(k)), \tag{10}$$

i.e.,

$$\langle A_i,(a'_\xi\varepsilon_i)_{\xi<\gamma}\rangle \models \sim P(a_{\lambda+n}\varepsilon_i).$$

Since $|\gamma| \leqslant \alpha$ and the cardinal of the set of all formulae does not exceed α, it follows that $|\Gamma| \leqslant \alpha$. For each $P \in \Gamma$ put

$$X_P = \{i \in I \mid \langle B_i,b'\zeta_i\rangle \vdash \bigvee_k P(k)\}. \tag{11}$$

By (c), with $\bigvee_k P(k)$ in place of P, we see that $|X_P| = \alpha$ whenever $P \in \Gamma$. We now use Lemma 6.1 to replace X_P by a subset X'_P such that $|X'_P| = \alpha$ and the X'_P are pairwise disjoint. Then by (11), there is an element $f \in B$ such that

$$\langle B_i,b'\zeta_i\rangle \models P(f\zeta_i) \qquad \text{for } i \in X'_P. \tag{12}$$

If we define $b'_{\lambda+2n}$ by the equation

$$b'_{\lambda+2n} = f,$$

then (c) holds for $\gamma = \lambda + 2n$. For if $\sim P$ holds in fewer than α structures $\langle A_i,(a'_\xi\varepsilon_i)_{\xi<\gamma+1}\rangle$ and if ρ_γ does not occur in P, then by the induction hypothesis (using (c)), P holds in α structures $\langle B_i,(b'_\xi\zeta_i)_{\xi<\gamma+1}\rangle$, while if ρ_γ does occur, say $P = P(\rho_\gamma)$, then by (10), $P(k) \in \Gamma$, and so (12) shows that P holds for α structures $\langle B_i,(b'_\xi\zeta_i)_{\xi<\gamma+1}\rangle$.

(b) $\gamma = \lambda + 2n + 1$. Here we merely repeat the construction with the roles of a' and b' interchanged.

For $\gamma = 1$ the hypothesis (c) just reduces to condition (ii), which is assumed to hold, and we therefore obtain structures $A_i^* = \langle A_i,a'\varepsilon_i\rangle$ and

$B_i^* = \langle B_i, b'\zeta_i \rangle$ over Ω^*, a unary enlargement of Ω by α unary predicates. For each elementary sentence P over Ω^* we define two subsets of I:

$$Y_P = \{i \in I \mid A_i^* \vdash P\},$$
$$Z_P = \{i \in I \mid B_i^* \vdash P\}.$$

It is clear that $Y_{\sim P} = I \setminus Y_P$, $Y_{P \wedge Q} = Y_P \cap Y_Q$, and similarly for Z. Moreover, by (c), either P holds for α structures A_i^* or $\sim P$ holds for α structures B_i^*; thus if $\Phi(I)$ denotes the minimal Fréchet filter on I, either $Y_{\sim P} \notin \Phi(I)$ or $Z_P \notin \Phi(I)$. If we replace P by $\sim P$, this states that

(13) $$\text{if } Y_P \in \Phi(I), \text{ then } Z_{\sim P} \notin \Phi(I).$$

Let $\mathscr{E}_0$ be the set of all Z_P such that $Y_P \in \Phi(I)$. Then $\mathscr{E}_0$ is closed under finite intersections, and by (13), every $\mathscr{E}_0$-set has cardinal α. Hence by Proposition V.2.10, there is a Fréchet ultrafilter $\mathscr{E}$ containing $\mathscr{E}_0$. Next let $\mathscr{D}_0$ be the set of all Y_P such that $Z_P \in \mathscr{E}$; then $\mathscr{D}_0$ again is a filter base and every $\mathscr{D}_0$-set has cardinal α, since any $J \in \mathscr{D}_0$ has the form $J = Y_P$, where $Z_P \in \mathscr{E}$, hence $Z_{\sim P} \notin \mathscr{E}_0$, and so by the definition of $\mathscr{E}_0$, $Y_{\sim P} \notin \Phi(I)$, i.e., $|J| = \alpha$. It follows that there is a Fréchet ultrafilter $\mathscr{D}$ containing $\mathscr{D}_0$.

If P is any sentence over Ω^*, then

$$Y_P \in \mathscr{D} \text{ if and only if } Z_P \in \mathscr{E}.$$

For if $Y_P \in \mathscr{D}$, then $Y_{\sim P} \notin \mathscr{D}$, hence $Y_{\sim P} \notin \mathscr{D}_0$ and so $Z_{\sim P} \notin \mathscr{E}$; therefore $Z_P \in \mathscr{E}$. Conversely, if $Z_P \in \mathscr{E}$, then $Y_P \in \mathscr{D}_0$ and so $Y_P \in \mathscr{D}$. Consider now the correspondence $a'_\xi \leftrightarrow b'_\xi$ between A and B. If $a'_\xi \equiv a'_\eta (\text{mod } \mathscr{D})$, then the sentence

$$\bigvee_k (\rho_\xi(k) \wedge \rho_\eta(k))$$

holds on a $\mathscr{D}$-set of structures A_i^*, and therefore it holds on an $\mathscr{E}$-set of structures B_i^*, i.e. $b'_\xi \equiv b'_\eta (\text{mod } \mathscr{E})$; by symmetry, the converse also holds. This means that the correspondence $a'_\xi \to b'_\xi$ induces a bijection between $A/\mathscr{D}$ and $B/\mathscr{E}$. Moreover, a sentence P holds in $\langle A/\mathscr{D}, a'/\mathscr{D} \rangle$ if and only if $Y_P \in \mathscr{D}$, i.e. $Z_P \in \mathscr{E}$, i.e. if and only if P holds in $\langle B/\mathscr{E}, b'/\mathscr{E} \rangle$. Therefore the correspondence $a'_\xi/\mathscr{D} \leftrightarrow b'_\xi/\mathscr{E}$ is an isomorphism, i.e. $A/\mathscr{D} \cong B/\mathscr{E}$. ∎

For a proof of Theorem 6.2 without GCH see Shelah [71].

Theorem 6.2 has the following consequences (Keisler [61]):

Theorem 6.3 (GCH)

Two Ω-structures M and N are indiscernible if and only if there exist ultrapowers $M^I/\mathscr{D}$ and $N^I/\mathscr{E}$ of M and N respectively which are isomorphic.

Here I can be taken to be any set of cardinal α, *where* $\alpha \geqslant \max\{|\Omega|, \aleph_0\}$ *and* $\alpha^+ \geqslant \max\{|M|, |N|\}$.

This follows by taking $A_i = M$, $B_i = N$ in Theorem 6.2. ∎

In the results which follow, we denote for any model class $\mathscr{K}$, by $\mathscr{K}'$ the class of all structures not in $\mathscr{K}$.

Corollary 6.4 (GCH)

The model class $\mathscr{K}$ *is a type class if and only if both* $\mathscr{K}$ *and* $\mathscr{K}'$ *admit ultrapowers.*

For if $\mathscr{K}$ is a type class, then so is $\mathscr{K}'$, and any type class admits ultrapowers, by Corollary V.6.5. Conversely, if both $\mathscr{K}$ and $\mathscr{K}'$ admit ultrapowers, then by Theorem 6.3, every $\mathscr{K}$-model is discernible from every $\mathscr{K}'$-model, hence $\mathscr{K}$ is a type class. ∎

Corollary 6.5 (GCH)

The model class $\mathscr{K}$ *is axiomatic if and only if* $\mathscr{K}$ *admits ultraproducts and* $\mathscr{K}'$ *admits ultrapowers.*

For the conditions are necessary by Corollary V.6.4; and when they are satisfied, $\mathscr{K}$ is a type class by Corollary 6.4 and hence is axiomatic by Corollary V.6.4. ∎

Corollary 6.6 (GCH)

The model class $\mathscr{K}$ *is elementary if and only if both* $\mathscr{K}$ *and* $\mathscr{K}'$ *admit ultraproducts.*

This follows from the preceding result and Corollary V.6.2. ∎

As another consequence of Theorem 6.3 we note that for any infinite Ω-structure M and any given cardinal α, there exist ultrapowers of M whose cardinal exceeds α. For by Corollary 3.7, there exists an Ω-structure N of cardinal exceeding α and indiscernible from M. Now any ultrapower of N has cardinal at least $|N|$ (cf. Exercise 3.1) and so the result follows by Theorem 6.3. The information so conveyed is not very precise, and in any case it is of interest to have a more direct estimate of the cardinal of an ultrapower. In the simplest case one has the following result, due to Halmos and Kochen (Kochen [61]; for related results see Frayne, Morel, & Scott [62]). We remark that the proof does not depend on any of the preceding theorems.

Theorem 6.7

Let $(M_n)_{n \in N}$ be a sequence of finite sets such that for any finite number k, only finitely many sets M_n are of cardinal k. Then if $\mathscr{D}$ is any nonprincipal ultrafilter on the set of positive integers N, $M_{\mathscr{D}} = \Pi M_n/\mathscr{D}$ has cardinal $2^{\aleph_0}$.

Proof:

Clearly,

$$|M_{\mathscr{D}}| \leqslant |\Pi M_n| = 2^{\aleph_0},$$

so to complete the proof we need only construct an injection $2^N \to M_{\mathscr{D}}$. We may assume that the M_n are ordered by size, i.e. the numbering is so arranged that $m \leqslant n$ implies $|M_m| \leqslant |M_n|$. Put $\Gamma = 2^N$; then Γ may be realized as the set of all sequences of 0's and 1's; we denote by Γ_k the set of finite sequences of length k. Each Γ_k is finite, and for a given n we shall denote by k_n the greatest integer k such that $|\Gamma_k| \leqslant |M_n|$. It follows that there is an injection

$$\phi_n : \Gamma_{k_n} \to M_n$$

and moreover

$$|\Gamma_{k_n}| \leqslant |M_{n'}| \qquad \text{for all } n' \geqslant n.$$

Now define $\phi : \Gamma \to \Pi M_n$ as follows. If $\gamma \in \Gamma$ and γ_k is the section consisting of the first k terms of γ, then we put

$$(\gamma\phi)_n = \gamma_{k_n}\phi_n.$$

If $h : \Pi M_n \to \Pi M_n/\mathscr{D}$ is the natural mapping nat $\mathscr{D}$, we assert that $\phi h : \Gamma \to M_{\mathscr{D}}$ is injective. For if $\gamma, \delta \in \Gamma$ and $\gamma \neq \delta$, then for some n_0, $\gamma_n \neq \delta_n$ for all $n \geqslant n_0$, and since the complement of any finite set lies in $\mathscr{D}$, $\gamma\phi h \neq \delta\phi h$. ∎

Given two families of sets (M_i), (N_i) over the same index set I, and an ultrafilter $\mathscr{D}$ on I, we can form the ultraproducts $M_{\mathscr{D}} = \Pi M_i/\mathscr{D}$ and $N_{\mathscr{D}} = \Pi N_i/\mathscr{D}$. Clearly any family of mappings $\phi_i : M_i \to N_i$ extends to a mapping of the products $\phi : \Pi M_i \to \Pi N_i$; moreover, elements which are congruent to each other mod $\mathscr{D}$ are mapped to elements congruent mod $\mathscr{D}$, so that we obtain a mapping $\phi_{\mathscr{D}} : M_{\mathscr{D}} \to N_{\mathscr{D}}$. We note that $\phi_{\mathscr{D}}$ is injective if and only if ϕ_i is injective for all i belonging to some $\mathscr{D}$-set; in particular, if each ϕ_i is injective, then so is $\phi_{\mathscr{D}}$. It follows that the cardinal $|M_{\mathscr{D}}|$ is an increasing function of the cardinal of each factor. Taking I and each M_i to be countable, and applying Theorem 6.7, we obtain

Corollary 6.8

Any ultraproduct of a countable family of countable sets with respect to a nonprincipal ultrafilter has cardinal $2^{\aleph_0}$. ∎

This applies in particular to countable ultrapowers of a countable set.

EXERCISE

1. Show that Theorem 6.7 does not hold if there are infinitely many sets of a given (finite) cardinal k. (If N_0 is the set of suffixes for which $|M_n| = k$, construct a Fréchet ultrafilter containing N_0.)

Chapter VII

Applications

In the course of the first six chapters a number of applications to particular situations were obtained by specializing the general theory. However, it is much more common for specific problems to fall into two parts, of which one, involving universal algebra, is relatively simple, while the more substantial piece of work still remains to be done. A typical instance is the universal mapping problem, in which the proof that the universal functor is injective is often the most difficult part, and one where universal algebra has less to contribute. Nevertheless, the use of universal algebra often helps to simplify the proof by setting out the precise nature of what has to be proved. These points are illustrated by the applications given in this chapter.

1. THE NATURAL NUMBERS

The natural numbers have already been discussed briefly in Chapter I, where a way was sketched of defining them within the framework of general set theory. Besides this method, which goes back to Frege, there is the axiomatic approach due to Peano. This is rather closer to the spirit of abstract algebra, and we shall now look at the natural numbers from this point of view. In the main we shall follow the very lucid account by Henkin [60], to which the reader is referred for further details and references.

Peano takes as his starting point the following five axioms:

P.1. 0 is a number.
P.2. Every number x has a uniquely determined successor x'.
P.3. For all x, $x' \neq 0$.
P.4. If $x' = y'$, then $x = y$.
P.5 (principle of induction)
Any set of numbers which includes 0, and with each number x includes its successor x', includes all numbers.

The first four of these axioms may be expressed as elementary sentences; we take a unary predicate ν (to represent 0) and a binary predicate σ (to represent the successor function). Then the axioms become

P′.1. $\bigvee_i \bigwedge_j [\nu(i) \wedge (\nu(j) \Rightarrow (i = j))]$.
P′.2. $\bigwedge_i \bigvee_j \bigwedge_k [\sigma(i,j) \wedge (\sigma(i,k) \Rightarrow (j = k))]$.
P′.3. $\bigwedge [\sigma(i,j) \Rightarrow \sim \nu(j)]$.
P′.4. $\bigwedge [(\sigma(i,k) \wedge \sigma(j,k)) \Rightarrow (i = j)]$.

For the induction principle (P.5) no such translation is possible, for as we shall see later, there exist systems which satisfy all elementary sentences holding for the natural numbers, but not the induction principle.

We shall not be concerned here with the question whether systems satisfying P.1–5 exist; for us, this is settled affirmatively by the axioms of set theory in Chapter I. Let N be any system satisfying P.1–5; then P.2 states that there is a unary operation defined on N, while P.1 asserts the existence of a distinguished element of N, i.e., a constant operator. We may therefore consider N as an algebra with a 0-ary operator, zero, and a unary operator, the successor function. Now P.5 may be expressed by saying that N has no proper subalgebras, or equivalently, that N is generated by the empty set.

Guided by these considerations, we define an *induction algebra* as an algebra with a constant operator (written 0) and a unary operator (written $x \to x'$), with the empty set as generating set. An induction algebra is said to be *numeral* if it satisfies P.3–4 as well. Now the usual development of the integers from Peano's axioms shows that there is in effect only one numeral algebra. We shall obtain this result as a consequence of

Theorem 1.1

A numeral algebra is the free induction algebra on $\emptyset$ as free generating set.

We shall prove this theorem by the following

Lemma. *Every element* $\neq 0$ *of an induction algebra* A *is a successor of an element of* A.

For the set of successors of elements of A, together with 0, is a subalgebra and hence coincides with A. ∎

To prove the theorem, let N be a numeral algebra and I any induction algebra. We have to find a homomorphism $N \to I$; this will necessarily be an epimorphism, and from this the result will follow. We form the direct product $N \times I$; the subalgebra H generated by $\emptyset$ is again an induction algebra. Now let ε be the restriction to H of the projection $N \times I \to N$. The image of H is a subalgebra of N, and hence must be N itself. We assert that ε is injective; this amounts to saying that for each $x \in N$,

(1) there exists a unique $y \in I$ such that $(x,y) \in H$.

Let S be the subset of N consisting of all elements $x \in N$ satisfying (1); then we must show that $S = N$, and this will follow if we show that S is a subalgebra of N.

(i) $0 = (0,0) \in H$; if $(0,y) \in H$ for some $y \neq 0$, then $(0,y) \neq 0$, and hence by the lemma, $(0,y) = (u,v)' = (u',v')$, i.e. $u' = 0$. But this contradicts P.3; therefore $(0,y) \in H$ only when $y = 0$, which shows that $0 \in S$.

(ii) If $x \in S$, then there exists y such that $(x,y) \in H$, and hence $(x',y') \in H$. Assume that $(x',y_1), (x',y_2) \in H$; then since $x' \neq 0$, it follows that $(x',y_i) \neq 0$, and so (x',y_i) is a successor in H, say $(x',y_i) = (u_i,v_i)' = (u_i',v_i')$, where $(u_i,v_i) \in H$. Now $u_1' = x' = u_2'$, and by P.4, $u_1 = u_2 = x$. Since $x \in S$ and $(x, v_i) \in H\,(i = 1, 2)$, it follows that $v_1 = v_2$; therefore $v_1' = v_2'$, i.e. $y_1 = y_2$. Thus x' satisfies (1), and so $x' \in S$.

We have now shown that $S = N$, i.e. (1) holds for all $x \in N$. Let $x\theta$ be the unique element y of I defined in this way; then $\theta : N \to I$ is the required homomorphism. ∎

Since numeral algebras exist, it follows that the free induction algebra (which is unique up to isomorphism) satisfies P.3–4 and we obtain

Corollary 1.2

The numeral algebra is unique up to isomorphism. ∎

Theorem 1.1 forms the basis for the notion of definition by induction. This is most clearly seen in the case of addition. Let N be a numeral algebra, and for any $a \in N$, denote by N_a the subset of N generated from

a by the successor function. This set N_a may itself be regarded as an induction algebra, with a as zero element. Hence by Theorem 1.1 there is a homomorphism $\alpha_a : N \to N_a$. In detail this means that

$$0\alpha_a = a, \quad x'\alpha_a = (x\alpha_a)'.$$

If we denote the homomorphism α_a by $+a$, these equations take on the familiar form

$$0 + a = a, \quad x' + a = (x + a)'.$$

It is important to note that whereas the method of proof by induction requires only P.5 and so holds in any induction algebra, a definition by induction usually needs further justification. For example, in the case of addition in N we must show that there exists a homomorphism of N into N_a; of course, with Theorem 1.1 at our disposal this follows immediately (in most accounts a direct justification is given, based on P.3–4).

Consider now the definition of multiplication in N; this is a mapping of N into itself, μ_a (for each $a \in N$), satisfying

$$0\mu_a = 0, \quad x'\mu_a = x\mu_a + a.$$

Since N is in fact a word algebra, there exists a unique mapping $\mu_a : N \to N$ satisfying these equations (cf. Exercise III.7.6). A similar definition by induction can be applied to any function on N. The details may be left to the reader.

We conclude with an example of a structure which is indiscernible from, but not isomorphic to, the natural numbers. Such a structure is called a *nonstandard model* of the natural numbers. Let N be the set of natural numbers, regarded as a numeral algebra, and let $M = N^N/\mathscr{D}$ be an ultrapower with respect to a nonprincipal ultrafilter $\mathscr{D}$ on N. Then M is indiscernible from N, but not isomorphic to N, since by Corollary VI.6.8 M is uncountable and so is not even equipotent with N. It may also be verified directly that M does not satisfy P.5; the subalgebra of M generated by $\emptyset$ consists of all mappings $N \to N$ which are constant on a $\mathscr{D}$-set. Now $\mathscr{D}$ is a nonprincipal ultrafilter and so contains no finite sets; therefore the identity mapping on N is not constant on any $\mathscr{D}$-set, and so does not belong to the minimal subalgebra of M; hence M has proper subalgebras, in contradiction with P.5.

In a similar way one can construct nonstandard models of set theory by taking sets to be defined axiomatically, in terms of the binary relation $\in$

(and equality). E.g., if $\mathcal{O}$ is the class of all ordinal numbers in a given universe, then $\mathcal{O}$ is a model of set theory in this sense, and any ultrapower of $\mathcal{O}$ will again be a model, but in this ultrapower the ordering will in general not be a well-ordering (cf. Vopenka [62]; also Exercise 8 below). In Chapter I we bypassed these difficulties by adopting a relatively naïve point of view, from which the axioms are regarded as self-evident assertions about intuitively known concepts. A glance at the axioms will convince the reader that they certainly cannot all be translated into elementary sentences; the study of theories defined by more general sorts of sentences, and their interpretations, lies outside the scope of this book. See Barwise [77] for a comprehensive survey.

It should be observed that one can in a sense 'approximate' P. 5 by elementary sentences, by replacing it by an axiom scheme. If L is a language referring to the natural numbers (N, 0, ′, +, ×), we may for each unary predicate α in L introduce the axiom $(\alpha(0) \wedge \bigwedge_x[\alpha(x) \Rightarrow \alpha(x')]) \Rightarrow \bigwedge_x \alpha(x)$. This infinite family of elementary sentences still is weaker than P. 5, since a nonstandard model of N will satisfy all of them but not P. 5. Nevertheless it will yield a large class of the consequences of P. 5 that are expressible in elementary sentences. However, no recursively enumerable axiom scheme can yield all first order theorems of arithmetic, by Gödel's incompleteness theorem.

EXERCISES

1. Show that each nonfree induction algebra I is determined up to isomorphism by two integers r,n ($r \geqslant 0$, $n \geqslant 1$), such that if $\theta: N \to I$ is the unique homomorphism of Theorem 1.1, then $x\theta = y\theta$ if and only if $x \geqslant r$, $y \geqslant r$, and $x \equiv y \pmod{n}$ or $x = y$. In the exercises which follow, the induction algebra here determined is denoted by $I_{r,n}$.

2. Show that every induction algebra I is relatively free; deduce that the definition by induction may be applied to endomorphisms of I, and use this to define addition in I. Can multiplication be defined in I? Can exponentiation be defined in I?

3. Give a direct proof that every induction algebra satisfies either P.3 or P.4.

4. Determine all induction algebras satisfying P.6: $x' \neq x$ for all x.

5. Prove in detail the validity of the inductive definition for free induction algebras; i.e., show that if N is the free induction algebra, then for every $a \in N$ and $f \in I^N$ (where I is some induction algebra), there exists a unique mapping $\theta: N \to I$ such that $0\theta = a$, $x'\theta = f(x)$.

6. Give a direct proof that a free induction algebra satisfies P.3–4. (Consider suitable mappings from the algebra to $I_{0,2}$.)

7. Express the fact that the natural numbers form a totally ordered set in terms of elementary sentences, using only zero, the successor function and addition. Deduce that every ultrapower of the natural numbers is totally ordered.

8. Prove that the well-ordering of the natural numbers cannot be expressed in terms of elementary sentences. (Construct an ultrapower which is not well-ordered.)

9. Show that in any model of the natural numbers any nonzero element is a successor.

10. (Henkin.) Show that there are countable nonstandard models of the natural numbers. (Apply Theorem VI.3.4 to a suitable ultrapower of the integers.)

11. Let Q be the set of nonnegative rational numbers and Z the set of all integers, both in their natural ordering, and consider the subset A of $Q \times Z$ consisting of all (x,y) except those for which $x = 0$, $y < 0$. If A is ordered lexicographically, i.e., $(x,y) \leqslant (z,t)$ if and only if $x < z$ or $x = z$ and $y \leqslant t$, then each element of A has an immediate successor. Show that A is a nonstandard model of the integers, and that every countable nonstandard model is isomorphic to A. (Use Exercises 10 and I.5.9.)

2. ABSTRACT DEPENDENCE RELATIONS

The theory of linear dependence in a vector space over a field presents so many similarities with the theory of algebraic dependence in a field over a given ground field that the general theorems are usually deduced from a few axioms common to both theories (v.d.Waerden [37], Zariski & Samuel [58]). It is therefore natural to formulate these axioms in the context of universal algebra. This makes a comprehensive treatment possible, including the notion of algebraic closure.

An *abstract dependence relation* on a set S is given by a system $\mathscr{D}$ of subsets of S such that

(1) any subset X of S belongs to $\mathscr{D}$ if and only if some finite nonempty subset of X belongs to $\mathscr{D}$.

A subset of S is called *dependent* if it belongs to $\mathscr{D}$, and *independent* otherwise. It is clear from (1) that any subset of S containing a dependent set is itself dependent; in other words, any subset of an independent set is independent. Further, the empty set is independent.

Let S be a set with a dependence relation; an element a of S is said to *depend on* a subset X of S, if $a \in X$ or if there is an independent subset X' of X such that $X' \cup \{a\}$ is dependent. The set of all elements which depend on X is called the *span* of X and is denoted by $\langle X \rangle$. If $\langle X \rangle = S$, X is called a *spanning set* of S; an independent spanning set of S is called a *basis* of S.

An obvious example of a dependence relation is provided by the notion of linear dependence in a vector space over a field (not necessarily commutative). Secondly, if E is a commutative field, then the usual notion of algebraic dependence (over a fixed subfield F of E) is another instance of a dependence relation. Both types of dependence have the property that

$$\langle\langle X \rangle\rangle = \langle X \rangle. \tag{2}$$

When (2) holds, the dependence relation is said to be *transitive*. Such dependence relations can also be described by means of a certain type of algebraic closure operator:

Proposition 2.1

For any transitive dependence relation, the mapping $X \to \langle X \rangle$ is an algebraic closure operator with the following exchange property:

$$\textit{If } y \notin \langle X \rangle \textit{ and } y \in \langle X \cup \{z\} \rangle, \textit{ then } z \in \langle X \cup \{y\} \rangle. \tag{3}$$

Conversely, any algebraic closure operator with the exchange property (3) *arises in this way from a transitive dependence relation.*

Proof:

For the moment let us call a subset T of S *closed*, if $\langle T \rangle = T$. We first show that the closed sets form a closure system: indeed, if $B = \bigcap C_\lambda$, where (C_λ) is a family of closed sets, let B_0 be an independent subset of B such that $B_0 \cup \{y\}$ is dependent; since $B_0 \subseteq C_\lambda$ for all λ, we have $y \in \langle C_\lambda \rangle = C_\lambda$, hence $y \in \bigcap C_\lambda = B$, which shows B to be closed. Next, if $y \in \langle X \rangle$, then by definition there is an independent subset X' of X such that $X' \cup \{y\}$ is dependent. By hypothesis, $X' \cup \{y\}$ must have a finite dependent subset Y, say. If $y \notin Y$, then $Y \subseteq X'$, which contradicts the independence of X'. Hence $y \in Y$, i.e. Y is of the form $Y' \cup \{y\}$, where $Y' \subseteq X'$, and therefore Y' is independent. Thus $y \in \langle Y' \rangle$, where Y' is a finite subset of X. This proves that the closed sets form an algebraic closure system, and since $\langle X \rangle$ is closed by (2), we have an algebraic closure operator. To verify (3) we note that as an immediate consequence of the definitions,

(4) if X is independent and $X \cup \{y\}$ is dependent, then $y \in \langle X \rangle$.

Now assume that $y \notin \langle X \rangle$, $y \in \langle X \cup \{z\} \rangle$. Since y is dependent on $X \cup \{z\}$, it is dependent on a finite subset Y of $X \cup \{z\}$, and by taking Y to be minimal with this property, we ensure that Y is independent. Now if $z \notin Y$, then Y would be a subset of X, and so $y \in \langle X \rangle$, contradicting our assumption. Therefore $z \in Y$, say $Y = Y' \cup \{z\}$, where $Y' \subseteq X$. Now $Y' \cup \{y\}$ is independent, because $y \notin \langle Y' \rangle$, and $Y' \cup \{y,z\} = Y \cup \{y\}$ is dependent, hence $z \in \langle Y' \cup \{y\} \rangle \subseteq \langle X \cup \{y\} \rangle$.

Conversely, let $X \to [X]$ be any algebraic closure operator with the exchange property, and define X to be *dependent* if, for some $y \in X$, $y \in [X \backslash \{y\}]$, and independent otherwise. Since the operator is algebraic, it follows that any dependent set has a finite dependent subset, and clearly any set containing a dependent set is itself dependent so (1) holds. Condition (2) holds by definition, and this shows that we have a transitive dependence relation. Now, for any $X \subseteq S$, $y \in S$, we have $y \in [X]$ if and only if $y \in [X']$ for some finite subset X' of X; taking X' minimal we may assume that X' is independent. It follows that $y \in \langle X' \rangle \subseteq \langle X \rangle$, and hence $[X] \subseteq \langle X \rangle$. Conversely, if $y \in \langle X \rangle$, then again $y \in \langle X' \rangle$ for a finite independent subset X' of X. This means that $X' \cup \{y\}$ is dependent, i.e. for some $z \in X' \cup \{y\}$, $z \in [X' \cup \{y\} \backslash \{z\}]$; by the exchange property it follows that $y \in [X']$ and $[X'] \subseteq [X]$; therefore $[X] = \langle X \rangle$. ▮

When we are dealing with dependence relations on Ω-algebras, the relation will usually be defined not for a single algebra, but for a whole class of algebras. A dependence relation on the algebras of a certain category $\mathscr{K}$ of Ω-algebras and homomorphisms is said to be *algebraic*, if it is transitive and

(i) every closed subset of a $\mathscr{K}$-algebra is again a $\mathscr{K}$-algebra,

(ii) for any $\mathscr{K}$-homomorphism $\theta : A \to B$ and any $X \subseteq A$, $\langle X \rangle \theta \subseteq \langle X\theta \rangle$.

We note that since $X \subseteq \langle X \rangle$, we have $X\theta \subseteq \langle X \rangle \theta$, and so when (ii) is satisfied, we have

$$\langle X \rangle \theta = \langle X\theta \rangle.$$

Suppose that $\mathscr{K}$ is a category with free algebras. Given any $\mathscr{K}$-algebra A and a subset X of A, let $\phi : F_X \to A$ be the homomorphism from the free $\mathscr{K}$-algebra F_X on X to A which extends the identity mapping on X. We define X to be *independent* if this homomorphism is injective, and *dependent* otherwise. In other words, X is independent if and only if the subalgebra of A generated by X is free on X. It is easily verified that the dependence thus defined satisfies (1), and in many cases also (2), though not always, e.g. not for groups (see below). The relation thus defined will be called the *standard dependence* on A (relative to $\mathscr{K}$). It always satisfies condition (ii) above, but not necessarily (i), although again this holds in many special cases. We now give some examples of the standard dependence in algebras:

(i) *The category of R-modules, where R is any ring.* The standard dependence reduces to linear dependence for R-modules, which generalizes the notion of linear dependence over a field. This dependence is algebraic if e.g. R is a commutative integral domain with 1, acting unitally.

(ii) *The category of linear K-algebras, where K is a commutative ring with* 1. The standard dependence generalizes the notion of algebraic dependence (with K as coefficient domain). If K is an integral domain and the K-algebras are taken to be commutative with injective mappings, then the dependence is algebraic.

(iii) *The category of extensions of a fixed $\mathscr{K}$-algebra.* Let $\mathscr{K}$ be any category of Ω-algebras and C a fixed $\mathscr{K}$-algebra; then the $\mathscr{K}$-algebras over C may be regarded as algebras over a unary enlargement of Ω (with a constant operator corresponding to each element of C); in this way we obtain a new category $\mathscr{K}_C$. The standard dependence in $\mathscr{K}_C$ may be defined whenever $\mathscr{K}$ admits free composition; this is also referred to as the standard dependence in $\mathscr{K}$ over C.

(iv) *The category of free groups and homomorphisms.* A subset X of a free group F is independent if and only if $\mathrm{Gp}(X)$ is free on X as free generating set. For any element x of F denote by $l(x)$ the length of x (in terms of some fixed free generating set of F) and more generally, for any finite subset $X = \{x_1, \cdots, x_k\}$ of F, write $l(X) = \Sigma l(x_i)$. Define a preordering on the finite subsets of F by putting $X \prec Y$ whenever $\mathrm{Gp}(X) = \mathrm{Gp}(Y)$ and $l(X) \leqslant l(Y)$. Then it may be shown that (i) the finite subsets of F satisfy the minimum condition with respect to this preordering, and (ii) any minimal set is independent. It follows that every finitely generated subgroup of a free group is free. This special case of Schreier's theorem (all subgroups of free groups are free) is due to Nielsen; for details of his method and other applications see M. Hall [59]. The same method may be applied in the case of free Lie algebras over a field (cf. Cohn [64]).

We now return to general transitive dependence relations. Our first task is to derive the existence of bases and the invariance of the dimension; the proofs follow a pattern which is familiar from linear algebra.

Lemma 2.2

Let S be a set with a transitive dependence relation and let X be a subset of S. Then the following three assertions are equivalent:

(i) *X is a maximal independent subset of S.*
(ii) *X is a minimal spanning set of S.*
(iii) *X is a basis of S.*

The proof is almost immediate, for a basis is both maximal independent and minimal spanning. Conversely, if X is maximal independent, then any element y of S either belongs to X or is such that $X \cup \{y\}$ is dependent, whence $y \in \langle X \rangle$ in either case. If X is minimal spanning, then it cannot be dependent, for otherwise it could be replaced by a proper subset which still spans S. ∎

Lemma 2.3 (exchange lemma)

Let S be a set with a transitive dependence relation. If X is an independent set and Y is a spanning set of S, then there is a subset Y' of Y such that $X \cap Y' = \emptyset$ and $X \cup Y'$ is a basis for S.

Proof:

Consider the system $\mathscr{I}$ of independent subsets Z of S such that

$$X \subseteq Z \subseteq X \cup Y. \tag{5}$$

Since X is independent, there are such sets; moreover, if (Z_λ) is any chain of sets in $\mathscr{I}$, then the union $Z = \bigcup Z_\lambda$ is again in $\mathscr{I}$, for it clearly satisfies (5), and if it were dependent, then some finite subset of Z would be dependent; this would be contained in some member of the chain (Z_λ), in contradiction with the fact that all the Z_λ are independent. By Zorn's lemma, $\mathscr{I}$ has a maximal element M; by the maximality, every element of Y either belongs to M or depends on M, whence $\langle M \rangle = \langle Y \rangle = S$. This proves that M is a basis of S. Since $X \subseteq M \subseteq X \cup Y$, M is of the form $M = X \cup Y'$, where $Y' = M \backslash X$ satisfies $Y' \subseteq Y$, $Y' \cap X = \emptyset$. ∎

The exchange lemma shows in particular that S has a basis:

Theorem 2.4

Let S be a set with a transitive dependence relation. Then S has a basis; more precisely, if X is any independent set and Y any spanning set of S such that $X \subseteq Y$, then there is a basis B of S such that

$$X \subseteq B \subseteq Y. \tag{6}$$

Moreover, any two bases have the same cardinal.

Proof:

Let X, Y be given as in the enunciation; then $X \cup Y = Y$, and so by Lemma 2.3 there is a basis satisfying (6). In particular, taking $X = \emptyset$, $Y = S$, we see that there always is a basis.

Now let B, C be any two bases of S. If one of B, C is infinite, then so is the other, and both have the same cardinal, by Proposition II.5.5, using Theorem II.5.2 to interpret the closure system on S as system of subalgebras. If B and C are both finite, let $B \cap C = \{a_1,\cdots,a_n\}$, $B = \{a_1,\cdots,a_n, b_1,\cdots,b_r\}$ and $C = \{a_1,\cdots,a_n,c_1,\cdots,c_s\}$, where distinct elements are denoted by distinct letters or suffixes. We shall use induction on $\max(r,s)$. If $r = 0$ or $s = 0$, then $B \subseteq C$ or $C \subseteq B$, and the result is clear. So we may assume $r \geqslant 1$, $s \geqslant 1$; further, $r > s$ without loss of generality, so that in fact $r > 1$. By Lemma 2.3, the set $\{a_1,\cdots,a_n,b_1\}$ may be completed to a basis D by using elements of C, say

$$D = \{a_1,\cdots,a_n,b_1,c_{i_1},\cdots,c_{i_t}\} \qquad t \leqslant s < r.$$

Now D has $n + 1$ elements in common with B and t ($<r$) elements besides, while B has $r - 1$ elements besides, so by the induction hypothesis, $|B| = |D|$, i.e.

$$r = t + 1.$$

Since $r > 1$, it follows that $t \geqslant 1$, and so D also has at least $n + 1$ elements in common with C. Using the induction hypothesis once again, we find that

$$s = t + 1,$$

and hence $r = s$. ∎

As an illustration, the standard dependence in groups does not satisfy the conclusion of Th. 2.4 and so cannot be transitive (i.e. (2) does not hold here).

In order to formulate a notion of algebraic closure one might proceed as follows: Let $\mathscr{K}$ be a category of Ω-algebras with an algebraic dependence relation. An extension E/A of $\mathscr{K}$-algebras is said to be *algebraic* if A spans E. Now it may seem natural to define a $\mathscr{K}$-algebra to be algebraically closed if it has no proper algebraic extension; such a definition would meet with difficulties because usually there will be no algebraically closed algebras in this sense, simply because we can always adjoin elements to A which are algebraic over $\emptyset$. The difficulty may be avoided by factoring out such elements and allowing only those extensions of A from which no elements can be factored out without collapsing part of A. Thus we make this

Definition

An extension E/A of $\mathscr{K}$-algebras is said to be *retractable* if there is a homomorphism $\theta: E/A \to F/A$ which is not injective; the image of θ is also called a *retraction* of E/A. If no such homomorphism exists, the extension is said to be *irretractable*. Further, E/A is an *algebraic* extension (relative to some algebraic dependence relation) if A spans E.

Proposition 2.5

Let $\mathscr{K}$ be a category of Ω-algebras admitting homomorphic images. Then any extension E/A of $\mathscr{K}$-algebras has a retraction which is itself irretractable.

Proof:

Consider the congruences on E which separate A; by Corollary II.6.4 there is a maximal such congruence, $\mathfrak{q}$ say. The natural homomorphism $E \to E/\mathfrak{q} = \bar{E}$ say, restricted to A, is injective and may therefore be used to embed A in $\bar{E}$. The resulting extension $\bar{E}/A$ is irretractable, for if not, then it would have a nontrivial congruence $\bar{\mathfrak{r}}$ separating A, which by Corollary II.3.12 corresponds to a congruence $\mathfrak{r}$ on E which properly contains $\mathfrak{q}$ and separates A, contradicting the definition of $\mathfrak{q}$. ∎

Let $\mathscr{K}$ be a category of algebras with an algebraic dependence relation. Then an extension E/A of $\mathscr{K}$-algebras is said to be *algebraically closed* if E/A is algebraic and irretractable, and given any algebraic irretractable extension C/B, any embedding $B \to A$ extends to an embedding $C \to E$. It follows immediately from this definition that if $\theta: B \to A$ is a monomorphism and C/B any algebraic extension, then θ may be extended to a homomorphism $\theta': C \to E$. For if C'/B is an irretractable retraction of C/B, this is still algebraic, and so may be embedded in E/A; combining this with the natural homomorphism $C \to C'$ we obtain the required homomorphism.

If we apply this definition to the case of the standard dependence in vector spaces (i.e. linear dependence), we see that no proper extension of a vector space can be algebraic; this explains why the notion of algebraic closure, as here defined, plays no role in the usual treatment of vector spaces. Taking next the case of standard dependence of field extensions (i.e. algebraic dependence in the usual sense), we see that every field extension E/F is irretractable, so that our definition of algebraically closed extension agrees here with the familiar notion.

Our object is to show that under suitable conditions on $\mathscr{K}$, every $\mathscr{K}$-algebra has an algebraically closed extension and that this is determined up to isomorphism by A. To establish this result we have to make another assumption, which is satisfied in most cases: We assume the dependence relation to be such that

I. Any algebraic irretractable extension of A has cardinal at most $\max(|A|,|\Omega|,\aleph_0)$.

Further, $\mathscr{K}$ is said to be *directed* if, for any two extensions of A, there exists a third containing them both. We can now state conditions for an algebraic closure to exist:

Theorem 2.6

Let $\mathscr{K}$ be a directed local category of Ω-algebras admitting homomorphic images, and suppose that an algebraic dependence relation is given on $\mathscr{K}$, satisfying condition I. *Then for every $\mathscr{K}$-algebra A, there exists an algebraically closed extension $\bar{A}/A$, and $\bar{A}$ is determined up to isomorphism by A.*

Proof:

Let $\gamma = \max(|A|,|\Omega|,\aleph_0)$; by assumption, any algebraic irretractable extension of A has cardinal at most γ. Further, since $\mathscr{K}$ is directed, any two extensions E_i/A $(i = 1,2)$ are contained in a third, E/A say. If E_i/A are algebraic, then on replacing E by the set of elements depending on A (in E), we may assume that E/A is algebraic too. If, in addition, the E_i/A are irretractable, then any congruence on E separating A will also separate E_i, and dividing out by a maximal such congruence we obtain an irretractable extension which is still algebraic and in which the E_i can again be embedded. Thus the family of algebraic irretractable extensions of A is directed by inclusion. Let (E_λ/A) be a family of algebraic irretractable extensions of A such that any algebraic irretractable extension of A is isomorphic to some E_λ/A, and define a preorder by putting $E_\lambda \prec E_\mu$ whenever there is a monomorphism $E_\lambda \to E_\mu$ (over A); then the E_λ form a directed system whose limit E may be defined as an Ω-algebra (Exercise III.1.5), and this is in fact a $\mathscr{K}$-algebra containing A, because $\mathscr{K}$ is local. Each element of E lies in some E_λ and so is dependent on A; moreover, E/A is irretractable, because any homomorphism θ of E/A which is not injective must identify a pair of distinct elements x,y say of E, and if E_λ contains x and y, then $\theta \mid E_\lambda$ is not injective, which contradicts the irretractability of E_λ. This proves that E/A is algebraic irretractable. To prove that E/A is algebraically closed, we must show that for any algebraic irretractable extension C/B and any embedding $\theta : B \to A$ there is an embedding $\theta' : C \to E$ extending θ. By identifying B with its image under θ we may regard E as an extension of B. Let F/B be an extension containing C/B and E/B, then F contains A and so may be regarded as an extension of A. Since E/A and C/B are algebraic, we may take F/A to be algebraic too.

If $\mathfrak{q}$ is any congruence on F which separates A, then $\mathfrak{q}$ also separates B and so it separates C. Therefore an irretractable retraction of F/A still contains C. But E/A was maximal algebraic irrectractable, therefore F is embeddable in E and hence so is C.

In order to show that E is determined up to isomorphism by A, we first show that E/A is a minimal algebraically closed extension of A. For suppose that F/A is a proper subextension of E/A which is also algebraically closed. Then E/A may be embedded in F/A, and replacing F by the image of E in the embedding, we may assume F/A to be isomorphic to E/A. It thus follows that E/A has a proper extension E'/A isomorphic to itself, and so algebraic irretractable. Repeating this process, we obtain an ascending well-ordered sequence of extensions of A, all algebraic irretractable:

$$E/A \subset E'/A \subset E''/A \subset \cdots \subset E^{(\alpha)}/A \subset \cdots$$

At a limit ordinal we have an algebraic irretractable extension of A, which can again be embedded in E/A, and so the process can be continued. If we continue this sequence to an ordinal whose cardinal exceeds γ, we obtain an extension $E^{(\alpha)}/A$ which is algebraic irretractable but has cardinal greater than γ, which is a contradiction. Therefore no proper subextension of E/A can be algebraically closed.

Now let E/A, F/A be any two algebraically closed extensions of A. Then F/A may be embedded in E/A; the image is again algebraically closed and hence by the minimality of E it coincides with E. This shows that $E/A \cong F/A$. ∎

The algebraically closed extension of A determined up to isomorphism by A is called the *algebraic closure* of A. From the proof we obtain

Corollary 2.7

If E/A is any extension of A containing an algebraic closure $\bar{A}$ of A, then $\bar{A}$ is uniquely determined within E as the set of elements of E which depend on A. ∎

For example, the category of (commutative) fields of a given characteristic is local and directed and admits homomorphic images (the only homomorphic images here are isomorphic images; in other words, every field is an irretractable extension of its prime field, and hence every field extension is irretractable). Since an equation of degree n cannot have more than n roots, and over a field F of infinite cardinal γ there are $\aleph_0\gamma = \gamma$ equations, it follows that an algebraic extension of F (necessarily irretractable) has cardinal at most γ. Thus condition I is satisfied, and Theorem

2.6 may be applied to deduce the existence of an algebraic closure of a given field.

For skew fields there is no corresponding result because the standard dependence is not algebraic (it does not even satisfy Th. 2.4, cf. Cohn [77']). However it can be shown that the class of skew fields is directed; based on this fact there is a construction possessing some of the properties of an algebraic closure, namely the existential closure (cf. Ch. IX).

The proof of Theorem 2.6 may be simplified if A has an extension I/A such that I is algebraically closed, as extension of itself. An algebra I with this property is said to be *injective*; thus an algebra I is injective if any embedding $B \to I$ may be extended to any algebraic irretractable extension C of B.

Proposition 2.8

If A is contained in an injective algebra I, then any maximal algebraic irretractable subextension of I/A is an algebraic closure of A.

Proof:

The algebraic irretractable subextensions of I/A form an inductive system, hence there is a maximal such subextension, E/A say. Now, if B/A is any algebraic irretractable subextension of I/A, then there is some algebraic subextension C/A of I/A containing both E/A and B/A. Let C'/A be an irretractable retraction of C/A; then C'/A is again algebraic and B/A, E/A may be embedded in C'/A. Now the inclusion mapping $E \to I$ may be extended to an embedding of C' in I, because I is injective. By the maximality of E it follows that $C' = E$; this means that B/A can be embedded in E/A. If C/B is any algebraic irretractable extension, then any embedding $B \to A$ may be extended to a homomorphism $C \to I$, which is necessarily an embedding, and hence C can be embedded in E. This proves E/A to be algebraically closed. Now the uniqueness of E is proved as follows. We observe that E is in fact injective, since any algebraic irretractable extension of E can be embedded in I and then embedded in E. As in the proof of Theorem 2.6, the result will follow if we can show that E is a minimal injective subalgebra of I containing A. Thus let F/A be any subextension of E/A which is also injective; then the identity on F extends to an embedding $E \to F$, which is possible only if $F = E$, hence E is minimal, as asserted. ∎

To illustrate Proposition 2.8, let K be the category of unital R-modules (over an associative ring R with 1) and take all nonempty subsets of a module to be dependent. Then every module can be embedded in an injective module (cf. e.g. MacLane [63]), and therefore every module has an algebraic closure which is uniquely determined (up to isomorphism) as the maximal irretractable extension, or equivalently, the minimal injective extension. This is usually known as the *injective hull* of the given module.

Still other definitions of algebraic closure are possible. In particular, by making more specific assumptions it is possible to deduce condition I (or to avoid it altogether). E.g., W. R. Scott [51] defines an algebra A to be weakly algebraically closed if (in effect) every finitely presented extension of A has a retraction onto A; with this definition a hypothesis of the type of condition I is not required. Another type of condition which entails I is considered by Jónsson [62] (cf. Exercise 6).

Dependence relations occur (in one form or another) in may different contexts and they have been studied intensively under the name matroid theory; for a recent account see Welsh [76].

EXERCISES

1. Let R be a noncommutative integral domain; show that the standard dependence in R-modules is algebraic if and only if any two nonzero right ideals of R have a nonzero intersection.

2. (H. Whitney.) Let Γ be a graph; a subgraph Γ_0 of Γ is said to be *dependent* if Γ_0 contains a circuit. Verify that this is a transitive dependence relation.

3. Show that the convex subsets of the plane form a closure system which does not possess the exchange property.

4. For any R-module M (over a fixed ring R) define a family of submodules to be dependent if their sum is not direct. Show that this is a dependence relation on the set of all submodules of M which is not transitive in general.

5. (A. Kertész.) Given a group G, denote by G_0 the set of all elements a of G such that the normal subgroup of G generated by a is minimal (among the nontrivial normal subgroups). If the span of $X \subseteq G_0$ is defined as the normal subgroup generated by X, show that this defines a transitive dependence relation on G_0. Deduce that any two decompositions of G into a direct product of simple groups (if any exist) have the same number of factors.

6. Two elements x and y of an extension E/A are said to be *A-isomorphic* if the mapping $x \to y$ extends to an isomorphism $A(x) \to A(y)$ leaving A elementwise fixed. If, in the standard dependence relation, any element algebraic over A is A-isomorphic to only a finite number of elements, show that this dependence satisfies condition I (cf. Jónsson [62]).

3. THE DIVISION PROBLEM FOR SEMIGROUPS AND RINGS

A group may be regarded as a semigroup admitting division. Hence, for a given semigroup, one may ask whether there exists a group containing it. In more formal terms, the variety Gp (of groups) is subordinate to the variety Sg (of semigroups), and therefore Sg may be represented in Gp. Moreover, by Corollary IV.4.3, this representation has a universal functor. This means that with every semigroup S there is associated a group $U(S)$ and a homomorphism

$$u : S \to U(S), \tag{1}$$

such that every homomorphism ϕ of S into a group G can be factored uniquely by u to give a homomorphism $\bar{\phi} : U(S) \to G$. We shall call $U(S)$ the *universal group* of the semigroup S.

Clearly, S can be embedded in a group if and only if (1) is injective; the division problem for semigroups consists in finding practical criteria for a semigroup to be embeddable in a group. Although the finding of such criteria may take one outside the domain of universal algebra, the abstract setting is quite likely to simplify the proof of such a criterion, once it has been found. We shall illustrate this by two examples: the first is a proof of Malcev's necessary and sufficient conditions for embeddability; in the second we consider semigroups of endomorphisms of an algebra A and obtain conditions under which A has an extension A^* with automorphisms inducing the given endomorphisms.

Let S be any semigroup and G a group containing S and generated by it; then it is clear that G is a homomorphic image of $U(S)$ by a congruence separating S, i.e. by a normal subgroup of $U(S)$ which meets the set $SS^{-1} = \{st^{-1} \mid s, t \in S\}$ in the unit element only. Conversely, if G is any homomorphic image of $U(S)$ by a congruence separating S (assuming S to be embedded in $U(S)$), then u may be used to embed S in G in such a way that S generates G. Any such group G will be called a *group of fractions* of S. Unlike the universal group, a group of fractions does not always exist, and when it does exist, it need not be unique (cf. Exercise 3).

The first step in embedding a semigroup in a group is to adjoin a unit element, if this is not there already. This can always be done; we take an element 1 and define a multiplication on the set $S^1 = S \sqcup \{1\}$ by keeping the multiplication on S as before and writing

$$a1 = 1a = a \qquad (a \in S^1).$$

Now it is clear that if S can be embedded in a group then either S has a unit element or S^1 can be embedded in a group. We may therefore restrict our attention to semigroups with 1. Thus instead of Sg we shall only be dealing with Sg*, the variety of semigroups with 1 (regarded as a constant operator), and all semigroups occurring in the sequel are assumed to belong to Sg*.

If S is any semigroup, then a subset P of S is said to be *potentially invertible*, if there is a monomorphism of S into a semigroup T in which the elements of P have inverses. Thus, S has a group of fractions if and only if S is itself potentially invertible. By the localization principle we can state:

Proposition 3.1

A semigroup S has a group of fractions whenever any finite subset of S is potentially invertible. ∎

Now suppose that we are given a semigroup S and a subset P of S, and we wish to test whether P is potentially invertible. For each element p of P we adjoin an indeterminate p^- to S and denote the semigroup so obtained by $S(P)$. Let $\mathfrak{q}$ be the congruence on $S(P)$ generated by all the pairs $(pp^-, 1)$, $(p^-p, 1)$, where $p \in P$; then the quotient $S(P)/\mathfrak{q}$ is the universal semigroup for S with inverses for the elements of P, and we have to find conditions under which the mapping nat $\mathfrak{q}$, restricted to S, is injective. This is of the type of the word problem (III.9), and no criterion is known which is generally applicable, practically useful, *and* simple. However, the general conditions have been put into a very striking form, which although not simple, is useful in some cases. Our description follows the paper by Malcev [39], with some simplifications.

The key step is to modify the construction of $S(P)$ above. Given S and a subset P, we take, for each $p \in P$, two indeterminates p^- and p^+, and denote by $S(P)$ the semigroup obtained by adjoining all these elements to S. Let $\mathfrak{q}$ be the congruence on $S(P)$ generated by all the pairs $(p^-p,1)$, $(pp^+,1)$, where $p \in P$; then the quotient $S(P)/\mathfrak{q}$ is again the universal semigroup for S with inverses for the elements of P, for if the images of p, p^-, p^+ under nat $\mathfrak{q}$ are denoted by a, a^-, a^+, then $a^-a = aa^+ = 1$, and hence

$$a^- = a^-aa^+ = a^+.$$

To describe $\mathfrak{q}$ more closely, we represent the elements of $S(P)$ as the vertices of a graph, with segments defined by four kinds of moves, indicated by the following symbols:

(: insert p^-p [: insert pp^+;

) : delete p^-p,] : delete pp^+.

Thus e.g. (can be applied to any element w of $S(P)$, expressed as a product $w = uv$ in any way, and gives rise to a segment from w to up^-pv; similarly for [, while) ,] can only be applied to elements of $S(P)$ which contain p^-p, pp^+ respectively. This is just the graphical representation described in III.9, and it is clear that the different $\mathfrak{q}$-classes are the connected components of the graph. It follows that P is potentially invertible if and only if distinct elements of S lie in different components of the graph. To make this statement more explicit, let u,v be any two elements of $S(P)$ which lie in the same component, and consider a path from u to v. It is described by a chain of elements

(2) $$w_0 = u,\ w_1, \cdots, w_n = v,$$

of $S(P)$, where successive elements w_{i-1}, w_i are obtained from each other by a move. Since the p^- are indeterminates, each p^- occurring at a certain place in w_i either was introduced by a move in passing from w_{i-1} to w_i, or occurred already in w_{i-1}; similarly, it will occur in w_{i+1} unless it disappears by a move in passing from w_i to w_{i+1}. In any case, we can trace each occurrence of p^- from its first appearance in the chain (2) to its disappearance, and similarly for each occurrence of p^+. We now fix our attention on a particular occurrence of p^-. If in (2) we assume that $u,v \in S$, then p^- cannot occur in u or v, so it must enter at some stage, say at the ith move:

$$(\quad : w_{i-1} = ab, \quad w_i = ap^-pb \qquad (a,b \in S(P))$$

and exit at some later stage, say at the jth move:

$$) \quad : w_{j-1} = a'p^-pb', \quad w_j = a'b' \qquad (a',b' \in S(P)).$$

The moves from i to $j-1$ will transform a to a' and pb to pb', each move affecting either one or the other. Now the changes on pb cannot be postponed until after the jth move because pb' is then no longer present, but the changes on a *can* be so postponed, because a' has not been changed by the jth move. In particular, we may postpone all the moves by which a becomes a' until after the step which gets rid of p^-. Thus, when p^- exits, we have

$$ap^-pb' \to ab',$$

and this is followed by the series of moves which change ab' to $a'b'$.

To describe this situation let us call the part to the left of any occurrence of p^- (or to the right of any occurrence of p^+) the *passive part*, and the part to the right (or in the case of p^+, to the left) the *active part*. Then we have shown that any changes in the passive part of an occurrence of p^- can be postponed until after the exit of p^-. For the moment let us call an occurrence of p^- or p^+ in any term of the chain (2) *regular* if no changes take place in its passive part. We now show that a path connecting two elements of S can always be replaced by one in which all occurrences are regular.

Lemma 3.2

Given any chain

$$w_0 = u, \ w_1, \cdots, w_n = v \tag{2}$$

connecting two elements u and v of S, there exists a chain

$$w_0 = u, \ w_1', \cdots, w_{n-1}', \ w_n = v$$

of the same length connecting u and v in which all occurrences are regular. In particular, any two elements of S congruent mod $\mathfrak{q}$ *can be connected by such a chain.*

Proof:

Consider the first move by which an irregular occurrence enters; let this be p^-, entering at the ith move: $w_{i-1} \to w_i$, and leaving at the jth move. The part of the chain from i to j is

$$ab, ap^-pb, \cdots, a'p^-pb', a'b',$$

and this may be replaced by

$$ab,\ ap^{-}pb, \cdots, ap^{-}pb',\ ab', \cdots, a'b',$$

as we have seen. The new chain is again of length n and the given occurrence of p^{-} is now regular. Moreover, the number of irregular occurrences has not been increased, for any p^{+} occurring in a must have been irregular to begin with, and any p^{-} occurring regularly in a will occur regularly in the new chain. Thus, the number of irregular occurrences has been diminished by one at least, and the result follows by induction. ∎

Consider now a chain (2), all of whose occurrences are regular. It is clear that in any term w_i of such a chain, any element p^{-} will occur to the left of any element q^{+}; of two factors p^{-}, p'^{-}, the one lying further to the right will have been introduced later, while the reverse holds of the factors q^{+}. Let us call the part of any term w_i between the rightmost p^{-} and the leftmost q^{+} the *active region*. The change at each step takes place in this region: if the active region of w_i is f and the step $w_i \to w_{i+1}$ has the form (, then this step consists in choosing a factorization $f = ab$ and inserting $p^{-}p$ between a and b, so the active region of w_{i+1} will be pb. The p^{-} just introduced will leave the chain at some step), say in passing from w_j to w_{j+1}. Then the active region of w_j must have the form $f' = pb'$; that of w_{j+1} is obtained by removing p and reattaching the factor a split off at the ith step, giving as active region ab'. Likewise, at a step [the active region changes from cd to cq and at the corresponding step] it changes from $c'q$ to $c'd$. These changes may be summarized in the following table, where $R(*)$, $L(*)$ indicate the active region before and after the move * respectively:

	(	)	[	]
$R(*)$	ab	pb'	cd	$c'q$
$L(*)$	pb	ab'	cq	$c'd$.

Now, if the moves in a chain connecting u to v are $\alpha_1, \cdots, \alpha_n$, we have the equations

(3) $$L(\alpha_1) = R(\alpha_2), \quad L(\alpha_2) = R(\alpha_3), \cdots, \quad L(\alpha_{n-1}) = R(\alpha_n),$$

which express the fact that after the $(i-1)$th move has been carried out, the active region is the same as before the ith move is carried out. Since the active regions are the only part which changes, the equations (3) are precisely the expression of the fact that a chain (2) (with regular occurrences) exists, connecting u and v. At the beginning and end of the chain the active regions are u and v respectively, and for embeddability these must be equal; this is expressed by the equation

$$L(\alpha_n) = R(\alpha_1). \tag{4}$$

Thus with the sequence $\alpha_1,\cdots,\alpha_n$ of moves we associate the condition (3) $\Rightarrow$ (4). If all these conditions hold, then the congruence $\mathfrak{q}$ separates S and the embedding is possible. Clearly these conditions are also necessary.

To find all sequences $\alpha_1,\cdots,\alpha_n$ which can occur, we observe that the sequence must contain as many terms **(** as **)** . If the moves **(** are numbered from 1 to r say, in the order of their appearance, then of the corresponding occurrences p_k^-, if two appear in the same term w_i, the one with the higher suffix must be to the right of the other, for the left of the latter is a passive region. It follows that the one with the higher suffix must also exit first (otherwise, *its* passive region would be disturbed). Moreover, in any left segment $\alpha_1,\cdots,\alpha_\nu$ there must be at least as many **(** 's as **)** 's. Thus the terms **(** , **)** are matched in such a way that no pair separates another pair; let us call this a *bracket formation* for short (Tamari [62]). Similar remarks apply to **[** and **]** and we thus obtain

***Theorem 3.3** (Malcev [39].)*

Let S be a semigroup and P any subset of S. With any finite sequence $\alpha_1,\cdots,\alpha_n$ of pairs **(** , **)** *and* **[** , **]** *such that the round brackets form a bracket formation, and likewise the square brackets, we associate the condition*

$$L(\alpha_1) = R(\alpha_2),\cdots,L(\alpha_{n-1}) = R(\alpha_n) \Rightarrow L(\alpha_n) = R(\alpha_1), \tag{5}$$

for all $a,b,b',c,c',d \in S$ and $p,q \in P$, where these letters are given as label the suffix of the corresponding bracket and $L(*)$, $R(*)$ are taken from the above table. Then the conditions (5), for all such sequences, are necessary and sufficient for P to be potentially invertible. ∎

In particular, if we take P to be S itself, or more generally, a generating set of S, we obtain necessary and sufficient conditions for S to have a group of fractions.

As an example, consider the simplest condition, corresponding to the sequence **( )** . This reads

$$pb = pb' \Rightarrow ab' = ab,$$

and it is essentially the condition for left cancellation by elements of P (as one sees by taking $a = 1$). Similarly, **[ ]** corresponds to right cancellation. The next condition, corresponding to ([ )] , states that

(6) $$pb = cd, \quad cq = pb', \quad ab' = c'q$$

implies

(7) $$c'd = ab.$$

It is not hard to show that the semigroup with generators a, b, c, d, b', c', p, q and relations (6) satisfies cancellation but not (7); this provides an example of a semigroup admitting cancellation but not embeddable in a group (Malcev [37]). More generally, it can be shown (Malcev [40]) that for every integer n there exist conditions (5) of length n which are not implied by all the shorter conditions taken together. Thus the infinite set of conditions in Theorem 3.3 cannot in general be replaced by a finite subset.

Just as groups arise naturally as permutations of a set, so semigroups arise as mappings of a set into itself. It is therefore natural to ask the following: Given a set A and a semigroup Σ of mappings of A into itself, when does there exist a set A^* containing A, together with a group of permutations which induce the given mappings on A? An obvious necessary condition is that the given mappings on A be injective; once this holds, it is almost trivial to construct a set A^* with the required group of permutations. In particular, the given semigroup need not be embeddable in a group at all; that the construction is possible nevertheless, rests on the fact that different permutations of A^* may induce the same mapping on A. Thus the problem as stated is not very closely related to the division problem in semigroups unless we establish a bijection between the given semigroup of mappings of A and a permutation group of A^*. If we try to find necessary and sufficient conditions, we are again led to conditions of the Malcev type (Theorem 3.3); we therefore have to make further restrictions in order to find a suitable sufficient condition.

A natural restriction, which at the same time increases the scope of the method, is to put a structure on A and to consider only mappings compatible with this structure; thus, we shall assume that A is an Ω-algebra with a semigroup of endomorphisms. Consider first the case of a single endomorphism θ, and assume that θ is injective. Then θ provides an isomorphism between A and a subalgebra A' of A. One can therefore define θ^{-1} on A', and since $A' \cong A$, the mapping θ^{-1} can be extended to an isomorphism of A onto an algebra A_1 containing A. Continuing in this way, one obtains an ascending sequence of algebras containing A (and, incidentally, all isomorphic to A); their union A^* is the required extension.

This construction can be generalized by replacing the ascending sequence by a directed system. Correspondingly, this will allow us to apply the method to semigroups which are directed by left-divisibility, i.e. semigroups Σ such that

(8) given $\alpha,\beta \in \Sigma$ there exist $\alpha',\beta' \in \Sigma$ such that $\alpha\beta' = \beta\alpha'$.

A semigroup satisfying this condition is said to be *directed.* We note that any commutative semigroup, and in particular, any semigroup generated by a single element, is always directed.

Let A be a $\mathscr{K}$-algebra, where $\mathscr{K}$ is some category of Ω-algebras. Then a semigroup Σ is said to *act on A by injections,* if with each $\alpha \in \Sigma$ an injective endomorphism θ_α of A is associated such that

$$\theta_{\alpha\beta} = \theta_\alpha\theta_\beta, \qquad \theta_1 = \iota,$$

where ι denotes the identity mapping on A. For brevity we shall simply write $x\alpha$ instead of $x\theta_\alpha$. Now the result to be proved may be stated as follows:

Theorem 3.4

Let A be a $\mathscr{K}$-algebra and let Σ be a semigroup acting on A by injections. If Σ satisfies cancellation on both sides and is directed, then there is a locally $\mathscr{K}$ algebra A^* with the following properties:

(i) *A is a subalgebra of A^*.*

(ii) *Σ acts on A^* by automorphisms which extend the action of Σ on A.*

(iii) *Every element of A^* is of the form $a\theta_\alpha^{-1}$, for some $a \in A$ and $\alpha \in \Sigma$.*

Moreover, A^* is determined up to isomorphism by (i)–(iii).

Proof:

Let us write $\alpha|\beta$ to indicate that $\alpha\lambda = \beta$ for some $\lambda \in \Sigma$; then '$|$' is a preordering of Σ, for which Σ is directed, by hypothesis. We refine this relation by writing $\alpha \leqslant \beta$ whenever there exist α', $\beta' \in \Sigma$ such that

$$\alpha\beta' = \beta\alpha' \quad \text{and} \quad A\beta' \subseteq A\alpha'.$$

Clearly, if $\alpha|\beta$, then $\alpha \leqslant \beta$, hence the relation $\leqslant$ is reflexive. We assert that it is again a preordering on Σ which directs Σ. For suppose that

$\alpha \leqslant \beta$, $\beta \leqslant \gamma$ and take $\lambda,\mu,\lambda',\mu' \in \Sigma$ such that $\alpha\lambda = \beta\mu$, $A\lambda \subseteq A\mu$, $\beta\lambda' = \gamma\mu'$, $A\lambda' \subseteq A\mu'$. Let $\mu,\lambda' \mid \nu$ say, $\nu = \mu\kappa' = \lambda'\kappa$; then $\alpha\lambda\kappa' = \beta\nu = \gamma\mu'\kappa$, and

$$A\lambda\kappa' \subseteq A\mu\kappa' = A\nu \subseteq A\mu'\kappa.$$

Therefore $\alpha \leqslant \gamma$, i.e., $\leqslant$ is a preordering of Σ. Now if $\alpha,\beta \in \Sigma$, then $\alpha,\beta \mid \gamma$ for some $\gamma \in \Sigma$; hence $\alpha,\beta \leqslant \gamma$ and so $\leqslant$ again directs Σ. We note also that like $\mid$ the relation $\leqslant$ is preserved by left multiplication.

Let $\lambda,\mu \in \Sigma$ be such that

(9) $$A\lambda \subseteq A\mu;$$

then there is a uniquely defined injection α of A such that

(10) $$x\alpha\mu = x\lambda \qquad (x \in A);$$

for, every element $x\lambda$ is of the form $y\mu$, with a uniquely determined element y. We denote this endomorphism α more briefly by $\lambda\mu^{-1}$, recalling however that this notation has only been justified under the assumption (9). Applying any $\nu \in \Sigma$ to (10), we obtain

(11) $$x\alpha\mu\nu = x\lambda\nu,$$

and conversely, (11) implies (10), because ν is injective. This shows that for the injection α defined by (10),

(12) $$\alpha = \lambda\mu^{-1} = (\lambda\nu)(\mu\nu)^{-1}.$$

Now for each $\alpha \in \Sigma$ take a copy A_α of A with an isomorphism $\eta_\alpha : A \to A_\alpha$ and embed A in A_α by a monomorphism ϕ_α:

(13) $$x\phi_\alpha = (x\alpha)\eta_\alpha \qquad (x \in A).$$

Given $\alpha,\beta \in \Sigma$ such that $\alpha \leqslant \beta$, we define a monomorphism

(14) $$\phi_{\alpha\beta} : A_\alpha \to A_\beta$$

as follows: let $\alpha\lambda = \beta\mu$, $A\lambda \subseteq A\mu$; then we put $x\phi_{\alpha\beta} = x\eta_\alpha^{-1}\lambda\mu^{-1}\eta_\beta$; this definition does not depend on the choice of λ,μ; for, if also $\alpha\lambda' = \beta\mu'$ and $A\lambda' \subseteq A\mu'$, let $\mu\nu' = \mu'\nu$; then $\alpha\lambda\nu' = \beta\mu\nu' = \beta\mu'\nu = \alpha\lambda'\nu$, hence $\lambda\nu' = \lambda'\nu$ by left cancellation, and so,

$$\lambda\mu^{-1} = (\lambda\nu')(\mu\nu')^{-1} = (\lambda'\nu)(\mu'\nu)^{-1} = \lambda'\mu'^{-1}.$$

We note that formally, $\phi_{\alpha\beta} = \phi_\alpha^{-1}\phi_\beta$; hence we have

$$\phi_{\alpha\beta}\phi_{\beta\gamma} = \phi_{\alpha\gamma} \qquad (\alpha \leqslant \beta \leqslant \gamma) \qquad \text{and}$$
$$\phi_{\alpha\alpha} = 1.$$

This shows that (14) is a directed system of $\mathscr{K}$-algebras and monomorphisms. Let A^* be the direct limit; this is locally $\mathscr{K}$, with monomorphisms

$$\varepsilon_\alpha : A_\alpha \to A^*, \tag{15}$$

such that

$$\varepsilon_\alpha = \phi_{\alpha\beta}\varepsilon_\beta \qquad (\alpha \leqslant \beta). \tag{16}$$

By (16) and (13) we have

$$\varepsilon_\alpha = \phi_{\alpha,\alpha\lambda}\varepsilon_{\alpha\lambda} = \eta_\alpha^{-1}\lambda\eta_{\alpha\lambda}\varepsilon_{\alpha\lambda},$$

hence

$$\eta_\alpha\varepsilon_\alpha = \lambda\eta_{\alpha\lambda}\varepsilon_{\alpha\lambda}.$$

Let us denote the monomorphism $\eta_\alpha\varepsilon_\alpha : A \to A^*$ by ζ_α; then the last equation reads $\zeta_\alpha = \lambda\zeta_{\alpha\lambda}$; it follows that for any $x \in A$ and any $\alpha,\beta,\lambda \in \Sigma$,

$$(x\alpha)\zeta_\beta = (x\alpha\lambda)\zeta_{\beta\lambda}. \tag{17}$$

To define the action of Σ on A^*, take any $\alpha \in \Sigma$ and any $x \in A^*$, say $x = y\zeta_\beta$ ($\beta \in \Sigma$). By hypothesis, $\alpha\lambda = \beta\mu$ for some $\lambda,\mu \in \Sigma$; we define $x\alpha$ by putting

$$x\alpha = (y\mu)\zeta_\lambda. \tag{18}$$

If we also had $\alpha\lambda' = \beta\mu'$ ($\lambda',\mu' \in \Sigma$), then $\mu\nu' = \mu'\nu$ for some $\nu,\nu' \in \Sigma$, hence $\alpha\lambda'\nu = \beta\mu'\nu = \beta\mu\nu' = \alpha\lambda\nu'$, i.e. $\lambda'\nu = \lambda\nu'$, and so by (17),

$$\begin{aligned}(y\mu)\zeta_\lambda = (y\mu\nu')\zeta_{\lambda\nu'} &= (y\mu'\nu)\zeta_{\lambda'\nu}\\ &= (y\mu')\zeta_{\lambda'}.\end{aligned}$$

This shows the right-hand side of (18) to be independent of the choice of λ and μ. Now (18) defines an action of Σ on A^*, for if $\alpha,\beta \in \Sigma$ and $x = y\zeta_\gamma$, let $\gamma\lambda = \alpha\mu$, $\mu\rho = \beta\nu$; then $\gamma\lambda\rho = \alpha\mu\rho = \alpha\beta\nu$, and hence

$$\begin{aligned}&(x\alpha)\beta = (y\zeta_\gamma\alpha)\beta = (y\lambda\zeta_\mu)\beta = y\lambda\rho\zeta_\nu = (y\zeta_\gamma)\alpha\beta = x(\alpha\beta),\\ &x\cdot 1 = y\zeta_\gamma = x.\end{aligned}$$

Further, if x is of the form $y\zeta_1$, then (18) reads

$$(y\zeta_1)\alpha = (y\alpha)\zeta_1.$$

Therefore if we embed A in A^* by identifying $x \in A$ with $x\zeta_1$, then the action of Σ on A^*, when restricted to A, corresponds to the given action on A. Moreover, since

$$(x\zeta_\alpha)\alpha = (x\alpha)\zeta_\alpha = x\zeta_1 = x,$$

it follows that to every $x \in A^*$ there corresponds an $\alpha \in \Sigma$ such that $x\alpha \in A$, i.e. every element of A^* is of the form $a\alpha^{-1}$, where $a \in A$. The same equation shows that ζ_α is the inverse of α, which is therefore an automorphism.

To establish the uniqueness of A^*, let U be the universal Ω-algebra with the property (ii) of the theorem (with U for A^*); clearly there is such an algebra (by Corollary IV.4.3, taking $\mathscr{L}$ to be the category consisting of all algebras isomorphic to A and monomorphisms between them), and since A^* has been found satisfying (ii) and (i), the canonical homomorphism $A \to U$ is injective; since A^* satisfies (iii), it is generated by A (under the action of the elements of Σ and their inverses), and therefore A^* is a homomorphic image of U. If it is a proper homomorphic image, assume that in A^*,

(19) $$x\alpha^{-1} = y\beta^{-1}$$

and let $\alpha\lambda = \beta\mu$; then

$$x\lambda = x\alpha^{-1}\cdot\alpha\lambda = y\beta^{-1}\cdot\beta\mu = y\mu;$$

it follows that $x\lambda = y\mu$ in A, and hence in U; applying $(\alpha\lambda)^{-1}$, we see that (19) already holds in U; therefore A^* is an isomorphic image of U and so is unique up to isomorphism. ∎

For abstract semigroups, this result provides a simple sufficient condition for a group of fractions to exist (cf. Dubreil [43]):

Proposition 3.5

Let S be any cancellation semigroup which is directed (by left-divisibility); then S has a group of fractions G; moreover, G is determined up to isomorphism by S, and is such that $G = SS^{-1}$. Conversely, any semigroup S with a group of fractions $G = SS^{-1}$ is directed.

To prove this result we need only let S act on itself by right multiplications and apply the theorem. The last part is obtained as follows: If S has a group of fractions of the form SS^{-1}, then for any $a,b \in S$, $a^{-1}b$ is of the form uv^{-1}, i.e. $a^{-1}b = uv^{-1}$, and multiplying up we obtain $bv = au$, whence S is directed. ∎

Returning to the situation of Theorem 3.4, we note that any constant element of A, i.e. the value of a 0-ary operator in A, is necessarily left fixed by every endomorphism of A. Let us say that Σ acts *regularly* on A if any two distinct elements of Σ have different effects on every non-constant element of A. Then we have

Corollary 3.6

If Σ, A, *and* A^* *are as in Theorem* 3.4 *and* Σ *acts regularly on* A, *then it also acts regularly on* A^*.

For assume that $x \in A$ and $\alpha,\beta,\gamma \in \Sigma$ are such that

(20) $$x\gamma^{-1}\alpha = x\gamma^{-1}\beta, \text{ where } x\gamma^{-1} \text{ is not constant;}$$

then either $x\gamma \neq x$ or $\gamma = 1$ and $x \in A$. In the latter case, $\alpha = \beta$ by hypothesis. Otherwise there exist $\lambda,\mu,\lambda',\mu' \in \Sigma$ such that $\alpha\lambda = \gamma\mu$, $\beta\lambda' = \gamma\mu'$ and hence there are $\nu,\nu' \in \Sigma$ such that $\lambda\nu' = \lambda'\nu = \rho$ say. It follows that

(21) $$\gamma\mu\nu' = \alpha\rho, \quad \gamma\mu'\nu = \beta\rho,$$

and multiplying (20) by ρ, we obtain

$$x\mu\nu' = x\gamma^{-1}\alpha\rho = x\gamma^{-1}\beta\rho = x\mu'\nu.$$

Since $x\gamma \neq x$, x is not constant; therefore $\mu\nu' = \mu'\nu$, and by (21),

$$\alpha\rho = \gamma\mu\nu' = \gamma\mu'\nu = \beta\rho;$$

therefore $\alpha = \beta$ as asserted. ∎

The division problem for rings, namely, to obtain conditions for an associative ring to be embeddable in a field (not necessarily commutative), is very much more difficult than the corresponding problem for semigroups. Again there is no loss of generality if we limit ourselves to rings with a unit element 1, and we may further assume that $1 \neq 0$. Any field containing R as a subring and generated by R will be called a *field of fractions* for R.

In the first place, we cannot assert the existence of a universal field for a given ring (even when a field of fractions exists) by appealing to universal algebra, since fields do not form a variety. In fact, taking fields as they are, we can show that there is no universal field in general. In the special case of commutative rings and fields, the existence of a universal field for a given ring can be established if one operates not in the category of fields and homomorphisms but in the category of fields and *places*, i.e., generalized homomorphisms which map some of the field elements to infinity (cf. e.g. Lang [58]). There is no difficulty in defining places for skew fields, but at present it is not even known whether the resulting category has anything corresponding to free fields.

A second difficulty, which also did not arise for semigroups, is that if we adjoin the inverses of all the nonzero elements to a ring R (assuming R to have a field of fractions), then the ring generated need not be a field; e.g. if $a,b \in R$, $ab \neq 0$, then $ab^{-1} + ba^{-1}$ may not have an inverse of the form uv^{-1} with $u,v \in R$. However, in the special case when R has a field of fractions whose elements are all of the form uv^{-1} ($u,v \in R$, $v \neq 0$), all these difficulties disappear, and one has the following elegant result, due to Ore [31]:

Theorem 3.7

Let R be an associative ring (*with a unit element* $1 \neq 0$), *whose nonzero elements form a semigroup under multiplication which is directed by left-divisibility; then there exists a skew field K containing R as subring, such that every element of K is of the form ab^{-1}* ($a,b \in R$). *Moreover, K is determined up to isomorphism by R.*

A ring R satisfying the conditions of Theorem 3.7 is called a (*right*) *Ore domain.*

Proof:

Denote the set of nonzero elements of R by R^*. By hypothesis, $R^* \neq \emptyset$, and it is a semigroup under multiplication if and only if R has no zero divisors; when this is so, R^* necessarily satisfies cancellation: $ab = ac$ implies that $a(b - c) = 0$, and if $a,b,c \in R^*$, this means that $b = c$; right cancellation follows similarly. By Proposition 3.5, R^* may be embedded in a group, whose elements are all of the form $ab^{-1}(a, b \in R^*)$. We shall write K for the set consisting of this group together with 0. Then K is closed under multiplication and every non-zero element has an inverse. Moreover, K is closed under the operation $x \to x + 1$, for if $x = ab^{-1}$, then $x + 1 = (a + b)b^{-1}$. To show that K is a field we need only show that it admits addition: Let $x, y \in K$; then

$$x + y = \begin{cases} x & \text{if } y = 0, \\ (xy^{-1} + 1)y & \text{if } y \neq 0. \end{cases}$$

Thus, R has been embedded in a field of the required form. Moreover, K^*, the set of nonzero elements of K, is uniquely determined as the group of fractions of R^*, and hence K is unique (up to isomorphism). ∎

The construction of Theorem 3.7 has been generalized to rings whose multiplicative semigroup satisfies the conditions of Theorem 3.7 residually

(Cohn [61]). This enables one to embed free associative algebras in fields. Another method of embedding free associative algebras in fields, due to Malcev [48] and Neumann [49], is based on the fact that free groups may be totally ordered (Neumann [49′]) and the group algebra (over a commutative field F) of the free group on X contains the free associative algebra on X over F as subalgebra. The embedding is now accomplished by using

Theorem 3.8 (Malcev, Neumann)

The group algebra over a commutative field F of a totally ordered group G can be embedded in a field.

Proof (Higman [52]):

Consider the direct power F^G, regarded as a vector space over F. With every element $f \in F^G$ we associate a subset $D(f)$ of G, its *support*, defined by

$$D(f) = \{s \in G \mid f(s) \neq 0\}.$$

Let A be the subset of F^G consisting of all elements with well-ordered support. Formally, each element of A may be written as a power series: $f = \Sigma f(s)s$. The sum of two such power series again belongs to A, because the union of two well-ordered sets is again well ordered. We define products in A by setting

$$fg = (\Sigma f(s)s)(\Sigma g(t)t) = \Sigma\Big(\sum_{st=u} f(s)g(t)\Big)u. \tag{22}$$

Given $u \in G$, the equation $st = u$ has only finitely many solutions (s,t) in $D(f) \times D(g)$, because both supports are well-ordered. This shows the inner sum on the right of (22) to be finite; now $D(fg)$ is the image of $D(f) \times D(g)$ under the mapping $(s,t) \to st$, and is well-ordered (as the image of a partly well-ordered set, cf. III.2). Hence the element fg defined by (22) lies in A, and it is easily verified that with these definitions A forms an algebra over F containing the group algebra of G as subalgebra (namely, the subalgebra of elements of finite support).

Now consider an element of A of the form $1 - f$, where $f(s) = 0$ for $s < 1$. We assert that $1 + f + f^2 + \cdots$ can be rearranged as a power series. For the free semigroup on $D(f)$ with the divisibility ordering is partly well-ordered by Theorem III.2.9; it follows that $\bigcup D(f^n)$ is partly well-ordered, and moreover, no element of G can belong to infinitely many of the $D(f^n)$, because the solutions $(s_1,\cdots,s_n)$ of

$$s_1\cdots s_n = u$$

for fixed u, are pairwise incomparable elements of a partly well-ordered set. This shows $1 + f + f^2 + \cdots$ to be an element of A, and it is easily verified that it is in fact the inverse of $1 - f$. Now every nonzero element of A is of the form $\alpha u(1 - f)$, where $\alpha \in F$, $\alpha \neq 0$, $u \in G$, and $f(s) = 0$ for $s < 1$. Thus it has an inverse in A, and this shows A to be a field. ∎

None of these methods of embedding a free associative algebra in a field is purely algebraic, in that they all involve limiting processes, and this makes it difficult to decide whether the fields obtained are 'universal' in any sense. However, by comparing these constructions with an earlier algebraic construction by R. Moufang [37], one obtains two nonisomorphic fields of fractions of a given free associative algebra.

In the 1930's Malcev raised the question whether a ring R exists whose non-zero elements form a semigroup which is embeddable in a group, without R being embeddable in a field. Such examples have recently been found; they are briefly discussed on p. 342f.

EXERCISES

1. Show that any idempotent element of a cancellation semigroup S is necessarily the unit element of S, and deduce that a cancellation semigroup has at most one idempotent.

2. Verify that the semigroup defined by the relations (6) admits two-sided cancellation. (Use Theorem III.9.3 to obtain a normal form.)

3. (Malcev.) Show that the free semigroup on two free generators a,b can be embedded in $G = \mathrm{Gp}\{a,b \mid (ab^{-1})^2 = 1\}$ and in $H = \mathrm{Gp}\{a,b \mid (ab^{-1})^3 = 1\}$, and that G, H are groups of fractions for S which are nonisomorphic. Show that S has nonisomorphic groups of fractions which are irretractable as extensions of S. (Apply Proposition VII.2.5 to G and H.)

4. Show that a semigroup S is embeddable in a group if and only if, for each $p \in S$ there is an embedding of S in a cancellation semigroup which maps p to an invertible element.

5. Define a sequence of moves to be *admissible* if it is of the form described in Theorem 3.3. Show that in Theorem 3.3 it is enough to take those sequences which do not contain any admissible proper subsequences.

6. Verify directly the conditions of Theorem 3.3 for directed cancellation semigroups.

7. Let S be a cancellation semigroup in which any two elements with a common right-multiple have a least common right-multiple (unique up to right multiplication by invertible elements). Show that in applying Theorem 3.3 one need only consider admissible sequences not containing ([ ) as a subsequence. (Write down the condition for a sequence ⋯ ([ ) ⋯ and show that it can be replaced by the condition for the sequence ⋯ [⋯ .)

8. Let S be a semigroup and p an element of S such that $ap = pa'$ for any $a,a' \in S$ implies that either $a = a' = 1$ or $a = pb$, $a' = bp$. Show that p is potentially invertible.

9. Let S be a semigroup and C a subset of the centre of S such that S admits cancellation by every element of C. Show that C is potentially invertible.

10. Show that the semigroup $S = \mathrm{Sg}^*\{a,b \mid ba = ab^r\}$ (where r is a positive integer) satisfies cancellation and is directed. (Obtain a normal form for the elements of S.)

11. Show that a subdirect product of two directed semigroups need not be directed. (Take the semigroup of Exercise 10, for different integers r.)

12. Show that any (associative) ring may be embedded in a ring with 1. If R^1 denotes the universal 'ring with 1' for R, obtain necessary and sufficient conditions on R for R^1 to have no zero divisors.

13. (Goldie.) Show that every integral domain with maximum condition on right ideals is a right Ore domain. (If a,b have no common right-multiple, show that the right ideals generated by $b,ab,\cdots,a^nb$ form a strictly ascending chain.)

14. (Albert.) Show that any totally ordered ring has no zero divisors. If such a ring is an Ore domain, show that the ordering has a unique extension to the field of fractions.

15. (Amitsur.) Let $\mathscr{K}$ be the class of rings of a given characteristic (0 or p). Show that any integral domain (not necessarily commutative) in $\mathscr{K}$ which satisfies any law not holding throughout $\mathscr{K}$ is embeddable in a (skew) field.

16. Let R be any ring with 1, and M a multiplicatively closed subset of R which contains 1 but no zero divisors, and which is directed by left-divisibility. If, for each $a \in M$, there exists $b \in M$ such that $Rb \subseteq aR$, and moreover, $Mx \cap M \neq \emptyset$ implies $x \in M$, show that R may be embedded in a ring S such that the elements of M are invertible in S and each element of S has the form am^{-1} ($a \in R$, $m \in M$).

17. (H. Friedman.) Consider the group G of permutations of the real line **R** generated by $\alpha: x \to x + 1$ and $\beta: x \to x^3$. Is G free on α, β?

18. Let $F(x)$ be the field of rational functions in x over a commutative field F, and denote by α, β, γ the endomorphisms generated by $x \to -x$, $x \to x^2$, $x \to x^3$ respectively. Then $F(x)$ can be embedded in a field E which has automorphisms α', β', γ' inducing α, β, γ respectively on $F(x)$, but the semigroup Σ generated by the endomorphisms α, β, γ of $F(x)$ cannot be extended to act on E (note that Σ does not satisfy cancellation).

4. THE DIVISION PROBLEM FOR GROUPOIDS

The division problem discussed in VII.3 becomes easier, oddly enough, if one drops the associativity condition. The problem now is to embed a groupoid in a quasigroup. Since a quasigroup admits unique left and right division, it is obviously necessary, for a groupoid G to be embeddable in a quasigroup, that G possess left and right cancellation. This condition is also sufficient for the embedding to be possible (Bates [47], Evans [51]). We shall use the semigroup representation of algebras (Proposition IV.4.6) to establish this fact. In the proof it is convenient to deal with left and right division separately; thus, we define a groupoid G to be a *right quasigroup*, if the equation

$$xa = b \tag{1}$$

has a unique solution in G, for all $a,b \in G$. Clearly a right quasigroup, and more generally, any groupoid contained in a right quasigroup as subgroupoid, possesses right cancellation:

$$\text{If } ac = bc, \quad \text{then } a = b. \tag{2}$$

Lemma 4.1

Let G be a groupoid with right cancellation; then G may be embedded in a right quasigroup G. Moreover, if G satisfies left cancellation, then so does G*.*

Proof:

We first construct an extension G_1 of G in which the equation (1) can be solved for all $a \in G$ and all $b \in G_1$. Let S be the free semigroup (with 1) on the set G and two further elements μ,ρ (distinct from the elements of G)

as free generating set. Denote by $\mathfrak{q}$ the congruence on S generated by the pairs $(a\mu a\rho,1)$, $(a\rho a\mu,1)$, $(ab\mu,c)$, $(cb\rho,a)$, where a,b,c range over all elements of G such that $ab = c$ in G. Writing $S' = S/\mathfrak{q}$, we see that the natural homomorphism $x \to x'$ of S onto S' has the properties

$$(ab)' = a'b' \qquad a,b \in G,$$
$$u'a'\mu'a'\rho' = u'a'\rho'a'\mu' = u' \qquad u \in S.$$

Interpreting μ and ρ as the operators of multiplication and right division, we thus obtain an embedding of the required kind provided we can show that $\mathfrak{q}$ separates G and that the cancellation conditions hold. Consider the graph defined by the presentation of S'; its segments are obtained by taking the moves

$$ua\mu a\rho v \to uv, \tag{3}$$

$$ua\rho a\mu v \to uv, \tag{4}$$

$$uab\mu v \to ucv, \tag{5}$$

$$ucb\rho v \to uav, \tag{6}$$

and their inverses, where $u,v \in S$, $a,b,c \in G$, and $ab = c$ in G. We note that by right cancellation a is uniquely determined by b and c, so that the move (6), when possible, is completely determined by its initial vertex $ucb\rho v$.

Our object is to show that the conditions of Theorem III.9.3 are verified. Clearly, each move (3)–(6) decreases the length of any word w to which it is applied; therefore we reach a reduced word after a finite number of steps which is bounded by the length of w. Suppose now that two words w_1, w_2 have each been obtained by a move (3)–(6) from the same word w; we must show that by applying further moves to w_1 and w_2 we can reach the same word from both. This is clear if the parts of w affected by the moves leading to w_1 and w_2 do not overlap. Now an overlap can only occur for the following pairs of moves: (3) and (4), (3) and (5), (4) and (6). If (3) and (4) overlap, w must have the form $w = ua\mu a\rho a\mu v$ or $w = ua\rho a\mu a\rho v$, say the former. Applying either (3) or (4) we obtain $ua\mu v$, and so $w_1 = w_2$ in this case; the same conclusion holds if $w = ua\rho a\mu a\rho v$. Next, let (3) and (5) overlap; then $w = uab\mu b\rho v$; applying (3) and (5) we obtain $w_1 = uav$ and $w_2 = ucb\rho v$ respectively, where $c = ab$. If we apply (6) to w_2, we obtain $uav = w_1$. The case where (4) and (6) overlap is treated similarly.

Thus the conditions of Theorem III.9.3 are satisfied, and we conclude that the reduced words in S form a transversal for S'. Since each element of G is clearly reduced, this shows that G is embedded in S'. Let T be the subalgebra of S with respect to the operators μ and ρ, generated by G; its image T' under nat $\mathfrak{q}$ is a groupoid containing G which admits right division by all the elements of G, i.e. the equation

$$xa\mu = g \qquad (a \in G, g \in T')$$

has a unique solution in T', namely

$$x = ga\rho.$$

Here we have identified the elements of G and μ,ρ with their images under nat $\mathfrak{q}$. To verify right cancellation in T', let $u,v,w \in T$ be such that

$$uw\mu \equiv vw\mu \pmod{\mathfrak{q}}. \tag{7}$$

We may take u, v, and w to be reduced words, without loss of generality; if both sides of (7) are reduced, then they must be equal, and so $u = v$, by cancellation in the free semigroup S. This leaves the case where either side of (7) is reducible. Now T is just the $\{\mu,\rho\}$-word algebra on G; therefore any proper subword of $uw\mu$ must occur in u or w (Corollary III.2.5); but each move (3)–(6) has the effect of replacing a certain word by another word. Therefore any reduction on a proper subword of $uw\mu$ takes place entirely within u or w and so can already be carried out in u or w respectively. Since u and w are reduced, it follows that $uw\mu$ is reduced except when $u,w \in G$, in which case $uw\mu = c \in G$. In that case $vw\mu$ must also be reducible, for if not, we should obtain an equality $vw\mu = c$ between reduced words, which leads to a contradiction (e.g. by comparing lengths). Therefore v also belongs to G; by (7) we have $uw = vw$ in G and by cancellation in G we find that $u = v$.

Assume now that G satisfies left cancellation and let

$$uv\mu \equiv uw\mu \pmod{\mathfrak{q}}, \tag{8}$$

where u, v, w are again reduced elements of T. The same argument as before shows that if either side can be reduced, then so can the other, and u, v, $w \in G$; now $v = w$ follows by cancellation in G.

We have now constructed a groupoid T' containing G, in which the equation (1) has a unique solution for any $a \in G$, $b \in T'$, and T' has (left or right) cancellation whenever G does. Let us put $G_1 = T'$ and apply the same process to G_1; repeating the construction we obtain an ascending chain

$$G \subseteq G_1 \subseteq G_2 \subseteq \cdots$$

of groupoids, all with right cancellation; their direct limit G^* is again a groupoid with right cancellation. Moreover, given any equation (1) in G^*, we can find n such that $a,b \in G_n$, and by construction, (1) has a solution in G_{n+1}, and therefore in G^*. Owing to right cancellation, the solution is unique. Thus G^* is a right quasigroup containing G. If, further, G has left cancellation, so do $G_1, G_2, \cdots$, and therefore their direct limit G^* also has left cancellation (because this is a local property). ∎

It is clear by symmetry that a corresponding lemma holds with right and left interchanged throughout (with an obvious definition of a left quasigroup). Therefore, if we take a groupoid G with two-sided cancellation, we can embed G in a right quasigroup with left cancellation and embed this in turn in a left quasigroup with right cancellation, G' say. Then any equation

$$xa = b \quad \text{or} \quad ay = b, \tag{9}$$

where $a,b \in G$, has a solution in G'. If we repeat this construction, we again obtain an ascending chain

$$G \subseteq G' \subseteq G'' \subseteq \cdots$$

whose direct limit H is a quasigroup containing G. For, given any equation of the form (9) in H, we can take n such that $a,b \in G^{(n)}$ and then solve the equation in $G^{(n+1)}$; thus the equation has a solution in H, and this is unique because H has two-sided cancellation. This proves

Theorem 4.2

Any groupoid with two-sided cancellation can be embedded in a quasigroup. ∎

We remark that a result similar to Lemma 4.1, involving one-sided division, can be proved for semigroups by the same method (Cohn [56]), but the resulting semigroup does not possess cancellation even when we start with a cancellation semigroup. For this reason, the method cannot be extended to two-sided division, a limitation which is to be expected in view of Malcev's theorem (Theorem 3.3).

In the case of nonassociative rings (NA-rings for short), there are also simple necessary and sufficient conditions for embeddability in a NA-field (i.e. a ring whose nonzero elements form a quasigroup under multiplication) (Neumann [51]). As in the commutative case, a category of NA-fields and places can be defined, and the representation of NA-rings in this category can be shown to have a universal functor (Skornyakov [57]).

EXERCISES

1. Show that every cancellation groupoid with a unit element can be embedded in a loop.

2. Show that any groupoid can be embedded in a groupoid with division, i.e. a groupoid in which every equation $xa = b$ or $ay = b$ has at least one solution.

3. Obtain a normal form for the elements of a free quasigroup on a single free generator. (Use Theorem III.2.3 and Proposition IV.4.6.)

5. LINEAR ALGEBRAS

One of the main applications of the representation theory of IV.4 is to the representation of Lie and Jordan algebras in associative algebras. Although the basic situation is the same in the Lie and in the Jordan case, there are many differences of detail, and each theory has its own peculiar difficulties, which are chiefly due to the presence of the associative law. We shall therefore devote this section to the case of not necessarily associative linear algebras, and in the following sections we shall consider the effect of associativity.

Let K be any commutative and associative ring with 1, and denote by (K) the category of all linear K-algebras (not necessarily associative). When $K = Z$, the ring of integers, (Z) is essentially the category of all rings. The multiplication in a K-algebra may be denoted by an operator symbol μ,

(1) $$ab\mu,$$

in the notation of Chapter II; it is more usual to omit the symbol altogether, but since the associative law need not hold, it is now necessary in repeated products to use parentheses, to distinguish e.g. $(ab)c$ and $a(bc)$. We shall adopt a compromise notation by using parentheses, but leaving out the left-hand parenthesis. Thus the product of a by b will be written

(2) $$ab).$$

This is in effect the notation (1), with ')' in place of 'μ'. Thus the associative law would now read: $ab)c) = abc))$.

Now consider any law in a K-algebra; such a law may be brought to the form

(3) $$f = 0,$$

where f is a word in the free K-algebra, K_X say, on some alphabet X. As in IV.2, we may show that K_X is the groupoid algebra over K on Γ_X, the free groupoid on X. Thus every element of K_X is uniquely expressible as

$$f = \Sigma f(s)s,$$

where $f(s) \in K$, $s \in \Gamma_X$, and $f(s) = 0$ for all but a finite number of elements s. With each element s of Γ_X, a positive integer $l(s)$ is associated, its *length*, equal to the number of factors (or equivalently: one greater than the number of parentheses). With its help, we define the *degree* of f as

$$d(f) = \max\{l(s) \mid f(s) \neq 0\}.$$

For $f = 0$, this is interpreted as $-\infty$.

Any systematic study of varieties of linear K-algebras would proceed by considering the possible sets of laws. Now the laws (3) may be classified by their degree. For general degree n this is a formidable task, involving the representation theory of the symmetric group on n letters and the general linear group over K (cf. Malcev [50], Specht [50], Cohn [52]), and unless a definite problem is kept in mind, such a classification would probably not yield a good return on the effort involved. We shall therefore confine ourselves to the simplest case, namely, when $n \leqslant 2$ and K is a field. The complete result is then given by

Theorem 5.1

Let K be any (commutative) field; then any law for linear K-algebras, of degree at most two, and not holding identically, is equivalent to one of the following: (i) $x = 0$; (ii) $xy) = 0$; (iii) $x^2) - x = 0$ (*idempotent law*); (iv) $x^2) = 0$ (*alternating law*); (v) $xy) - yx) = 0$ (*commutative law*). *Moreover, case* (iii) *is possible only if K is the field of two elements.*

Proof:

Any law involves only a finite number of distinct indeterminates; now the most general law of degree two in $x_1,\cdots,x_n$ is

(4) $$\Sigma x_i\alpha_i + \Sigma x_ix_j)\beta_{ij} = 0.$$

If we fix k and put $x_i = 0$ for $i \neq k$ in (4), we obtain

(5) $$x_k\alpha_k + x_k^2)\beta_{kk} = 0 \qquad (k = 1,\cdots,n);$$

subtracting the equations (5) from (4), we get

(6) $$\sum_{i \neq j} x_ix_j)\beta_{ij} = 0.$$

Thus, (4) entails (5) and (6), and conversely, when (5) and (6) hold, we can deduce (4), so that (4) is equivalent to (5) and (6).

Suppose now that not all the α_i are zero, say $\alpha_1 \neq 0$. Then dividing by α_1, we obtain from (5)

$$x + x^2)\gamma = 0, \tag{7}$$

where we have written x for x_1. If $\gamma = 0$, we thus obtain the law $x = 0$, and this clearly implies (4), so that we have case (i). In any case, if we replace x in (7) by $x\lambda$ ($\lambda \in K$) and subtract the result from (7) multiplied by λ^2, we get $x(\lambda^2 - \lambda) = 0$; this again leads to case (i), unless $\lambda^2 = \lambda$ for all elements λ of K, which can only happen when K consists of two elements. When this is so, and case (i) does not hold, we therefore have

$$x^2) - x = 0. \tag{8}$$

If in (8) we replace x by $x + y$ and then simplify, using (8) (as in V.2), we obtain

$$xy) + yx) = 0. \tag{9}$$

Thus, using (8) alone we can reduce (4) to the form

$$\sum_{i<j} x_ix_j)\gamma_{ij} = 0. \tag{10}$$

If this is not identically zero, let $\gamma_{ij} \neq 0$; then putting $x_i = x_j = x$, $x_k = 0$ ($k \neq i,j$), we obtain $x^2) = 0$, which together with (8) shows that $x = 0$, i.e. case (i). In the alternative case (10) vanishes identically, i.e. (8) implies (6), and only (5) is left. But we have seen that this can only give case (i) or case (iii), and in the latter case K is the field of two elements.

It remains to discuss the case when all the α_i in (4) vanish. Putting all except two of the variables equal to zero, we obtain from (6) the laws

$$x_ix_j)\beta_{ij} + x_jx_i)\beta_{ji} = 0 \qquad (i,j = 1,\cdots,n), \tag{11}$$

which together are equivalent to (6). If $\beta_{kk} \neq 0$ for some k, or if $\beta_{ij} + \beta_{ji} \neq 0$ for some i,j, we can deduce the law

$$x^2) = 0 \tag{12}$$

from (5) or (11) respectively (remembering that now $\alpha_k = 0$ in (5)). From (12) we obtain (9) (by replacing x by $x + y$ and using (12) to simplify the result), and this may again be used to reduce (4) to the form (10). If this is not identically zero, say if $\gamma_{ij} \neq 0$, then putting all except x_i and x_j equal to zero, we derive the law

$$xy) = 0, \tag{13}$$

and conversely, (13) implies (4) (under our assumption that all the α_k vanish). Thus we have case (ii); on the other hand, if the form (10) obtained is identically zero, this means that (12) entails (4), and is therefore equivalent to (4), i.e., case (iv).

There remains only the case where $\beta_{kk} = 0$ for all k in (5), so that (5) holds identically, and $\beta_{ij} + \beta_{ji} = 0$ for all i,j in (11). Since (4) is not identically satisfied, some β_{ij} is different from zero, and we thus obtain

$$xy) - yx) = 0;$$

conversely, this may be used to reduce (6) to zero, and so is equivalent to our original law (4), i.e., we have case (v). ∎

For $\alpha =$ i,...,v, denote by $(K)_\alpha$ the variety of linear K-algebras defined by law (α) of Theorem 5.1. Clearly $(K)_{\text{i}}$ is the trivial variety and $(K)_{\text{ii}}$ is the variety of 'zero algebras', i.e., K-modules regarded as linear algebras with multiplication identically zero. $(K)_{\text{iii}}$ is the variety of *idempotent* algebras; in the presence of the associative law, these are just the Boolean algebras (cf. V.2), but it is easy to construct nonassociative algebras in $(K)_{\text{iii}}$ (cf. Exercise 2). In fact, general (i.e., nonlinear) algebras have been constructed, all of whose two-generator subalgebras are Boolean algebras, but which are not themselves Boolean algebras (Diamond & McKinsey [47]). Such algebras cannot be linear, by the result of Exercise 1 below. They show that Boolean algebras have axiom rank at least three (cf. IV.3); since Boolean algebras may be defined by laws in three variables, the axiom rank is actually equal to three.

The algebras of $(K)_{\text{iv}}$ are called *anticommutative*[1] or *(−)-algebras*. There is a natural way of representing (−)-algebras in linear K-algebras, which we shall now describe. With each K-algebra A we may associate another K-algebra A^-, whose K-module structure is the same as that of A, but whose multiplication is given by

$$xy] = xy) - yx). \tag{14}$$

We shall call A^- the (−)-algebra associated with A; clearly, it is in fact a (−)-algebra, whose operations are derived from those of A. In this way

[1] From some points of view it would be more natural to call these algebras *alternating*; we have avoided this name because of possible confusion with alternative algebras (cf. Exercise 6).

the category (K) of all linear K-algebras becomes subordinate to $(K)_{iv}$. Now, by Corollary IV.4.3 the representation (14) of $(-)$-algebras in K-algebras has a universal functor, associating with each $(-)$-algebra B a linear K-algebra $U(B)$ and a homomorphism

$$u: B \to U(B)^-, \tag{15}$$

which is universal for homomorphisms of B into an algebra of the form A^-. Assume now that K is a field of characteristic not two; then on any $(-)$-algebra B, with multiplication $xy]$, a second multiplication $xy)$ may be defined by setting

$$xy) = \tfrac{1}{2}xy]. \tag{16}$$

If the algebra so obtained is denoted by $B^\natural$, then the identity mapping provides a representation of B in $B^\natural$, because

$$xy) - yx) = \tfrac{1}{2}(xy] - yx]) = xy].$$

Thus every $(-)$-algebra has a faithful representation (14), and it follows that (15) is an embedding. For this reason, $U(B)$ is called the *universal linear K-envelope* of B, even in the case of a general ground ring, when (15) need not be injective. We may sum up our result as

Theorem 5.2

Let K be any field of characteristic not two; then any anticommutative K-algebra may be embedded in the $(-)$-algebra of a suitable linear K-algebra, and there is a universal K-algebra, the universal linear K-envelope of B, for all such embeddings. ∎

The result holds for any field, although not for any ring K (cf. Exercises 3 and 4). It may be used e.g. to prove that subalgebras of free $(-)$-algebras (over a field) are again free (Širšov [54]), using the corresponding result for linear algebras (Kuroš [47], Witt [53]); but we shall not enter into the details.

The algebras of $(K)_v$ are just the commutative (not necessarily associative) algebras. With each K-algebra A we associate a commutative algebra A^+, its $(+)$-algebra, by taking the K-module A with the multiplication

$$xy\} = xy) + yx). \tag{17}$$

As before this gives rise to a universal functor which associates with each commutative algebra C a linear K-algebra $V(C)$ and a homomorphism

$$v: C \to V(C)^+, \tag{18}$$

which is universal for homomorphisms of C into algebras of the form A^+, where A is a linear K-algebra. When K is a field of characteristic not two, the identity mapping of C into $C^\natural$, defined as in (16), again provides a faithful representation, and we thus obtain

Theorem 5.3

Let K be any field of characteristic not two; then any commutative K-algebra may be embedded in the (+)-algebra of a suitable linear K-algebra, and there is a universal K-algebra for all such embeddings. ▮

EXERCISES

1. Show that any ring with 1 (but not assumed to be associative) which satisfies the law $xy)y) = xy)$, is a Boolean algebra.

2. Show that the commutative algebra with unit element 1, over the ground field of two elements, with basis 1, a, b, c, and relations $a^2) = a$, $b^2) = b$, $ab) = ac)$ $= c^2) = c$, $bc) = 0$, is nonassociative and belongs to $(K)_{iii}$.

3. Show that for a ground field of characteristic two, the universal functors U and V for the representation of (−)- and (+)-algebras are still injective. (Use a totally ordered basis to construct a faithful representation.)

4. Let F be a field of characteristic two and K the associative and commutative F-algebra generated by α, β, γ with the relations $\alpha^2 = \beta^2 = \gamma^2 = 0$. Show that the universal functor U is not injective for free algebras over this ring K. (Take the (−)-algebra with generators a, b, c and relation $a\alpha + b\beta = c\gamma$, and show that every representation maps $ab]\,\alpha\beta$ to zero.)

5. (Kaplansky.) Show that in a linear algebra over a field of characteristic zero, every law is equivalent to laws which are linear homogeneous in each variable.

6. Show that any variety of linear algebras with 1 over a field of characteristic zero, which is defined by laws of degree at most three, is defined by one or more of the following laws, where $A(x,y,z) = xy)z) - xyz))$, $xy] = xy) - yx)$: (i) $xy) = yx)$, (ii) $x^2)x) = xx^2))$, (iii) $xy]z] + yz]x] + zx]y] = 0$ (Jacobi identity), (iv) $A(x,y,z) = 0$ (associative law), (v) $A(x,y,x) = 0$ (flexible law), (vi) $A(x,y,y)$ $= 0$ (right-alternative law), (vii) $xy]y] = 0$, (viii) $xy]z] = xz]y]$, (ix) $A(x,y,z) =$ $A(z,y,x)$, (x) $xy)z) - xz)y) - yz)x) + zy)x) = \lambda\{xyz)) - xzy)) - yzx)) + zyx))\}$, and the laws obtained by reversing the order of the factors (i.e. taking the anti-isomorphs) of (vi) and (x).

7. Show that in an associative algebra satisfying the law $xy]z] = 0$, the derived operation $\alpha xy) + \beta yx)$ defines an associative multiplication for all α, β.

8. Let A be a $(+)$-algebra (with coefficient ring K); then a K-module M with a mapping $v: A \times M \to M$, which is linear in each argument, is called an *A-module*. If the multiplication in A is denoted by μ, show that the direct product $E = A \times M$, with the multiplication

$$(a,x) \cdot (b,y) = (ab\mu,\ ayv + bxv),$$

is a $(+)$-algebra containing A as subalgebra and M as ideal. (This is called the *split null extension* of M by A.)

9. (Tamari [62].) If nonassociative products are distinguished by enclosing the second factor in brackets, show that each bracket formation of r pairs (,) gives rise to a different way of bracketing a product of $r + 1$ factors, and that all possible ways are thus accounted for. Deduce the formula

$$b_r = \binom{2r}{r} - \binom{2r}{r-1}$$

for the number b_r of bracketing a product of $r + 1$ factors. (If s_k^h denotes the number of all sequences of h terms '(' and k terms ')', show that $s_r^r - b_r = s_{r-1}^{r+1}$, by changing in any sequence the last unmatched ')' into '('.)

6. LIE ALGEBRAS

Let K be any commutative and associative ring with 1 and denote by As(K) the category of associative K-algebras. Since As(K) is a (full) subcategory of (K), it is again subordinate to the category of anticommutative algebras, which we shall here denote by $(K)^-$ (instead of $(K)_{\text{iv}}$). We therefore have a natural representation of $(K)^-$ in As(K). An admissible mapping is just a K-linear mapping μ such that

$$\mu(\,xy]\,) = \mu(x)\mu(y) - \mu(y)\mu(x). \tag{1}$$

The admissible mappings are therefore defined by identities, and it follows (by Corollary IV.4.2) that there is a universal functor, associating with every $(-)$-algebra X a unique associative algebra $U(X)$, the *universal associative envelope* of X, with a canonical homomorphism

$$u: X \to U(X). \tag{2}$$

This homomorphism is not usually injective; in fact, it is easily verified that any element of the form

(3) $$xy]z] + yz]x] + zx]y]$$

is mapped to zero. We therefore confine our attention to $(-)$-algebras in which this expression vanishes identically, and define:

A *Lie algebra* is a $(-)$-algebra satisfying the law

(4) $$xy]z] + yz]x] + zx]y] = 0 \qquad \text{(Jacobi identity).}$$

The previous remark shows that X must be a Lie algebra if (2) is to be injective; a Lie algebra for which (2) is injective is said to be *special*. Over a field as coefficient domain, every Lie algebra is special. This follows from the Birkhoff-Witt theorem, which, moreover, gives a basis for the universal associative envelope of a Lie algebra L, in terms of a basis of L. Below we shall give a proof of this result, but first we consider the special case of a free Lie algebra. Here we shall prove rather more, namely we shall give a basis of the algebra in terms of a free generating set. In particular, this provides a solution of the word problem for free Lie algebras.

Let X be any totally ordered set and consider the free semigroup Φ_X on X. The elements of Φ_X, called the *words* in X, are the finite rows of elements of X and may be ordered lexicographically. More precisely, if $u = u_1 \cdots u_r$, $v = v_1 \cdots v_s$ $(u_i, v_j \in X)$, then we put $u < v$ whenever $u_i = v_i$ for $i = 1, \cdots, k-1$ and $u_k < v_k$, or $u_i = v_i$ for $i = 1, \cdots, s$ and $r > s$. With this definition, Φ_X is totally ordered; if e.g. $x_1 < x_2 < x_3$, then $x_1x_3x_2^2x_1 < x_1x_3x_2$. Two words u, v are said to be *cyclic conjugates* if $u = ab$, $v = ba$. A word u is said to be *regular* if, for any factorization $u = ab$ $(a,b \neq 1)$, we have $u > ba$; thus a word is regular if it comes after all its cyclic conjugates. We note that $u = a^k$ with $k > 1$ is impossible for a regular word; further, we remark that

(5) if u is regular and $u < v$, then $uv < vu$.

For if $u < v$ and $uv \geqslant vu$, then $u = vz$, and so $vzv \geqslant vvz$, i.e. $zv \geqslant vz$, which contradicts the regularity of u.

We now use the regular words to define nonassociative 'basic products', which are to serve as a basis in our free Lie algebra. Let Γ_X be the free groupoid on X; its elements are rows of elements of X, bracketed in some way. We indicate such products by writing $[u]$, where there are $n-1$ pairs of brackets if u has degree n (or $n-2$ pairs if we omit the outermost pair, as we shall do occasionally).

Definition

A product $[u]$ is called *basic* if either $u \in X$, or the following conditions hold:

(i) if $[u] = [v][w]$, then $[v]$, $[w]$ are basic and $v > w$,
(ii) if $[u] = [[v_1][v_2]][w]$, then $v_2 \leqslant w$.

This defines basic products by induction on the degree. E.g., if $X = \{x,y\}$ with $x < y$, then the basic products include

$$x,\ y,\ [yx],\ [[yx]x],\ \cdots,\ [[yx]\,[[yx]x]],\ \cdots.$$

The notation soon becomes unwieldy, but the situation is saved by the following

Lemma 6.1

Every word u in X can be bracketed in just one way so as to be of the form

$$u = [b_1][b_2]\cdots[b_r], \quad b_1 \leqslant b_2 \leqslant \cdots \leqslant b_r, \tag{6}$$

where each $[b_i]$ *is basic.*

Moreover, if in any basic product $[u]$ *the brackets are removed, the resulting word is regular, and conversely, each regular word u can be bracketed in just one way as a basic product.*

Proof:

We use induction on the length (= degree) of u. Let x_1 be the first element of X (in the ordering) which occurs in u. If x_1 occurs in the first place in u, we have

$$u = x_1 v,$$

where v has shorter length, and by the induction hypothesis, $v = [b_2]\ldots[b_r]$, where $b_2 \leqslant \ldots \leqslant b_r$; hence

$$u = [x_1][b_2]\cdots[b_r]. \tag{7}$$

Now the only basic product beginning with x_1 is x_1 itself. Hence $r > 1$ except when $u = x_1$; moreover, $x_1 \leqslant b_2$. This shows that (7) is of the required form and is unique.

Next, suppose that x_1 does not occur in the first place in u, so that

$$u = x_i \cdots x_1 \cdots \qquad (x_i > x_1)$$

In any division of u into basic products, the only one beginning with x_1 is x_1 itself, as we have just seen. But the first basic product begins with x_i

and so comes after x_1. Thus, there is no division of the form (6) with x_1 as factor. Now take any x_1 in u which is preceded by $x_j \neq x_1$. Since x_1 is not the initial letter of a basic product (in any division of the form (6)), it can only be the final letter of some part, itself basic, of a basic product: $[[w]x_1]$, say. But if w has degree greater than one, say $w = w_1w_2$, then $w_2 > x_1$, and this contradicts the fact that $[w]x_1$ is basic. Thus, $w = x_j$, and in any division of u into basic products, x_jx_1 must occur as a factor. By bracketing x_jx_1 together and regarding it as a single letter we obtain a shorter word, and using the induction hypothesis we obtain a unique factorization (6) of u.

If in (6), $r > 1$, then using (5), we have $b_i \leqslant b_j$ for $i < j$, and hence $b_ib_j \leqslant b_jb_i$. This shows that

$$b_1b_2\cdots b_r \leqslant b_2b_1b_3\cdots b_r \leqslant \cdots \leqslant b_2b_3\ldots b_rb_1;$$

hence, if $r > 1$, u cannot be regular. Thus if u is regular, there is just one way of bracketing u to obtain a basic product $[u]$. Conversely, if $[u]$ is a basic product, then either u has degree 1, and so is regular, or u has degree greater than 1, and x_1 is the earliest letter occurring in u; then the first occurrence of x_1 in $[u]$ is in the combination $[x_ix_1]$ with $x_i > x_1$. Regarding this as a new letter and using induction on the degree, we see that any cyclic conjugate of u must precede u, except possibly one starting with x_1 itself. But this also precedes u because u clearly cannot start with x_1. Thus u is regular, as asserted. ∎

We can now prove a theorem giving a canonical basis for free Lie algebras. The proof (as well as that of Lemma 6.1) is based on that of Širšov [58] (cf. also M. Hall [50] and P. Hall [58]).

Theorem 6.2

Let L be the free Lie algebra on a totally ordered set X. Then the basic products in X form a basis of L, as free K-module.

Proof:

We first show that the basic products in X span L. Any element of degree greater than one in L is a linear combination of terms $[v][w]$, where $[v]$ and $[w]$ may be taken to be basic, by the induction hypothesis, and we may assume that $v > w$, by anticommutativity. If $[v]$ has degree 1 or if $[v] = [v_1][v_2]$, where $v_2 \leqslant w$, then $[v][w]$ is also basic. Otherwise $v_2 > w$, and we have by the Jacobi identity

(8) $$[v][w] = -[[v_2][w]][v_1] + [[v_1][w]][v_2].$$

Now v_1, v_2, and w are regular, and $v_1 > v_2 > w$; hence $v_1v_2 > v_2v_1$ and $v_1w > wv_1$; therefore

$$v_1v_2w > v_2v_1w > v_2wv_1.$$

Secondly, $v_2w > wv_2$, and so $v_1v_2w > v_1wv_2$. By induction on the order (of terms of the same degree), the terms on the right of (8) can be expressed in terms of basic products, and hence so can $[v][w]$. This shows that the basic products span L.

To prove the independence, let A be the free associative algebra on X and let $[X]$ be the subalgebra of A^- generated by X. Then $[X]$ is a Lie algebra on X, and since L is free on X, the identity mapping on X can be extended to a homomorphism

$$L \to [X]. \tag{9}$$

We shall complete the proof of the theorem by showing that the basic products in $[X]$ are linearly independent over K. This will show incidentally that (9) is an isomorphism. Any basic product has the form $[u]$, where u is regular. We assert that when $[u]$ is written out in terms of associative words in A, then

$$[u] = u + \text{words preceding } u \text{ in the ordering.} \tag{10}$$

For let $[u] = [v][w]$; then by induction on the degree, since $[v]$ and $[w]$ may be taken to be basic,

$$[v] = v + v^*, \quad [w] = w + w^*,$$

where v^* denotes terms preceding v and w^* denotes terms preceding w. Now

$$\begin{aligned}[u] = [v + v^*, w + w^*] = vw + v^*w + vw^* + v^*w^* \\ -wv - wv^* - w^*v - w^*v^*.\end{aligned}$$

On the right, all the terms in the first line precede vw, and those in the second line precede wv, which itself precedes vw, by the regularity of u. This proves (10). If we now have a relation

$$\Sigma\alpha_iu_i = 0 \qquad (\alpha_i \in K), \tag{11}$$

between distinct basic products, let $[u_k]$ be the last basic product (in the ordering) occurring in (11) with a nonzero coefficient. Writing out (11) in terms of associative words, we find that

$$\alpha_ku_k + \text{earlier words} = 0,$$

which is a contradiction. Hence the basic products are linearly independent. ▮

Theorem 6.3 (Birkhoff [37], Witt [37].)

Let L be any Lie algebra over a field K, with basis B. If B is totally ordered in any way, then the universal associative envelope U(L) of L has the basis of ascending products

(12) $$b_1 b_2 \cdots b_r \qquad (b_i \in B, \quad b_1 \leqslant \cdots \leqslant b_r).$$

In particular, L is special.

Proof:

Suppose first that L is free on X; then we have a homomorphism

(13) $$\phi : L \to [X]$$

of L into the Lie algebra generated by X in A^-, where A is the free associative algebra on X. Now, any admissible mapping α of L into an associative algebra C (i.e., any homomorphism $\alpha : L \to C^-$) induces a mapping $\alpha_0 : X \to C$. This can be extended to a homomorphism $\alpha' : A \to C$, because A is free. Now $\phi\alpha'$ and α are two homomorphisms of L into C^- which agree on X and hence on L. Thus, α may be factored by ϕ, and the mapping α', restricted to $[X]$, is unique because it is determined by its values on X. This shows that $[X]$, with the canonical mapping (13), is the universal associative envelope of L; we shall show that the products

(14) $$[b_1] \cdots [b_r] \qquad ([b_i] \text{ basic}, \quad b_1 \leqslant \cdots \leqslant b_r)$$

form a basis of A. In the proof of Theorem 6.2 we saw that any basic product $[b]$ has the form

$$[b] = b + \text{terms preceding } b.$$

Hence, if u is any word in X, then by Lemma 6.1,

$$u = b_1 b_2 \cdots b_r \qquad (b_1 \leqslant \cdots \leqslant b_r);$$

thus,

(15) $$u = [b_1][b_2] \cdots [b_r] + \text{earlier terms}.$$

Now the distinct words u form a basis of A, and by (15) these basic elements can be expressed recursively in terms of the products (14); therefore the latter again form a basis. Taking $r = 1$, we see that the mapping (13) is injective, so L may be identified with a subspace of A.

Let L^r be the subspace of A spanned by all products (14) with at most r factors. Since for any two basic products b_i, b_j,

$$b_i b_j = b_j b_i + \sum \gamma_{ijk} b_k \qquad (\gamma_{ijk} \in K),$$

it follows that in any product of r basic products, the factors commute (mod L^{r-1}), so that L^r contains all expressions of degree r in the basic products, taken in any order. In particular, if B is any basis of L, ordered in some way, then the ascending products of degree r in B form a basis of L^r (mod L^{r-1}).

Now let L be any Lie algebra and write $L = F/N$, where F is a free Lie algebra and N is an ideal of F. If B is any basis of F of the form $B = B' \cup B''$, where B' is a basis of N and B'' is a basis for a complement of N in F, then $U(F)$ has a basis of elements

$$v'v'', \tag{16}$$

when v', v'' run over the ascending products in B', B'' respectively. Since N is an ideal in F, the subspace V of $U(F)$ spanned by the elements (16) with $v' \neq 1$ forms an ideal in $U(F)$. For if $v'v''$ is of degree r, say $v' = b_1 b_2 \cdots b_s$, $v'' = b_{s+1} \cdots b_r$ $(s \geqslant 1)$, and $b \in B$, suppose that $b_{i-1} \leqslant b \leqslant b_i$; then, by induction on r, we have

$$b_1 \cdots b_r b \equiv b_1 \cdots b_{i-1} b b_i \cdots b_r \qquad (\text{mod } V \cap F^{r-1}).$$

Here the right-hand side belongs to V, whence $b_1 \cdots b_r b \in V$, and similarly, $b b_1 \cdots b_r \in V$. This shows that V admits multiplication by all elements of B, and hence by all elements of F, i.e. it is an ideal. The quotient $U(F)/V$ has the basis of elements v'', i.e. all ascending products in B'', and since $V \supseteq N$, the natural mapping $F \to U(F)$ induces a mapping $L \to U(F)/V$ such that the diagram

$$\begin{array}{ccc} F & \longrightarrow & U(F) \\ \downarrow & & \downarrow \\ L & \longrightarrow & U(F)/V \end{array}$$

commutes. Thus we have a homomorphism $U(L) \to U(F)/V$, and since the ascending products in B'' are linearly independent in $U(F)/V$, they are also independent in $U(L)$; clearly, they span $U(L)$, and hence they form a basis. ∎

We remark that it is also possible to prove this theorem without using Theorem 6.2, by a direct, though somewhat lengthy, computation (cf. e.g.

Cartan & Eilenberg [56], ch. 13). That method has the advantage of applying to any Lie algebra which is free as K-module, so that it is not necessary to assume K to be a field. The Birkhoff-Witt theorem has been formulated for arbitrary Lie K-algebras, and has been established for any Lie algebras over a principal ideal domain (Lazard [54]), and more generally, any Dedekind domain (Cartier [58]), but it does not hold in all cases, since there exist Lie algebras which are not special (Cartier [58], Širšov [53]). In all examples of nonspecial Lie algebras the additive group of L has torsion elements, and in fact it has been shown that a Lie algebra without torsion elements $\neq 0$ is necessarily special (Cohn[63]).

In conclusion we return to the free Lie algebra F and compute the dimension ψ_n of F_n, the component of degree n of F, in case F is of rank q. Given any word u of length n, either all the cyclic conjugates of u are distinct, in which case u is conjugate to exactly one regular word, or u has the form v^k, where v has length $d = n/k$ and is conjugate to just one regular word. Therefore if we enumerate all the q^n words of length n we get, for each factor d of n, all the regular words of length d, each repeated d times. Thus

$$q^n = \sum_{d|n} d\psi_d;$$

to solve this for ψ_n, we write this down for all divisors of a given n and solve recursively for $n\psi_n$:

$$n\psi_n = q^n - \sum_{p_1|n} q^{n/p_1} + \sum_{p_1p_2|n} q^{n/p_1p_2} - \cdots, \tag{17}$$

where $p_1,p_2,\cdots$ run over the distinct primes dividing n. To write this more concisely, we introduce the Möbius function $\mu(r)$:

$$\mu(r) = \begin{cases} (-1)^k & \text{if } r = p_1p_2\cdots p_k \text{ (distinct primes),} \\ 0 & \text{otherwise.} \end{cases}$$

Then (17) yields

Theorem 6.4 (Witt's formula)

Let F be the free Lie algebra on $x_1,\cdots,x_q$ and F_n the component of degree n of F; then the dimension of F_n is given by

$$\psi_n = \frac{1}{n}\sum_{d|n} \mu(d)q^{n/d}. \quad \blacksquare$$

EXERCISES

1. Let L be any Lie algebra and denote by ρ_a the right multiplication $x \to xa]$. Show that $a \to \rho_a$ is an admissible mapping from L to $\mathscr{L}(L)$, the ring of K-linear mappings on L.

2. Show that a nonzero Lie algebra cannot have a unit element.

3. Show that if X is a generating set for a Lie algebra L, then L is spanned by the left-normed products in X, where a product is said to be left-normed if any factor is of degree 1 or is a product in which the left-hand factor is left normed and the right-hand factor is of degree 1. (Note that by an application of Th. 6.2, the basic products in X with respect to any ordering also span L. However in general, basic products are not left normed, nor vice versa.)

4. In the free Lie algebra on $\{x\} \cup Y$, show that the left-normed products obtained by bracketing the elements $xy_1 \cdots y_n$ $(y_i \in Y)$ appropriately form a basis for the elements of degree $n + 1$ linear in x.

5. In the free Lie algebra on $x_1, \ldots, x_q$, show that the dimension of the space of elements of degree n_i in x_i is

$$\frac{1}{n}\sum_{d|n_i} \mu(d) \frac{\left(\frac{n}{d}\right)!}{\left(\frac{n_1}{d}\right)! \cdots \left(\frac{n_q}{d}\right)!},$$

where $n = \Sigma n_i$.

7. JORDAN ALGEBRAS

We now consider the natural representation of the category $(K)_v$ of commutative algebras in As(K). For brevity, we denote the product in $(+)$-algebras by $x \cdot y$ and in associative algebras by xy (as before). Moreover, we shall assume that K contains an element $\frac{1}{2}$ satisfying the equation

$$\tfrac{1}{2} + \tfrac{1}{2} = 1.$$

This is satisfied e.g. if K is a field of characteristic not two. An admissible mapping is a K-linear mapping μ satisfying

$$\mu(a \cdot a) = 2\mu(a)^2. \tag{1}$$

By linearity and commutativity we have

$$a \cdot b = \tfrac{1}{2}\{(a + b) \cdot (a + b) - a \cdot a - b \cdot b\}.$$

Substituting into (1) and using the linearity of μ, we obtain

$$\mu(a \cdot b) = \mu(a)\mu(b) + \mu(b)\mu(a). \tag{2}$$

The process by which (2) was obtained from (1) is called *linearization*; we shall also say that (2) was obtained by *linearizing* (1).

Again there is a universal functor associating with every (+)-algebra X an associative algebra which we denote by $V'(X)$, and a canonical mapping

$$v' : X \to V'(X). \tag{3}$$

To determine when v' is injective, we again look for elements in the free (+)-algebra which are mapped to zero by v'. It turns out that there are no such elements of degree less than four. The most general element[1] of degree four which is mapped to zero by v' is

$$(x \cdot y) \cdot y^{\cdot 2} - (x \cdot y^{\cdot 2}) \cdot y. \tag{4}$$

The corresponding law

$$(x \cdot y) \cdot y^{\cdot 2} - (x \cdot y^{\cdot 2}) \cdot y = 0 \tag{5}$$

is called the *Jordan identity*, and any (+)-algebra satisfying the law (5) is called a *Jordan algebra*. What has been said shows that if v' is injective, then X must be a Jordan algebra; such a Jordan algebra is said to be *special*. A K-linear mapping satisfying (1) will be called a *special representation* or *associative specialization*.

The reader may now expect a treatment of Jordan algebras which runs parallel to the theory of Lie algebras. It turns out, however, that there are fundamental differences which lead to a completely different development of Jordan algebras. The first of these differences is the seemingly insignificant one, that whereas the Jacobi identity is of degree three (which is also the degree of the associative law), the Jordan identity is of degree four. This already makes it unreasonable to expect the Jordan identity to be a full substitute for the associative law. Thus, even when K is a field of characteristic zero, there are Jordan algebras which are not special. The best known of these 'exceptional' Jordan algebras is the algebra M_3^8 of Hermitian 3×3 matrices over the Cayley-Dickson algebra (Albert [34], [50]). The class of special Jordan algebras clearly admits subalgebras and direct products, but it is not a variety, as will be shown in Proposition 7.9 by an example of an exceptional Jordan algebra which is a homomorphic

[1] In the sense that the fully invariant ideal generated by the word (4) contains every element of degree four mapped to zero by v'.

image of a special Jordan algebra (Cohn [54]). Thus if $\mathscr{J}$ is the variety of all Jordan algebras and $\mathscr{J}'$ the class of special Jordan algebras, we have

$$\mathscr{J}' \subseteq \mathrm{v}\mathscr{J}' \subseteq \mathscr{J}. \tag{6}$$

Here the first inequality is strict, by what has been said. Now Albert & Paige [59] have shown that the exceptional Jordan algebra M_3^8 is not a homomorphic image of a special Jordan algebra, and therefore does not belong to $\mathrm{v}\mathscr{J}'$. This shows the second inequality in (6) to be strict. It means in effect that there are laws which hold in all special Jordan algebras but not in all Jordan algebras. The proof by Albert and Paige gave no explicit identities, although it appeared from the proof that there should be such identities of degree not exceeding 45. In fact some identities (of degrees 8 and 9) holding in special Jordan algebras, but not in all Jordan algebras, have now been found by C. M. Glennie [63].

In any Jordan algebra J, we denote by R_a the right multiplication

$$R_a : x \to x \cdot a. \tag{7}$$

This is also called the *regular representation* of J; in contradistinction to the Lie case, this is not in general an admissible mapping of J into $\mathscr{L}(J)$, the ring of K-linear transformations of J. However, it does satisfy certain identities which follow from the Jordan identity. In the first place, we have from (5)

$$R_y R_{y \cdot 2} = R_{y \cdot 2} R_y; \tag{8}$$

secondly, we can linearize (5) and express it as an operator equation acting on one of the variables replacing y. Thus, denoting the left-hand side of (5) by $f(y)$, consider the expression $f(u + v + w) - f(u + v) - f(v + w) - f(w + u) + f(u) + f(v) + f(w)$. Its vanishing is expressed by the equation

$$\begin{aligned} x \cdot u) \cdot v \cdot w)) + x \cdot v) \cdot w \cdot u)) + x \cdot w) \cdot u \cdot v)) \\ = u \cdot v) \cdot x) \cdot w) + v \cdot w) \cdot x) \cdot u) + w \cdot u) \cdot x) \cdot v). \end{aligned}$$

Writing this as an operator equation for w, we obtain

$$R_v R_{x \cdot u} + R_u R_{x \cdot v} + R_x R_{u \cdot v} = R_{u \cdot v) \cdot x)} + R_v R_x R_u + R_u R_x R_v. \tag{9}$$

Any K-linear mapping μ of a Jordan algebra into an associative algebra satisfying the identities (8), (9) (with R replaced by μ) is called a *Jordan representation* or *multiplication specialization* of J. It is easily verified that a special representation of a Jordan algebra is also a Jordan representation (cf. Exercise 2).

Consider a special Jordan algebra B, contained in an associative algebra A. If we denote right and left multiplication by $a \in A$ by ρ_a and λ_a, respectively, and the regular representation in B by R_a, then, since $x \cdot y = xy + yx$, it follows that

$$R_a = \lambda_a + \rho_a. \tag{10}$$

Now it is clear that both λ_a and ρ_a, restricted to B, are special representations, and moreover, the associative law in A: $(ax)b = a(xb)$ may be stated as

$$\lambda_a \rho_b = \rho_b \lambda_a. \tag{11}$$

This expresses the fact that the representations λ and ρ commute, and it shows that the regular representation $a \to R_a$ in B is the sum of two commuting special representations. We shall generally call a K-linear mapping of a Jordan algebra a *semispecial representation* if it can be expressed as a sum of two commuting special representations. Then what we have shown may be expressed as

Proposition 7.1
The regular representation of a special Jordan algebra is semispecial. ∎

Since the regular representation of any Jordan algebra is a Jordan representation, Proposition 7.1 makes it seem plausible that any semispecial representation is a Jordan representation, and this is easily seen to be the case, by a direct verification, which may be left to the reader. Moreover, a special representation μ may be trivially expressed as a sum of two commuting special representations:

$$\mu(a) = \mu(a) + 0,$$

and is therefore semispecial. We thus have three types of representations of Jordan algebras, in decreasing order of generality:

(i) general representations,
(ii) semispecial representations,
(iii) special representations.

Of these (i) and (ii) are defined by identities and hence possess universal functors V, V' with canonical mappings v, v' respectively. Likewise semispecial representations have a universal functor (V'', v''), where $v'' = v_1'' + v_2''$, the v_i'' being special representations. More precisely, given any semispecial representation μ: $J \to A$ and a decomposition into commuting special representations $\mu = \alpha + \beta$, there is a unique semispecial representation μ^*: $V'' \to A$ such that $\alpha = v_1''\mu^*$, $\beta = v_2''\mu^*$. We need only put $V'' = V' \otimes V'$, $v_1'' = v' \otimes 1$, $v_2'' = 1 \otimes v'$, $v'' = v_1'' + v_2''$; it is easily verified that (V'', v'') has the desired universal property. We note explicitly, if $v'(a) = a'$ for short, then

$$\mu^*(\Sigma x_i' \otimes y_i') = \Sigma\alpha(x_i) \otimes \beta(y_i) \qquad (x_i, y_i \in J), \tag{12}$$

where the right-hand side is well-defined because $x' = 0$ implies $\alpha(x) = \beta(x) = 0$.

The canonical mappings v, v'', v' are also called the *universal* general, semispecial and special representation respectively. Here (V', v') agrees with the functor on $(+)$-algebras introduced earlier. Since the representations are in decreasing order of generality, we have natural homomorphisms $\sigma\colon V(J) \to V''(J)$, $\tau\colon V''(J) \to V'(J)$, where τ corresponds to the decomposition $v' = v' + 0$. These mappings can be combined with the canonical mappings v, v'', v' into a commutative diagram

$$\begin{array}{ccccc} J & \longrightarrow & J & \longrightarrow & J \\ \downarrow{\scriptstyle v} & & \downarrow{\scriptstyle v''} & & \downarrow{\scriptstyle v'} \\ V(J) & \xrightarrow[\sigma]{} & V''(J) & \xrightarrow[\tau]{} & V'(J) \end{array}$$

Since each of V, V'', V' is generated by the image of J, the mappings σ, τ are both surjective. Whether they are injective depends on whether the canonical mappings are injective, and we therefore begin by considering these.

The universal special representation v' is injective if and only if J is special, by the definition of a special Jordan algebra. It follows immediately that for special Jordan algebras, v'' is also injective; conversely, when v'' is injective, then J has a faithful semispecial representation, and hence a faithful special representation in a direct product of two associative algebras, replacing the representation $a \to \mu(a)$, where $\mu(a) = \alpha(a) + \beta(a)$ with α, β special, by: $a \to (\alpha(a), \beta(a))$. Thus J is again special. Turning now to v, we show that this is always injective, by constructing, for any Jordan algebra J, a faithful general representation. Let $J^1 = J \times (u)$ be the direct product of J and a free K-module generated by u, and define a multiplication in J^1 by

$$(a,\alpha u)\cdot(b,\beta u) = (a\cdot b + \alpha b + \beta a, \alpha\beta u) \qquad (a,b \in J;\ \alpha,\beta \in K).$$

Thus J^1 is just the algebra obtained by adjoining a unit element u to J. It is easily verified that J^1 is again a Jordan algebra; hence its regular representation is a Jordan representation. Now $a \to R_a$ is a faithful mapping from J to $\mathscr{L}(J^1)$, because $R_a = R_b$ implies that $a = u \cdot a = u \cdot b = b$; thus R_a is a faithful Jordan representation of J. Summing up, we have

Proposition 7.2

The universal Jordan representation v is always injective, while the universal semispecial representation v″ and the universal special representation v′ are injective if and only if the Jordan algebra is special. ∎

Corollary 7.3

The mapping $\sigma : V(J) \to V''(J)$ *is not injective except possibly when* J *is special.*

For if σ is injective, then $v\sigma = v''$ is injective, and this can happen only if J is special. ∎

We remark that σ need not be injective, even when J is special, as may be seen by considering the Jordan algebra of 3×3 Hermitian matrices over the quaternions (cf. Jacobson [54]); another example will be indicated later.

By contrast τ is not injective, except in trivial cases. Thus, let J be a Jordan algebra over a field K, and assume that τ is injective.
Since $\tau(1 \otimes v'(a)) = 0$ for all $a \in J$, we find that $1 \otimes v'(a) = 0$ for all $a \in J$, hence $V'(J) = 0$.
We remark that J may well be nonzero even when $V'(J) = 0$. For example, the exceptional Jordan algebra M_3^8 mentioned earlier is simple; now the kernel of the universal special representation v' is clearly an ideal in M_3^8 and, being nonzero, must coincide with the whole algebra. Thus, im $v' = 0$, and since $V'(J)$ is generated by im v', we have $V'(M_3^8) = 0$.

As for associative and Lie algebras, we have a correspondence between representations and modules. Thus if J is any Jordan algebra, then by a Jordan module for J one understands a K-module M together with a Jordan representation $J \to \mathscr{L}(M)$. Given a Jordan module M for J, with representation: $a \to \mu(a)$, we may form the *split null extension* $E = J \times M$ with multiplication

$$(a,x)\cdot(b,y) = (a\cdot b, y\mu(a) + x\mu(b)) \qquad (a,b \in J;\ x,y \in M);$$

(cf. Exercise 5.8) and it is not hard to verify, using (8) and (9), that E is again a Jordan algebra. In terms of E there is a simple criterion for a representation of J to be semispecial (Jacobson [54]):

Proposition 7.4

Let J be a special Jordan algebra and M a J-module; then the representation defined by M is semispecial if and only if the split null extension J × M is a special Jordan algebra.

Proof:

Let $a \to \mu(a)$ be the representation defined by M; if this is semispecial, let $\mu(a) = \alpha(a) + \beta(a)$, where α,β are commuting special representations of J by linear transformations of M. Since J is special, we may take it to be embedded in $V'(J)$ (i.e. as subalgebra of $(V'(J))^+$). Now form the product $V'(J) \times M$ with the multiplication

$$(a, x)(b, y) = (ab, x\alpha(b) + y\beta(a)) \qquad (a, b \in J;\ x, y \in M).$$

Since $V'(J)$ is universal for special representations, this defines an associative algebra structure on $A = V'(J) \times M$, and it is easily seen that $E = J \times M$ is a subalgebra of A^+. Conversely, if E is embedded in the $(+)$-algebra of some associative algebra A, then we have for any $x \in M$, $a \in J$,

$$x\cdot a = xa + ax \qquad \text{in } A,$$

hence $\mu(a) = \rho_a + \lambda_a$, and this shows μ to be semispecial. ∎

So far, all associative algebras have been assumed to possess a unit element 1. Now the Jordan algebra may itself have a unit element, which we shall denote by u, to avoid confusion. This leads to a further subdivision of representations, according to their effect on u. Let μ be any Jordan representation and write $\mu_1 = \mu(u)$; then by (9),

$$\mu_1 + 2\mu_1^3 - 3\mu_1^2 = 0,$$

i.e.,

(13) $$\mu_1(\mu_1 - 1)(2\mu_1 - 1) = 0.$$

Moreover, μ_1 is central in $V(J)$. For by linearizing (8) and using x, u, u

as variables we obtain

$$\mu(x)\mu(u) + 2\mu(u)\mu(x) = 2\mu(x)\mu(u) + \mu(u)\mu(x),$$

hence $\mu_1\mu(x) = \mu(x)\mu_1$ for all $x \in J$. Since $V(J)$ is generated by $\mu(J)$, we see that μ_1 is central.

Thus we see that in any Jordan representation (over a field) the operator representing μ has eigenvalues 0, $\frac{1}{2}$, and 1. In particular, if 1 is the unit element of $V(J)$, then we have a decomposition

$$1 = 1_0 + 1_{\frac{1}{2}} + 1_1$$

of 1 into mutually orthogonal central idempotents, and a corresponding decomposition

$$V(J) = V_0(J) \oplus V_{\frac{1}{2}}(J) \oplus V_1(J) \tag{14}$$

of $V(J)$, where $V_i(J)$ is the universal associative envelope for representations $a \to \mu(a)$ such that

$$\mu(u) = i1 \qquad (i = 0,\tfrac{1}{2},1). \tag{15}$$

The representation μ is called *unital*, *special unital*, or a *zero representation*, according as (15) holds with $i = 1$, $\frac{1}{2}$, or 0. Clearly a zero representation of J is identically zero on J: applying (9) with $v = u$, $\mu(u) = 0$, we have

$$\mu(x) = \mu(u \cdot x) = 0.$$

If μ is special unital, then by (9),

$$\begin{aligned}\mu(a^{\cdot 2}) = \mu(a^{\cdot 2}\cdot u) &= 2\mu(a)\mu(a\cdot u) + \mu(u)\mu(a^{\cdot 2}) - 2\mu(a)\mu(u)\mu(a)\\ &= 2\mu(a)^2 + \tfrac{1}{2}u(a^{\cdot 2}) - \mu(a)^2,\end{aligned}$$

hence $\mu(a^{\cdot 2}) = 2\mu(a)^2$, i.e. μ is in fact special. Consider now $V'(J)$ and $V''(J)$; they have decompositions precisely analogous to (14). Since every zero representation and every special unital representation is special and a fortiori semispecial, we obtain the canonical isomorphisms

$$V_0'(J) \cong V_0''(J) \cong V_0(J); \quad V_{\frac{1}{2}}'(J) \cong V_{\frac{1}{2}}''(J) \cong V_{\frac{1}{2}}(J).$$

In particular, it follows that the canonical mapping

$$\sigma : V(J) \to V''(J)$$

is an isomorphism if and only if the induced mapping

$$\sigma_1 \colon V_1(J) \to V_1''(J)$$

is an isomorphism (Jacobson [54]). The result may be expressed as

Proposition 7.5

The universal general representation of a Jordan algebra J with unit element is semispecial if and only if the universal unital representation of J is semispecial. ∎

The structure theory of Jordan algebras is farthest advanced in the case of finite-dimensional Jordan algebras over a field (Albert [47]). We shall not enter into this theory (mainly because universal algebra has little bearing on it), but instead apply the theory of IV.4 to construct nonspecial Jordan algebras and non-semispecial representations. Henceforth all Jordan algebras are understood to be over a field K of characteristic not two.

Let A be the free associative algebra on X and denote by $(X)^+$ the subalgebra of A^+ generated by X; the elements of $(X)^+$ are also called *Jordan elements* in X. Our first task is to find a test for an element of A to be a Jordan element. Such a test is known only when X has at most three elements, and is obtained as follows. On A we define a linear mapping $a \to a^*$, the *reversal operator*, by the equation

$$(x_1 x_2 \cdots x_n)^* = x_n \cdots x_2 x_1 \qquad (x_i \in X) \tag{16}$$

together with linearity. This determines the mapping completely, because the monomials $x_1 x_2 \cdots x_n$ form a basis for A. We note that

$$(ab)^* = b^* a^*, \quad a^{**} = a \qquad (a,b \in A).$$

An element $a \in A$ satisfying the equation $a^* = a$ is said to be *reversible*. For any $a \in A$, we write

$$\{a\} = \tfrac{1}{2}(a + a^*);$$

from this definition it is clear that $\{a\}$ is reversible, for all elements a of A. The set of all reversible elements of A is denoted by H; it is a subalgebra of A^+ which contains $(X)^+$, but need not be equal to it. To verify this, let us assume that X has at least four distinct elements x_1,x_2,x_3,x_4, and consider the *tetrad*

$$\{x_1 x_2 x_3 x_4\}. \tag{17}$$

Clearly this lies in H, but it does not belong to $(X)^+$, for if we apply all 24 permutations π to the variables $x_1,\cdots,x_4$, multiply by the sign of the permutation π, and sum, we obtain from (17):

$$\Sigma(\text{sign } \pi) x_{1\pi} x_{2\pi} x_{3\pi} x_{4\pi},$$

because (1,2,3,4) differs from (4,3,2,1) by an even permutation. On the other hand, the same operation performed on a Jordan element linear in each of x_1,x_2,x_3,x_4 gives zero, because any such element is a sum of terms involving a factor $\{x_ix_j\}$. However, we obtain the whole of H once we include the tetrads in the generating set (Cohn [54]):

Theorem 7.6

The (+)*-algebra H of reversible elements of A is generated by X and all the tetrads* (17) *where* $x_i \in X$.

Proof:

Let H' be the subalgebra of A^+ generated by X and the tetrads; then it is clear that $H' \subseteq H$; to prove equality, we need only show that $p_n = \{x_1\cdots x_n\} \in H'$ for any $x_i \in X$. For $n = 0,1,2$ this is clear, so we may use induction on n. Thus for $n \geqslant 3$,

$$\{x_1x_2\cdots x_n\} + \{x_2\cdots x_nx_1\} = x_1 \cdot \{x_2\cdots x_n\} \equiv 0 \pmod{H'}, \tag{18}$$

by the induction hypothesis. Thus, mod H', the product p_n changes sign under the permutation $(1,\cdots,n)$; in particular, when n is odd, this proves that $p_n \in H'$. For even values of n greater than two, we have

$$\{x_1x_2\}\cdot\{x_3\cdots x_n\} \equiv 0 \pmod{H'},$$

whence

$$\{x_1x_2x_3\cdots x_n\} + \{x_3x_4\cdots x_nx_1x_2\} + \{x_2x_1x_3\cdots x_n\} + \{x_3\cdots x_nx_2x_1\} \equiv 0 \pmod{H'},$$

or, using (18),

$$\{x_1x_2\cdots x_n\} \equiv -\{x_2x_1x_3\cdots x_n\} \pmod{H'}. \tag{19}$$

Since the permutations $(12\cdots n)$ and (12) generate the symmetric group, we see by repeated application of (18) and (19) that p_n is skew-symmetric (mod H'). Hence $p_n \in H'$ unless $x_1,\cdots x_n$ are all distinct. For $n = 4$, this reduces all products to tetrads (17) with distinct arguments, so we may assume that $n \geqslant 6$ and n is even. Then,

$$\{x_1x_2x_3x_4\}\cdot\{x_5\cdots x_n\} \equiv 0 \pmod{H'},$$

i.e.

$$\{x_1x_2x_3x_4x_5\cdots x_n\} + \{x_4x_3x_2x_1x_5\cdots x_n\} \equiv 0 \pmod{H'},$$

and hence $p_n \in H'$; this shows that $H' = H$. ∎

When X has less than four elements, no tetrads occur, and we obtain

Corollary 7.7

In the free associative algebra on three free generators x,y,z, the set H of reversible elements is precisely the subalgebra generated by x,y,z. ▮

In other words, an expression in x,y,z is a Jordan element if and only if it is reversible. As a consequence we have

Theorem 7.8

Any homomorphic image of a two-generator special Jordan algebra is special.

Proof:

Let A be free associative on x and y, and denote by J the subalgebra of A^+ generated by x and y; thus J is the free special Jordan algebra on x and y. We have to show that any homomorphic image of J is special, and by Theorem IV.4.5, this will follow if we show that for any ideal N of J,

$$ANA \cap J = N. \tag{20}$$

Any element of $ANA \cap J$ is a sum of terms $aub + b^*ua^*$, where $a,b \in A$ and $u \in N$. Now consider the element $azb + b^*za^*$ in the free associative algebra on x,y,z. Clearly this is reversible, and by Corollary 7.7 it can be expressed as a Jordan element in x, y, z:

$$azb + b^*za^* = f(x,y,z).$$

Hence

$$aub + b^*ua^* = f(x,y,u) \in N.$$

Since $N \subseteq ANA \cap J$ in any case, this proves (20), and the assertion follows. ▮

This theorem, together with the result of Širšov [56] that the free Jordan algebra on two free generators is special, shows that in fact every two-generator Jordan algebra is special. In this context it is of interest to note that the exceptional Jordan algebra M_3^8 can be generated by three elements. We shall not prove that this algebra is exceptional, nor even that it is a Jordan algebra, but instead we shall construct a homomorphic image of a special Jordan algebra which is exceptional. Let A be the free associative algebra on x,y,z and J the Jordan algebra generated by x,y,z. Consider the ideal P in J generated by $u = x \cdot y (= \frac{1}{2}(xy + yx))$; if J/P were special, we should have $APA \cap J = P$, whence

$$\{uxyz\} \in P.$$

This means that there is a Jordan element $f(u,x,y,z)$ in four free variables which reduces to $\{uxyz\}$ when we put $u = x \cdot y$. Hence f must be a linear combination of reversible elements which are homogeneous linear in each of u,x,y,z. Taking only the terms ending in z, we see that the only contributions come from $v_1 = \{uxyz\}, v_2 = \{uyxz\}, v_3 = \{xyuz\}, v_4 = \{yxuz\}$, $v_5 = \{xuyz\}$, $v_6 = \{yuxz\}$. When we specialize to $u = x \cdot y$, only v_5 has a term in x^2y^2z and only v_6 has a term in y^2x^2z, so neither of these can occur Now xy^2xz occurs only in v_2 and v_3, so these must occur with opposite coefficients, α and $-\alpha$ respectively, say. Similarly, $yxyxz$ occurs only in v_2 and v_4, so v_4 has coefficient $-\alpha$. But the contributions to $xyxyz$ and yx^2yz must be equal, and this can only happen when $\alpha = 0$; this leaves only v_1, and we have seen that $v_1 = \{uxyz\}$ is not a Jordan element in u,x,y,z. This establishes

Proposition 7.9

The homomorphic image of the free special Jordan algebra on x,y,z obtained by adjoining the relation $x \cdot y = 0$ is exceptional. In particular, the class $\mathcal{J}'$ of special Jordan algebras is not a variety. ▮

In the same way, it may be shown that the Jordan algebra generated by x,y,z with relations $x \cdot y = 0$ and $Z^{\cdot 2} = 0$, where Z is the ideal generated by z, is exceptional. But this is just the split null extension of the algebra

$$\mathcal{J}\{x,y \mid x \cdot y = 0\} \tag{21}$$

by the universal general representation. Hence by Proposition 7.4 this representation is not semispecial; thus in (21) we have a special Jordan algebra for which the canonical mapping $\sigma : V(J) \to V''(J)$ is not an isomorphism. This result is of special interest in view of Macdonald's theorem (Macdonald [60] or also Jacobson [62]), which states that any identity in x,y,z and linear in z, which holds in all special Jordan algebras, holds in all Jordan algebras. An equivalent formulation of the theorem states that the universal general representation of the free Jordan algebra F_2 on two free generators is semispecial (in particular, Širšov's theorem quoted earlier follows from this). From this one may deduce, using the same technique as in the proof of Theorem 7.8, that *all* representations of F_2 are semispecial, in contrast to the representations of the algebra (21).

EXERCISES

1. Show that every Jordan algebra satisfies the law

$$(a^{\cdot 2}\cdot b)\cdot c-(a^{\cdot 2}\cdot c)\cdot b=2a\cdot b)\cdot c)\cdot a-2a\cdot c)\cdot b)\cdot a.$$

Is this law (in a (+)-algebra) equivalent to the Jordan identity?

2. Verify that every semispecial representation is a Jordan representation.

3. Verify that the split null extension of an A-module M, where A is a (+)-algebra (cf. Exercise 5.8) is a Jordan algebra if and only if A is a Jordan algebra and M defines a Jordan representation.

4. If N is the kernel of a general representation of a Jordan algebra J, verify that $A(k,a,b)\equiv A(a,k,b)\equiv A(a,b,k)\equiv 0 \pmod N$ for all $a,b\in A$, $k\in N$, where $A(a,b,c)=(a\cdot b)\cdot c-a\cdot(b\cdot c)$. When the representation is special, show that moreover N is an ideal of J.

5. If A is an associative algebra with 1 over a field of characteristic not two, show that an element $a\in A$ has an inverse b in A if and only if the equations $a\cdot b=2$, $a^{\cdot 2}\cdot b=4a$ hold in A^+.

6. Show that a Jordan algebra on a single generator is associative.

7. If J is a Jordan algebra on a single generator, show that $V'(J)$ is commutative; give an example of an associative Jordan algebra whose universal associative envelope for special representations is not commutative.

8. Let J be an n-dimensional vector space (over a field of characteristic not two), and define an algebra structure on J by putting $x\cdot y=0$ $(x,y\in J)$. Verify that J is a Jordan algebra and describe the associative algebras $V'(J)$ and $V(J)$.

9. (i) Let J be a Jordan algebra, X a generating set of J and c an element of J. If $R_xR_c=R_cR_x$ and $R_{x\cdot y}R_c=R_cR_{x\cdot y}$ for all $x,y\in X$, show that $R_cR_a=R_aR_c$ for all $a\in J$.

(ii) Deduce that for all positive integers i and j, every Jordan algebra satisfies the identity $R_{y\cdot i}R_{y\cdot j}=R_{y\cdot i}R_{y\cdot j}$.

(iii) Show that (i) does not hold if J is merely generated by $X\cup\{c\}$ or if we onit the second equation from the hypothesis.

10. Let V be the free Jordan algebra on two generators x, y. Show that the subalgebra of V generated by x, $x\cdot y$, $(x\cdot y)\cdot y$, ... is the free special Jordan algebra on this countable set.

Foreword to the supplements

Since the appearance of the first edition there has been much activity in universal algebra. Most of this has been the technical development of the subject, and so does not primarily concern us here, since we are more interested in the applications. But the power and scope of the subject has been greatly increased by two outside influences, namely category theory and logic, particularly model theory. In the supplementary chapters we can do no more than survey some of the salient features of the new development, and pick out one or two details of particular relevance to the topics in the main body of the book.

In quite another direction computer science ('information theory') has made a study of automata, which has revealed an underlying similarity with the kind of structure encountered in universal algebra; this has been brought out particularly clearly in algebraic language theory. It would take us too far afield to develop this ab ovo, but we include a survey article on the subject which appeared recently, as Chapter XI.

Chapter VIII

Category Theory and Universal Algebra

1. THE PRINCIPLE OF DUALITY

One of the more notable features of the axioms for an abstract category is the complete duality; this plays an important role in simplifying proofs. However, most of the concrete categories such as Sets, Groups, Rings, and Modules are not self-dual. This means that not all the important features of these categories can be described by a self-dual set of axioms, and it is an interesting task to push the description in terms of self-dual axioms as far as possible. There is also the practical advantage that any result obtained as a consequence of these axioms may immediately be dualized.

Perhaps the best illustration of this phenomencn is the categorical description of modules. What may be regarded as a particularly useful self-dual approximation to a category of modules is the notion of an 'abelian category', and the Mitchell–Freyd embedding theorem (cf. Mitchell [65], p. 104 or Gabriel [62]) makes it explicit what has to be added to obtain a category of modules over a ring. Most notable among the non-self-dual conditions holding in the category of modules over a ring is Grothendieck's axiom AB.5 (C.3 in Mitchell). This is essentially the relation at the foot of p. 85 in Ex. II.5.8.

For a simple example illustrating the lack of duality we consider St, the category of sets. This category has direct and inverse limits and every set is the direct limit of its finite subsets, but in general it will not be the inverse limit of its finite quotient sets, for the latter, as inverse limit of compact spaces (in the discrete topology) is a non-empty compact space (cf. Eilenberg–Steenrod [52], p. 217) and hence either finite or uncountable. For an even more concrete illustration of this lack of duality in St, we note that whereas the direct limit of nonempty sets with injective mappings is clearly nonempty, an inverse limit of nonempty sets with surjective mappings may well be empty (Higman–Stone [54]). Here is a simple example of this phenomenon, due to Douady. We denote by Map (A, B) the set of all injective mappings from A to B. Let X be an uncountable infinite set, then $X = \varinjlim F$, where F ranges over all finite subsets of X; as a consequence we have, for any set Y,

$$\mathrm{Map}(X, Y) = \varprojlim \mathrm{Map}(F, Y),$$

as is easily checked. If we take any countable set Y, then $\mathrm{Map}(F, Y)$ is non-empty for each finite F, but $\mathrm{Map}(X, Y) = \emptyset$.

2. ADJOINT PAIRS OF FUNCTORS

In this section we shall consider the notion of a pair of adjoint functors, which in the abstract is perfectly symmetric, and examine what asymmetries arise in concrete instances. This is of particular relevance to universal algebra, since it includes the notion of a free algebra, in a form which is general enough to extend to infinitary algebras.

In any category $\mathscr{A}$, an *initial* object is an object I such that for each $\mathscr{A}$-object A there is exactly one morphism $I \to A$. Dually, a *final* or *terminal* object in $\mathscr{A}$ is an initial object of $\mathscr{A}^0$, the opposite category. Thus if Z is final, there exists for each $A \in \mathrm{Ob}\mathscr{A}$, just one morphism $A \to Z$. A category may have more than one initial object, but they are all isomorphic, for if I, I' are both initial then there exist unique $\mathscr{A}$-morphisms $f\colon I \to I'$ and $g\colon I' \to I$, so $fg \in \mathscr{A}(I, I)$ $(= \mathrm{Hom}_{\mathscr{A}}(I, I))$. But $\mathscr{A}(I, I)$ has only one element, which must be the identity for I, so $fg = 1_I$ and similarly $gf = 1_{I'}$. Thus f is an isomorphism, and so we have shown that any two initial objects are isomorphic, by a unique isomorphism. Similarly for final objects. To illustrate these notions, consider the category St of all sets; this has ∅ as initial object and any 1-element set as final object. Likewise the category of all non-empty sets has no initial object and any 1-element set as final object. Thus neither of these categories is self-dual.

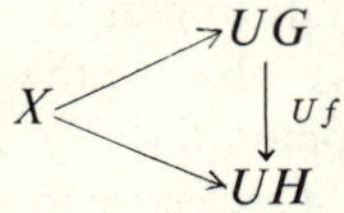

Initial objects may be used to describe universal functors (p. 111) as follows. Consider e.g. the category Gp of all groups, the category St of all sets and the forgetful functor U from Gp to St. Fix a set X and consider the category (X, U) whose objects are maps $X \to UG$ from X to the set underlying a group G, and whose morphisms are commutative triangles arising from a homomorphism $f\colon G \to H$. This is called the *comma category* based on X and U. Now the free group FX, with the canonical map $X \to UFX$, may be described as an initial object in the comma category (X, U). We note that this proves the uniqueness (up to a unique isomorphism) of the free group FX, without further work.

Another way of characterizing the universal functor F above is as left adjoint of the forgetful functor U. We have a natural equivalence (also called *natural isomorphism*)

$$\text{(1)} \qquad \mathrm{Hom}_{\mathrm{Gp}}(FX, G) \cong \mathrm{Hom}_{\mathrm{St}}(X, UG),$$

and it may be shown (cf. e.g. Cohn [77], p. 95) that each of U, F determines the other up to natural isomorphism by (1). The situation (1) is also described by saying that (F, U) is an adjoint pair, more precisely F is a *left adjoint* and U a *right adjoint*.

Let us now consider adjoints in a general setting. We have two functors $F\colon \mathscr{X} \to \mathscr{A}$, $G\colon \mathscr{A} \to \mathscr{X}$, such that for any $\mathscr{A}$-object A and $\mathscr{X}$-object X,

$$\text{(2)} \qquad \mathscr{A}(FX, A) \cong \mathscr{X}(X, GA),$$

where $\cong$ indicates an isomorphism of sets (i.e. bijection) which is natural in A and X. Of course for suitable (viz. additive) categories $\mathscr{A}$ and $\mathscr{X}$, (2) will be an isomorphism of abelian groups. If we put FX for A in (2), then the identity 1_{FX} on the left corresponds on the right to an $\mathscr{X}$-morphism (natural transformation) $\eta\colon X \to GFX$, called the *unit* of the adjunction. For any $\mathscr{A}$-object A, any morphism $f\colon X \to GA$ can be factored uniquely by η, to give $f = \eta \cdot Gf'$ (read left to right), where f':

$FX \to A$ corresponds to f under (2). Next take GA for X in (2), then 1_{GA} on the right corresponds on the left to an $\mathscr{A}$-morphism $\varepsilon: FGA \to A$; this is called the *counit* of the adjunction. For any $\mathscr{A}$-object A, any $\mathscr{A}$-morphism $g: FX \to A$ can be factored uniquely by ε to give $g = F('g)\cdot\varepsilon$, where $'g: X \to GA$ corresponds to g in (2). In all we speak of an adjunction $(F, G, \eta, \varepsilon): \mathscr{X} \to \mathscr{A}$; we observe that this determines the isomorphism (2). For $f: X \to GA$ determines $Ff: FX \to FGA$ and hence $Ff\cdot\varepsilon: FX \to A$; similarly $g: FX \to A$ gives rise to $\eta\cdot Gg: X \to GA$.

It is clear how any left adjoint may be defined as an initial object in a suitable comma category. Now we can of course dualize the above construction and find that each value of a right adjoint functor is obtained as a terminal object in an appropriate comma category.

It is often advantageous to regard a naturally occurring functor as a left or right adjoint, when possible; thus in an abelian category it may be shown that any left adjoint is right exact and any right adjoint is left exact (cf. e.g. Cohn [77], p. 96). Although there is complete symmetry between left and right adjoint at the formal level, the adjunctions encountered in practice by no means all show the same behaviour. Moreover, a forgetful functor on a variety of algebras always has a left adjoint, but only occasionally a right adjoint. In the cases most frequently encountered, the left adjoint of a forgetful functor turns out to be something like a free algebra, while the right adjoint (when it exists) usually describes a subset with some closure property. To give an example of the latter, consider the categories Gp (Groups) and Mon (Monoids, i.e. semigroups with neutral element, and homomorphisms preserving the neutral). We have a forgetful functor S from Gp to Mon, which forgets

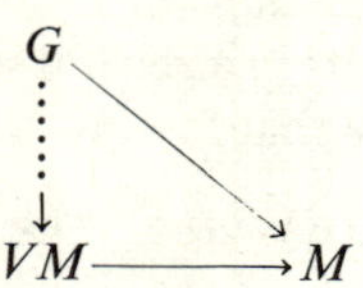

inverses. Its right adjoint V is a functor from Mon to Gp such that for any monoid M, VM is a group with a monoid homomorphism from VM to M which is terminal for homomorphisms from groups to M, i.e. for each group G and each monoid homomorphism $G \to M$ there is a unique group homomorphism $G \to VM$ such that the accompanying diagram commutes. It is easily seen that VM is just the group of units of M.

3. MONADS

Let (F, G) be a pair of adjoint functors; as we have seen, F may be the universal functor associated with a forgetful functor G, or G may be a type of closure functor associated with F, and these two situations show quite different behaviour. In order to take account of the differences of these various types of adjunctions we introduce another construction, from which certain adjunctions may be derived.

Let $\mathscr{X}$ be a category; by a *monad* in $\mathscr{X}$ we understand a functor $T\colon \mathscr{X} \to \mathscr{X}$ together with two natural transformations, the unit $\eta\colon I \to T$ (where I is the identity functor) and the multiplication $\mu\colon T^2 \to T$, such that the following diagrams (describing the unit law and associativity) commute:

$$
\begin{array}{ccccc}
 & & T^2 & & \\
 & \overset{\eta T}{\nearrow} & \downarrow{\scriptstyle \mu} & \overset{T\eta}{\nwarrow} & \\
IT & \xrightarrow{\ \cong\ } & T & \xrightarrow{\ \cong\ } & TI
\end{array}
\qquad\qquad
\begin{array}{ccc}
T^3 & \xrightarrow{\ T\mu\ } & T^2 \\
\downarrow{\scriptstyle \mu T} & & \downarrow{\scriptstyle \mu} \\
T^2 & \xrightarrow{\ \mu\ } & T
\end{array}
$$

The monad is written (T, η, μ) or simply T. Another name used is 'triple', but the actual choice of term is surrounded by a certain amount of controversy. We have adopted the former term since it has the advantage of not referring to any other mathematical concept (also the above diagrams are reminiscent of those used to define monoids). Of course one can also avoid the controversy altogether by using yet another term such as 'algebraic theory' (as is done by Manes [76]). There is a dual notion of *comonad* (cotriple), obtained by reversing all arrows, but we shall not have occasion to deal with this.

Any adjunction $(F, G, \eta, \varepsilon)\colon \mathscr{X} \to \mathscr{A}$ gives rise to a monad in $\mathscr{X}$: $(GF, \eta, G\varepsilon F)$. Here GF is a functor from $\mathscr{X}$ to $\mathscr{X}$, with unit η, and multiplication $G\varepsilon F$, in the sense that $\varepsilon\colon FG \to I$, hence $G\varepsilon F\colon GFGF \to GF$. The unit and associative laws are easily checked. Similarly we have a comonad in $\mathscr{A}$: $(FG, \varepsilon, F\eta G)$.

In general different adjunctions may give rise to the same monad, and for each such monad $T = (T, \eta, \mu)$ these adjunctions may be formed into a category in which the objects correspond to adjunctions $(F, G, \eta, \varepsilon)$ with $GF = T$, $G\varepsilon F = \mu$, represented as commutative triangles with the same base $T\colon \mathscr{X} \to \mathscr{X}$, while the morphisms are commutative 'tetra-

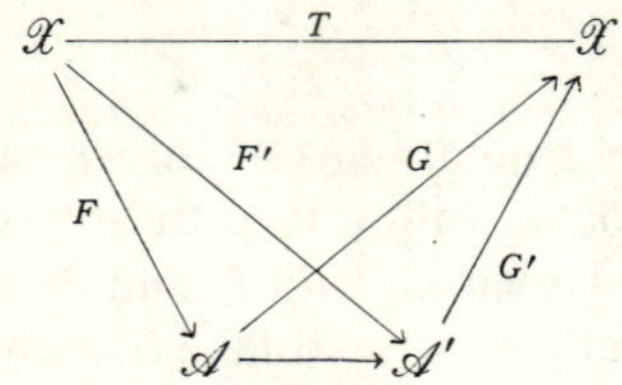

hedra' consisting of two such triangles with third vertex $\mathscr{A}$, $\mathscr{A}'$ respectively and a functor from $\mathscr{A}$ to $\mathscr{A}'$ making the whole commute. This may be described as the *adjunction category* of the given monad. We can now state the basic theorem which will show in particular that every monad arises from a suitable adjunction.

Theorem 3.1

Let (T, η, μ) *be a monad in* $\mathscr{X}$, *then the adjunction category of* T *has an initial object* $(F_T, G_T, \eta_T, \varepsilon_T)\colon \mathscr{X} \to \mathscr{X}_T$ *and a terminal object* $(F^T, G^T, \eta^T, \varepsilon^T)\colon \mathscr{X} \to \mathscr{X}^T$; *in particular, the category is not empty.*

We shall describe the initial and terminal objects but for the proof (which is largely formal) refer to the literature (e.g. Pareigis [69], p. 51 or Mac Lane [71], p. 136).

We begin with $\mathscr{X}_T$; its objects are the same as those of $\mathscr{X}$, while the morphisms from X to Y in $\mathscr{X}_T$ are those morphisms $f\colon TX \to TY$ for

$$\begin{array}{ccc} T^2X & \xrightarrow{Tf} & T^2Y \\ \downarrow{\scriptstyle \mu X} & & \downarrow{\scriptstyle \mu Y} \\ TX & \xrightarrow{f} & TY \end{array}$$

which the diagram shown commutes, with composition of morphisms as in $\mathscr{X}$. Now F_T, G_T are defined by $F_TX = X$, $F_Tf = Tf$, $G_TX = TX$, $G_Tf = f$.

Next $\mathscr{X}^T$ has as objects pairs (X, h), where $X \in Ob\mathscr{X}$ and $h\colon TX \to X$ is such that the diagrams below commute:

$$\begin{array}{ccc} & TX & \\ {\scriptstyle \varepsilon X}\nearrow & & \searrow{\scriptstyle h} \\ X & \xrightarrow{1} & X \end{array} \qquad\qquad \begin{array}{ccc} T^2X & \xrightarrow{Th} & XT \\ \downarrow{\scriptstyle \mu X} & & \downarrow{\scriptstyle h} \\ TX & \xrightarrow{h} & X \end{array}$$

The morphisms in $\mathscr{X}^T$ from (X, h) to (Y, k) are morphisms $f\colon X \to Y$ in $\mathscr{X}$ such that the square below commutes, with composition of morphisms as in $\mathscr{X}$. The initial category $\mathscr{X}_T$ is called the *Kleisli construction*

$$\begin{array}{ccc} TX & \longrightarrow & TY \\ \downarrow{\scriptstyle h} & & \downarrow{\scriptstyle k} \\ X & \longrightarrow & Y \end{array}$$

and the terminal category $\mathscr{X}^T$ is the *Eilenberg–Moore construction*; the objects in $\mathscr{X}^T$ are also called the *T-algebras*. We remark that for any monad (T, η, μ) on $\mathscr{X}$ and any $X \in Ob\mathscr{X}$, (TX, μ) is a T-algebra. Finally, an adjunction $(F, G, \eta, \varepsilon)\colon \mathscr{X} \to \mathscr{A}$ in which $\mathscr{A} \cong \mathscr{X}^T$ for some monad T of $\mathscr{X}$ is said to be *monadic* (tripleable).

To give an example, let G: Mon $\to$ St be the forgetful functor from monoids to sets and F: St $\to$ Mon its left adjoint (the free monoid functor). The monad $T = GF$: St $\to$ St associates with each set X the underlying set of the free monoid FX. A T-algebra consists of a set X and a map $h\colon GFX \to X$ such that the two diagrams above for T-algebras commute. The first means that h reduces to the identity on X, i.e. it is a retraction. To interpret the second, we observe that GFX, the underlying set of the free monoid on X, consists of words in X, i.e. strings of letters from X, and $(GF)^2X$ consists of words in GFX. Given an element w of $(GF)^2X$,

$$w = (x_{11} \dots x_{1i_1})(x_{21} \dots x_{2i_2}) \dots (x_{n1} \dots x_{ni_n}),$$

we can either apply h to each bracket, and then again to the resulting expression, or lift the brackets and then apply h, and according to the diagram the outcome is the same. Writing u^h for the image of $u \in GFX$ under h and 1 for the neutral of FX, we thus have

$$(1) \qquad (uv)^h = u^h v^h, \quad 1^h = 1;$$

thus Fh is a homomorphism from $FGFX$ to FX. Since X is a retraction of GFX, we can use h to define a monoid structure on X, and with this definition X^T is just the category Mon. In other words, the free monoid functor is monadic. In a similar way it may be shown that any free algebra functor is monadic.

As a second example, take F: Gp $\to$ Mon to be the forgetful functor and G: Mon → Gp its right adjoint, then $T = GF$: Gp → Gp consists in regarding a group as a monoid and then taking its group of units, i.e. the identity functor. Since Mon is not equivalent to Gp = Gp^T, this adjunction is not monadic.

It can be shown that any monad in St arises from an adjunction in a variety of Ω-algebras, provided that we interpret this term as including infinitary operations (cf. Manes [76], p. 66). In fact there is an abstract condition for monadicity in terms of coequalizers, the Beck tripleability theorem (Mac Lane [71], p. 147, Pareigis [69], p. 58, Manes [76], p. 165). It may be used e.g. to show that the forgetful functor from compact Hausdorff spaces to sets is monadic; its left adjoint is the Stone–Čech compactification (cf. Mac Lane [71], p. 153f.). On the other hand, the class of complete Boolean algebras can be regarded as a variety (allowing infinitary operations), but its forgetful functor is not monadic. This is reflected in the fact that the 'free complete Boolean algebra' of rank $\aleph_0$ say, does not exist. For a proof see Manes [76], p. 69; he shows there that a sufficient condition for monadicity is that the forgetful functor U: A → St be *tractable* at each cardinal ν, i.e. the class of all natural transformations from U^ν to U is a set. Now complete Boolean algebras are not tractable at $\aleph_0$. More surprisingly, complete lattices (with inf- and sup-preserving homomorphisms) are not tractable at 3. Intuitively this means that for any cardinal α we can define at least α ternary operations on a complete lattice, and for a suitable complete lattice these operations are distinct (cf. Hales [64], Manes [76], p. 68).

4. ALGEBRAIC THEORIES

In II.1 (p. 49) we defined an Ω-algebra as a set A together with a set of finitary operations on A indexed by Ω. But what matters, at least from a certain point of view, is not the set of operations corresponding to Ω but the subclone of $\mathcal{O}(A)$ generated by these operations. To obtain this subclone one can introduce the derived operators of Ω, essentially the elements of the Ω-word algebra (p. 145), or more generally, the elements of the free V-algebra in a given variety V of Ω-algebras, and now take a clone homomorphism to $\mathcal{O}(A)$.

The treatment sketched below, due to Lawvere [63], avoids the use of clones altogether. Let **N** be the category whose objects are all the non-negative integers, each occurring just once, while the morphisms $m \to n$

are the mappings from m to n, considered as finite sets. Thus we may regard $\mathbf{N}$ as a subcategory of St (it is the 'skeleton' of the category of finite sets and mappings). In $\mathbf{N}$ we have an operation of *coproduct* or *sum*, $m + n$, represented by the usual sum, i.e. the disjoint union of sets. In particular, each $n \in \mathbf{N}$ can be obtained as a sum of n terms: $n = 1 + 1 + \cdots + 1$. In any category $\mathscr{A}$ the coproduct $\amalg$, when it exists, satisfies the relation $\mathscr{A}(\amalg X_i, Y) \cong \Pi\mathscr{A}(X_i, Y)$, hence we have in $\mathbf{N}$,

$$\mathbf{N}(m, n) \cong \mathbf{N}(1, n)^m, \tag{1}$$

where of course $\mathbf{N}(1, n)$ consists of n injection maps. By an *algebraic theory* one understands a category $\mathscr{P}$ containing $\mathbf{N}$ as subcategory, with the same objects and coproduct maps as $\mathbf{N}$. Thus $\mathscr{P}$ differs from $\mathbf{N}$ only in having (possibly) more morphisms, but it will satisfy the analogue of (1):

$$\mathscr{P}(m, n) \cong \mathscr{P}(1, n)^m. \tag{2}$$

Now a $\mathscr{P}$-algebra is a contravariant functor F from $\mathscr{P}$ to St, which converts coproducts to products. Thus $Fn = A^n$, where A^n is the (set) product of n copies of $A = A^1$, and the $\mathscr{P}$-morphisms $1 \to n$ define mappings $A^n \to A$, i.e. n-ary operations on A, with the $\mathbf{N}$-morphisms being the unit operators $\delta_n^{(i)}$ (cf. p. 126). In this way F gives rise to the subclone of operations on A defined by the Ω-algebra structure.

E.g. for the theory of groups we need in $\mathscr{P}$ three morphisms, corresponding to the group operations, $\mu: 1 \to 2$, $\iota: 1 \to 1$ and $\varepsilon: 1 \to 0$. The category $\mathscr{P}$ will be generated (over $\mathbf{N}$) by these morphisms, and under the action of F the morphisms of $\mathscr{P}$ correspond to the derived group operations.

Now any identity in groups is represented by a pair of equal morphisms in $\mathscr{P}$, e.g. the associative law becomes $\mu(\mu + 1) = (1 + \mu)\mu$, where 1 is the identity morphism. This follows by drawing the appropriate diagram in $\mathscr{P}$ and applying our functor F.

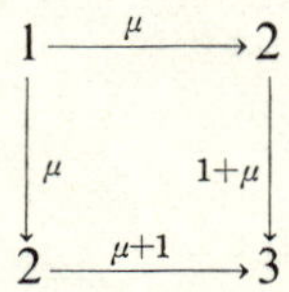

Given any algebraic theory $\mathscr{P}$, we can form the category Fun($\mathscr{P}^0$, St)

of all product-preserving functors from $\mathscr{P}^0$ (the opposite of $\mathscr{P}$) to St. Each object in this category is a functor F turning a certain set A_F into a $\mathscr{P}$-algebra, and it is not hard to see that the assignment $U: F \to A_F$ is a forgetful functor from Fun($\mathscr{P}^0$, St) to St. Now it may be shown that this functor U is monadic (cf. Pareigis [69], p. 101); its left adjoint provides the free $\mathscr{P}$-algebra on a given set.

Chapter IX

Model Theory and Universal Algebra

The relation between model theory and universal algebra has been summed up concisely by Chang and Keisler [73] in the equation

$$\text{universal algebra} + \text{logic} = \text{model theory}.$$

In this chapter we give some illustrations of the way in which the influence of model theory on our subject has grown in recent years; we shall be concerned particularly with various notions of algebraic closure. There are of course various other important aspects of the subject, such as stability theory (cf. Shelah [71′]), but this would require more tools from logic than we have at our disposal.

1. INDUCTIVE THEORIES

Given a class $\mathscr{X}$ of algebraic structures (in the sense of Chapter V), its theory Th($\mathscr{X}$) is the collection of all (first-order) sentences which hold in every $M \in \mathscr{X}$. If Σ is a collection of sentences, we write Mod(Σ) for the class of all models of Σ; thus for any collection Σ of sentences, Th(Mod(Σ)) is the collection of all sentences holding in all models of Σ.

Let M be any structure and M' a substructure of M, then the *diagram* of M with constants M', $\Delta(M, M')$, is the set of all open atomic and negated atomic formulae holding in M, with a constant c_p for each $p \in M'$, which is to be interpreted by p; we also put $\Delta_M = \Delta(M, M)$. The *full diagram* of M, Th(M), is the set of all sentences holding in M (with the elements of M as constants); thus the models of Th(M) are just the elementary extensions of M. For any set Σ of sentences and any sentence P we write $\Sigma \vdash P$ to indicate that P can be deduced from Σ. By the theory 'generated' by Σ we understand the set of all consequences of Σ.

The Gödel completeness theorem, which states that a set of formulae has a model if and only if f (= falsity, cf. p. 200) cannot be derived from the set, is essentially Th. V.5.4, p. 212. It means that to find out the consequences of a set of axioms we may proceed either syntactically, by testing which sentences are derivable, or semantically, by seeing which sentences hold in all models. However, it should be borne in mind that the models also contain information not expressible in terms of first-order sentences. An important consequence of the completeness theorem is the (Gödel–Malcev) compactness theorem (Cor. V.5.6, p. 213), which states that a set of formulae is consistent if and only if every finite subset is consistent.

In Chapter VI several results were proved giving syntactical descriptions of certain kinds of model classes, e.g. Prop. VI.1.1 or Th. VI.2.8. Below we give another such result, which will be used later. By a *universal-existential* sentence or briefly an $\forall\exists$-sentence we understand a sentence in prenex normal form in which the universal quantifiers precede all the existential ones. Intuitively an $\forall\exists$-sentence asserts the solubility of a given set of equations and inequations, possibly involving some free variables; when there are no free variables, we have an $\exists$-sentence. A theory $\mathscr{T}$ is said to be *inductive* if the class of all $\mathscr{T}$-models is inductive, i.e. the union of any chain of $\mathscr{T}$-models is again a $\mathscr{T}$-model.

***Theorem 1.1* (*Chang-Łos-Suszko*)**

A theory $\mathscr{T}$ is inductive if and only if it is equivalent to a set of $\forall\exists$-sentences.

Proof:

Suppose that $\mathscr{T}$ is equivalent to a set of $\forall\exists$-sentences, let (M_α) be a chain of $\mathscr{T}$-models and write $M = \bigcup M_\alpha$. The defining sentences of $\mathscr{T}$ may be taken to be of the form $S = \bigwedge_I \bigvee_J P(I, J)$, where $P(I, J)$ is an open formula in the variables $I = (i_1, \ldots, i_m)$ and $J = (j_1, \ldots, j_n)$. If $M \models \sim S$, then $\bigvee_I \bigwedge_J \sim P(I, J)$ holds in M, thus there exist $a_1, \ldots, a_m \in M$

such that $M \models \bigwedge_J \sim P(a_1, \ldots, a_m; J)$. Each a_μ lies in some M_β, and by taking the largest of them we obtain M_α to contain all the a_μ, but then, by the choice of S, $M_\alpha \models \bigvee_J P(a_1, \ldots, a_m; J)$, hence $M \models \bigvee_J P(a_1, \ldots, a_m; J)$, which is a contradiction.

Conversely, assume that $\mathscr{T}$ is inductive and let $\mathscr{T}'$ be the set of all consequences of $\mathscr{T}$ which have the form of ∀∃-sentences. Let M be a $\mathscr{T}'$-model and Δ' the set of all universal sentences holding in M (with elements of M as constants); we claim that $\mathscr{T} \cup \Delta'$ is consistent. For if this were not so, $\mathscr{T}$ would be inconsistent with a finite set of sentences of Δ', which are equivalent to some universal sentence holding in M:

(1) $$M \models \bigwedge_J P(a_1, \ldots, a_m; J),$$

where $a_1, \ldots, a_m$ are the constants in P representing elements of M. Since $a_1, \ldots, a_m$ do not occur in the sentences of $\mathscr{T}$, this inconsistency means that $\mathscr{T}$ implies $\sim \bigvee_I \bigwedge_J P(I, J)$ or equivalently,

$$\bigwedge_I \bigvee_J \sim P(I; J).$$

Since this is an ∀∃-sentence implied by $\mathscr{T}$, it is a member of $\mathscr{T}'$ and so is satisfied by M, in contradiction to (1).

Hence $\mathscr{T} \cup \Delta'$ is consistent, and it follows that it has a model N containing M. By definition of Δ', every existential sentence with constants from M holding in N also holds in M, so the diagram Δ_N of N is consistent with Th(M). By compactness, N has an extension M_1 which is an elementary extension of M. If we now repeat this construction with M_1 in place of M, and continue thus, we get a chain

(2) $$M \subseteq N \subseteq M_1 \subseteq N_1 \subseteq M_2 \subseteq \cdots$$

where each N_i is a $\mathscr{T}$-model, while the M_i form an elementary chain. The union is therefore both a $\mathscr{T}$-model and an elementary extension of M. It follows that M is a $\mathscr{T}$-model, so $\mathscr{T}$ is equivalent to $\mathscr{T}'$, and the conclusion follows. ∎

This type of argument, involving a chain of the form (2), is sometimes called an alternating chain argument. As a consequence of the proof we obtain

Corollary 1.2

For any theory $\mathscr{T}$, the theory generated by all ∀∃-sentences of $\mathscr{T}$ is the theory of all unions of chains of $\mathscr{T}$-models. ∎

2. COMPLETE THEORIES AND MODEL COMPLETE THEORIES

We recall that a theory $\mathscr{T}$ is called *complete* if it is maximal in the set of all theories (p. 212) or equivalently, if for every sentence P exactly one of $\mathscr{T} \vdash P, \mathscr{T} \vdash \sim P$ holds. Another equivalent condition is that $\mathscr{T}$ is the theory of a single structure, or also that all $\mathscr{T}$-models are elementarily equivalent. This last formulation shows that the notion of completeness is rather restrictive, and such concepts as that of an algebraically closed field cannot be axiomatized by a complete theory. We therefore introduce the following notion, due to A. Robinson. A class Σ of structures is said to be *model complete* if for any Σ-structures M, N such that $N \subseteq M$, we have $N \prec M$; thus every inclusion of Σ-structures is an elementary embedding. Now a theory $\mathscr{T}$ is called *model complete* when the class of all $\mathscr{T}$-models is so.

The two notions 'complete' and 'model complete' overlap, but neither includes the other. E.g. the theory of algebraically closed fields is model complete, but not complete (fields of different characteristics are not elementarily equivalent). On the other hand, Th(**Z**), where **Z** is considered as abelian group, is complete, by definition, but not model complete, since the enbedding $2\mathbf{Z} \subseteq \mathbf{Z}$ is clearly not elementary.

Let $\mathscr{T}$ be a model complete theory; given any chain of $\mathscr{T}$-models, their union is an elementary extension of each and hence is again a $\mathscr{T}$-model. By Th. 1.1 this shows the truth of

Proposition 2.1

Any model complete theory can be defined by a set of $\forall\exists$*-sentences.* ∎

In fact, 'model complete' may be thought of as 'complete with respect to every model'. This idea is made precise in part (b) of the next theorem, which lists some conditions for model completeness. We remark that a theory $\mathscr{T}$ which is model complete, remains so when constants are added to its language (this will not usually be true of a complete theory).

Theorem 2.2

For any theory $\mathscr{T}$, the following are equivalent:

(a) *$\mathscr{T}$ is model complete.*

(b) *For every $\mathscr{T}$-model M, the theory generated by $\mathscr{T} \cup \Delta_M$ is complete, where Δ_M is the diagram of M.*

(c) *If M, N are any $\mathscr{T}$-models such that $M \subseteq N$, then every existential sentence (with elements of M as constants) holding in N also holds in M.*

(d) *For every formula P there is a universal formula Q such that $\mathscr{T} \vdash P \Leftrightarrow Q$.*

Proof:

(a) $\Rightarrow$ (b). Assume (a) and let M be a $\mathscr{T}$-model. Since every $\mathscr{T}$-model extension of M is an elementary one, $\mathscr{T} \cup \Delta_M$ has the same models as Th(M), the theory of M with constants M, and so is complete.

(b) $\Rightarrow$ (c). Assume (b) and let M, N be $\mathscr{T}$-models such that $M \subseteq N$, then M, N are both models of the complete theory generated by $\mathscr{T} \cup \Delta_M$, hence any $\exists$-sentence holding in M also holds in N.

(c) $\Rightarrow$ (d). Suppose first that P is an existential sentence. Let Φ be the set of all universal sentences Q such that $\mathscr{T} \vdash P \Rightarrow Q$. Let M be a model of $\mathscr{T} \cup \Phi$, with diagram Δ_M. Every finite conjunction $S(a_1, \ldots, a_r)$ of formulae of Δ_M is consistent with $\mathscr{T} \cup \{P\}$, because $\bigwedge_I \sim S(i_1, \ldots, i_r)$ is false in M and so does not belong to Φ and is not a consequence of $\mathscr{T} \cup \{P\}$. It follows that $\mathscr{T} \cup \{P\} \cup \Delta_M$ has a model, necessarily an elementary extension of M, therefore P holds in M. Thus every model of $\mathscr{T} \cup \Phi$ is a model of P, hence $\mathscr{T} \cup \Phi \vdash P$. By compactness there exists $Q = Q_1 \wedge \cdots \wedge Q_n$, $Q_i \in \Phi$, such that $\mathscr{T} \vdash Q \Rightarrow P$, hence (by the definition of Φ) $\mathscr{T} \vdash P \Leftrightarrow Q$ and (d) is proved in this case. Next let $P(i_1, \ldots, i_n)$ be an existential formula; we adjoin constants $c_1, \ldots, c_n$ to the language of $\mathscr{T}$, then $\mathscr{T}$ is still model complete, and we obtain an existential sentence $P(c_1, \ldots, c_n)$. By what has been shown,

$$\mathscr{T} \vdash P(c_1, \ldots, c_n) \Leftrightarrow Q(c_1, \ldots, c_n)$$

for some universal formula $Q(i_1, \ldots, i_n)$. Replacing the c's by i's we find that $\mathscr{T} \vdash P \Leftrightarrow Q$.

Now let $P = F_1F_2 \cdots F_tP'$, where P' is an open formula and $F_1, \ldots, F_t$ are alternate groups of universal and existential quantifiers. We shall use induction on t; for $t = 0$ the result holds. If $t \geqslant 1$, $\mathscr{T} \vdash F_2 \cdots F_tP' \Leftrightarrow Q$ by induction, where Q is universal, hence $\mathscr{T} \vdash P \Leftrightarrow F_1Q$. Now either F_1 is universal, and the result follows, or F_1 is existential. In that case we can in this way prove $\mathscr{T} \vdash \sim P \Leftrightarrow Q$, where Q is universal, so $\mathscr{T} \vdash P \Leftrightarrow \sim Q$. Now $\sim Q$ is existential, so by the first part, $\mathscr{T} \vdash \sim Q \Leftrightarrow R$, where R is universal, thus $\mathscr{T} \vdash P \Leftrightarrow R$, as claimed.

(d) $\Rightarrow$ (a). Let M, N be $\mathscr{T}$-models such that $N \subseteq M$, take $a_1, \ldots, a_n$ in N and suppose that $M \models P(a_1, \ldots, a_n)$. By (d), $\mathscr{T} \vdash P \Leftrightarrow Q$, where Q is universal. Therefore $M \models Q(a_1, \ldots, a_n)$, hence $N \models Q(a_1, \ldots, a_n)$ and so

$N \models Q(a_1, \ldots, a_n)$. Since P was arbitrary, we have $N \prec M$ and it follows that $\mathscr{T}$ is model complete. ∎

A connexion with complete theories is given by the

Prime Model Test.

A model complete theory $\mathscr{T}$ is complete if there is a model M_0 of $\mathscr{T}$ which is contained in all $\mathscr{T}$-models.

For when such an M_0 exists, every $\mathscr{T}$-model is elementarily equivalent to M_0, so all $\mathscr{T}$-models are elementarily equivalent. ∎

3. MODEL COMPLETIONS

In practice we frequently need, not a model complete theory, but a model completion of a given theory. To define this notion, let Σ, Σ' be two classes of structures; we shall say that Σ' is *cofinal* in Σ if every Σ-structure is a substructure of some Σ'-structure. If Σ' is cofinal in Σ and Σ cofinal in Σ', we say that Σ and Σ' are *model consistent*. The same terms apply to theories, thus two theories $\mathscr{T}$, $\mathscr{T}'$ are model consistent if each $\mathscr{T}$-model is contained in a $\mathscr{T}'$-model and vice versa.

Let $\mathscr{T}$ be a theory, then a *model completion* of $\mathscr{T}$ is an extension theory $\mathscr{T}^*$ of $\mathscr{T}$ such that $\mathscr{T}^*$ is model consistent with $\mathscr{T}$ and any two $\mathscr{T}^*$-extensions of a $\mathscr{T}$-model M are elementarily equivalent, i.e. for any $M \models \mathscr{T}$, the theory generated by $\mathscr{T}^* \cup \Delta_M$ is complete. Since every $\mathscr{T}^*$-model is a $\mathscr{T}$-model, it is clear that the model completion of $\mathscr{T}$ is necessarily model complete. We shall see later that the model completion, if it exists, is unique.

Examples of model completions are: (1) $\mathscr{T}$ = fields, $\mathscr{T}^*$ = algebraically closed fields. (2) $\mathscr{T}$ = ordered fields, $\mathscr{T}^*$ = real closed ordered fields. (3) $\mathscr{T}$ = discretely valued fields, $\mathscr{T}^*$ = Henselian discretely valued fields. (4) In characteristic 0, $\mathscr{T}$ = differential fields, $\mathscr{T}^*$ = differentially closed fields (in characteristic $\neq 0$ one may take $\mathscr{T}$ = differentially perfect fields, cf. Wood [76]). If we examine how these completions are constructed, we find that e.g. in the case of fields one ensures that all consistent systems of equations can be solved; thus all existential sentences that are true in some extension are already true in the model itself. For fields the situation is simplified by the fact that the negation of an atomic sentence is itself a positive sentence, thus $\sim(f = g)$, i.e. $f \neq g$ is equivalent to $\bigvee_z [z(f - g) = 1]$.

We shall construct the model completion by forming extensions in which all existential sentences that can possibly be true are in fact true. Let M be a structure and M' a substructure of M; we shall say that M' is *existentially closed* in M, $M' \prec_\exists M$, if every existential sentence defined in M' and true in M is also true in M'. Given a class Σ of structures, if a Σ-structure M is existentially closed in every Σ-structure containing it, then we call M *existentially closed* in Σ.

To obtain existentially closed structures for Σ we need not assume that Σ is axiomatizable, but merely that it is inductive.

Proposition 3.1

Let Σ be any inductive class of structures, then each Σ-structure is contained in an existentially closed Σ-structure.

Proof:

Let $M \in \Sigma$ and let $\{P_\alpha \mid \alpha < \tau\}$ be the set of all existential sentences in the language Ω of M, where $\tau = \max\{|M|, |\Omega|, \aleph_0\}$. We define Σ-structures M_α as follows: $M_0 = M$; if $\alpha = \beta + 1$, and there exists $M' \supseteq M_\beta$ such that $M' \in \Sigma$ and $M' \models P_\beta$, then we put $M_{\beta+1} = M'$, otherwise $M'_{\beta+1} = M_\beta$; finally, if α is a limit ordinal, then $M_\alpha = \bigcup_{\beta<\alpha} M_\beta$. This lies in Σ because Σ is inductive.

Now put $M^1 = \bigcup_{\alpha<\tau} M_\alpha$, then $M^1 \in \Sigma$ and for any existential sentence P defined in M, if $N \models P$ for some extension N of M^1, it is already the case that $M^1 \models P$, by construction. We now repeat the construction on M^1 and obtain a countable chain

$$M \subseteq M^1 \subseteq M^2 \subseteq \cdots \quad .$$

Put $M^* = \bigcup_n M^n$, then $M^* \in \Sigma$ and any existential sentence P defined in M^* is defined in some M^n, because the number of terms occurring in P is finite. Now if $N \models P$ for some extension N of M^*, then by construction $M^{n+1} \models P$, hence $M^* \models P$, thus M^* is existentially closed in Σ. ∎

We denote by $\mathscr{E}_\Sigma$ the class of all existentially closed Σ-structures and remark that it may be uniquely described as follows:

Proposition 3.2

Let Σ be an inductive class and $\mathscr{E}_\Sigma$ the class of all existentially closed Σ-structures, then $\mathscr{E}_\Sigma$ is the unique class $\mathscr{C} \subseteq \Sigma$ with the properties:

(i) *$\mathscr{C}$ is cofinal in Σ*, (ii) $M \subseteq N,\ M, N \in \mathscr{C} \Rightarrow M \prec_\exists N$, (iii) *$\mathscr{C}$ is maximal subject to* (i), (ii).

For $\mathscr{E}_\Sigma$ clearly satisfies (i)–(iii); if $\mathscr{C}$ is a subclass of Σ satisfying (i)–(iii), then $\mathscr{C} \subseteq \mathscr{E}_\Sigma$. For let $M_0 \in \mathscr{C}$; since $\mathscr{C}$ and $\mathscr{E}_\Sigma$ are both cofinal in Σ, we can find a chain

$$M_0 \subseteq N_0 \subseteq M_1 \subseteq N_1 \subseteq \cdots$$

where $M_i \in \mathscr{C}$ and $N_i \in \mathscr{E}_\Sigma$. By (ii) the union of this chain is an elementary extension of both M_0 and N_0, so M_0 is elementarily equivalent to a member of $\mathscr{E}_\Sigma$ and hence itself belongs to $\mathscr{E}_\Sigma$. By (iii) we now deduce that $\mathscr{C} = \mathscr{E}_\Sigma$. ∎

We now add the assumption that Σ is axiomatizable. This means in effect that we have an inductive theory $\mathscr{T}$ such that $\Sigma = \text{Mod}(\mathscr{T})$. Then we have

Proposition 3.3

Let $\mathscr{T}$ be an inductive theory and suppose that $\mathscr{T}$ has a model completion $\mathscr{T}^*$, then a $\mathscr{T}$-model M is a $\mathscr{T}^*$-model if and only if it is existentially closed in every $\mathscr{T}$-model containing it.

Proof:

Let M be a $\mathscr{T}^*$-model, then for any $\mathscr{T}$-model N such that $M \subseteq N$, there is a $\mathscr{T}^*$-model N' such that $N \subseteq N'$. Thus $M \subseteq N'$, hence $M \prec N'$. Now any existential sentence holding in N also holds in N', hence in M, and this proves M to be existentially closed in N.

Conversely, let M be a $\mathscr{T}$-model and assume that whenever $M \subseteq N$ for some $N \in \text{Mod}(\mathscr{T})$, then $M \prec_\exists N$. Given $M \subseteq N_1 \subseteq N_2$, we have $M \prec_\exists N_1$, $M \prec_\exists N_2$, hence $N_1 \prec_\exists N_2$, thus $\mathscr{T} \cup \Delta_M$ satisfies (c) of Th. 2.2 and so the theory generated by $\mathscr{T} \cup \Delta_M$ is model complete. Now M may be embedded in a $\mathscr{T}^*$-model M_1; by model completeness $M \prec M_1$ and this shows M to be a $\mathscr{T}^*$-model, as claimed. ∎

4. THE FORCING COMPANION

Prop. 3.1 in conjunction with Prop. 3.3 allows us to form the model completion $\mathscr{T}^*$ of any inductive theory $\mathscr{T}$ having such a completion as the 'existential closure', but this fact is no guarantee that $\mathscr{T}^*$ exists. For example, if $\mathscr{T}$ is the theory of groups, we can form the class $\mathscr{E}_T$ of existentially closed groups, but this class is not axiomatizable, as we shall see later. As a matter of fact, for many algebraic purposes $\mathscr{E}_T$ fulfils the functions of an algebraic closure, even though it is not model complete.

But by using the method of forcing one can construct a smaller class which *is* model complete. This is the infinite forcing companion of A. Robinson; another form, called 'finite forcing' relates to finite sets of sentences (rather than to structures, as does the infinite kind. The original form, due to P. J. Cohen [63, 66] is essentially finite forcing). For more details we refer to Keisler's account in Morley [73] and for applications to set theory, to the chapter by J. P. Burgess in Barwise [77]. Since we shall not have to deal with finite forcing here, we shall frequently omit the qualifying adjective.

Let Σ be an inductive class of structures. An *infinite forcing companion* of Σ is a class Σ' such that

F.1. $\Sigma' \subseteq \Sigma$ and Σ, Σ' are model consistent,

F.2. Σ' is model complete,

F.3. Σ' is maximal, subject to **F.1, 2**.

Our aim will be to show that for any inductive class Σ of structures, every Σ-structure can be embedded in a member of Σ', where Σ' is a forcing companion. The proof parallels that of Prop. 3.1, but whereas in Prop. 3.1 we only had to deal with existential sentences, which by their nature are persistent (cf. p. 226), we now have to deal with any sentence, and so have to modify the notion of satisfaction: $M \models P$, so as to ensure persistence. This is accomplished by the notion of forcing, defined below. Given an inductive class Σ, let $M \in \Sigma$ and let P be any sentence defined in M, then we shall say that *M forces P*, in symbols: $M \Vdash P$, under the following circumstances:

1. If P is atomic, $M \Vdash P$ if and only if $M \models P$.
2. If $P = P_1 \vee P_2$, $M \Vdash P$ if and only if $M \Vdash P_1$ or $M \Vdash P_2$.
3. If $P = P_1 \wedge P_2$, $M \Vdash P$ if and only if $M \Vdash P_1$ and $M \Vdash P_2$.
4. If $P = \sim Q$, $M \Vdash P$ if and only if there is no extension M' of M in Σ such that $M' \Vdash Q$.
5. If $P = \bigvee_i Q(i)$, $M \Vdash P$ if and only if $M \Vdash Q(a)$ for some $a \in M$.
6. If $P = \bigwedge_i Q(i)$, $M \Vdash P$ if and only if $M \Vdash \sim \bigvee_i \sim Q(i)$.

We observe that forcing satisfies the following persistence condition: If $M \Vdash P$ and $M' \supseteq M$, $M' \in \Sigma$, then $M' \Vdash P$. Further we note that if $M \Vdash P$ then $M \Vdash \sim\sim P$, but not conversely, so the forcing concept does not respect logical equivalence. We also remark that for any P, M cannot force both P and $\sim P$, but it need not be the case that M forces either P or $\sim P$. *A* Σ-structure M is said to be (infinitely) *generic* if for each sentence P defined in M, either $M \Vdash P$ or $M \Vdash \sim P$. We can now prove the analogue to Prop. 3.1.

Theorem 4.1

Let Σ be an inductive class of structures, then every Σ-structure can be embedded in a generic Σ-structure.

The proof is word for word the same as that of Prop. 3.1, but taking all sentences instead of just existential ones, and replacing $\models$ by $\Vdash$. ∎

We shall also need the following alternative description of generic Σ-structures.

Proposition 4.2

Let Σ be an inductive class of structures and $M \in \Sigma$, then the following conditions are equivalent:

(a) *For each sentence P defined in M, $M \models P$ if and only if $M \Vdash P$,*

(b) *For each sentence of the form $\sim Q$ defined in M, $M \models \sim Q$ if and only if $M \Vdash \sim Q$,*

(c) *M is generic in Σ.*

Proof:

(a) $\Rightarrow$ (b), (c). Suppose that (a) holds, then clearly (b) holds too; further, for any P, either $M \models P$ or $M \models \sim P$, hence $M \Vdash P$ or $M \Vdash \sim P$, so M is generic, i.e. (c).

To prove that each of (b), (c) implies (a), we first note that the inductive definition of $M \Vdash P$ agrees with that of $M \models P$ for atomic formulae, conjunctions, disjunctions, existential and universal quantifications; hence the class of sentences P for which (a) holds if M satisfies (b) or (c), contains all atomic formulae and is closed under the above four operations. It only remains to prove its closure under negation. Let $P = \sim Q$, then in case (b), $M \models P \Leftrightarrow M \Vdash P$ by hypothesis, so assume (c). Then $M \models P \Leftrightarrow$ not $M \models Q \Leftrightarrow$ not $M \Vdash Q$; since M is generic, this holds if and only if $M \Vdash \sim Q$, i.e. $M \Vdash P$. ∎

We are now in a position to describe the forcing companion.

Theorem 4.3

Every inductive class Σ has a unique infinite forcing companion $\mathcal{G}_\Sigma$, and $\mathcal{G}_\Sigma$ is the class of all infinitely generic Σ-structures.

Proof:

We first prove the uniqueness. Let Σ be an inductive class and $\mathcal{G}$, $\mathcal{G}'$

any classes satisfying **F.1–2**; we claim that $\mathscr{G} \cup \mathscr{G}'$ is model complete. Clearly $\mathscr{G}$ and $\mathscr{G}'$ are mutually model consistent; now consider $M \in \mathscr{G}$, $N \in \mathscr{G}'$ such that $M \subseteq N$. We can find $M_1 \in \mathscr{G}$ such that $N \subseteq M_1$ and $N_1 \in \mathscr{G}'$ such that $M_1 \subseteq N_1$. Continuing this process, we obtain a chain

$$M \subseteq N \subseteq M_1 \subseteq N_1 \subseteq \cdots$$

Since the M_i form a chain of elementary extensions, the union $L = \bigcup M_i = \bigcup N_i$ is an elementary extension of M, likewise of N, therefore $M \prec N$ and this shows $\mathscr{G} \cup \mathscr{G}'$ to be model complete.

Now let $\mathscr{F}$ be a forcing companion of Σ and $\mathscr{G}$ any class satisfying **F.1–2**, then $\mathscr{F} \cup \mathscr{G}$ is model complete, by what has been shown, and it is clearly model consistent with Σ, hence by the maximality of $\mathscr{F}$, $\mathscr{F} = \mathscr{F} \cup \mathscr{G}$, i.e. $\mathscr{G} \subseteq \mathscr{F}$. Thus $\mathscr{F}$ is uniquely determined as the largest class satisfying **F.1, 2**.

Let $\mathscr{G} = \mathscr{G}_\Sigma$ be the class of all generic Σ-models, then $\mathscr{G} \subseteq \Sigma$ and by Th. 4.1, $\mathscr{G}$ is model consistent with Σ. If $M \subseteq N$ in $\mathscr{G}$, then for any sentence P, if $M \models P$, then $M \Vdash P$, hence $N \Vdash P$ (by persistence) so $N \models P$, and this shows $M \prec N$. Thus $\mathscr{G}$ is model complete, and so it satisfies **F.1–2**. To prove **F.3**, assume that $\mathscr{G} \subseteq \mathscr{G}'$, where $\mathscr{G}'$ satisfies **F.1–2**, and take $M \in \mathscr{G}'$. We have to show that $M \in \mathscr{G}$ and by Prop. 4.2(b) it will be enough to show that $M \models {\sim}Q \Leftrightarrow M \Vdash {\sim}Q$, for any sentence Q defined in M. Take Q and suppose that $M \Vdash {\sim}Q$; there exists M' in $\mathscr{G}$ extending M, hence $M' \Vdash {\sim}Q$ and so $M' \models {\sim}Q$, and since $M \prec M'$ (by model completeness of $\mathscr{G}'$) we have $M \models {\sim}Q$. Next suppose that M does not force ${\sim}Q$, thus there exists an extension M' of M in Σ with $M' \Vdash Q$. We take an extension M'' of M' in $\mathscr{G}$, then $M'' \Vdash Q$, hence $M'' \models Q$. As before, $M \prec M''$, therefore $M \models Q$, so it is not true that $M \models {\sim}Q$. This shows that $M \in \mathscr{G}$, so $\mathscr{G}$ is indeed maximal, and hence is the forcing companion of Σ. ∎

5. THE MODEL COMPANION

The last result (Th. 4.3) shows in particular that every inductive theory $\mathscr{T}$ has a forcing companion Mod($\mathscr{T}$)*. Moreover, if $\mathscr{T}$ has a model completion $\mathscr{T}^*$, then the forcing companion is in fact Mod($\mathscr{T}^*$), by Prop. 3.2 and 3.3. But even when there is no model completion of $\mathscr{T}$, it may happen that the forcing companion is axiomatizable. To describe this case, let us define the *model companion* of a theory $\mathscr{T}$ as a theory $\mathscr{T}^*$

extending $\mathcal{T}$ such that $\mathcal{T}^*$ is model consistent with $\mathcal{T}$ and model complete.

Proposition 5.1

An inductive theory $\mathcal{T}$ has a model companion $\mathcal{T}^*$ precisely when its forcing companion $\mathcal{G}$ is axiomatizable, and then $\mathcal{T}^* = \mathrm{Th}(\mathcal{G})$, $\mathcal{G} = \mathrm{Mod}(\mathcal{T}^*)$.

Proof:

Let $\mathcal{G}$ be the forcing companion of $\mathcal{T}$; if $\mathcal{G}$ is axiomatizable with theory $\mathcal{T}^*$, then by translating **F.1–3** we see that $\mathcal{T}^*$ is a model companion. Conversely, assume that $\mathcal{T}$ has a model companion $\mathcal{T}^*$ and put $\mathcal{F} = \mathrm{Mod}(\mathcal{T}^*)$, then $\mathcal{F}$ clearly satisfies **F.1–2** relative to $\mathrm{Mod}(\mathcal{T})$. We claim that it also satisfies **F.3**. For let $\mathcal{F}'$ be a class satisfying **F.1–2** and such that $\mathcal{F}' \supseteq \mathcal{F}$. Given $M \in \mathcal{F}'$, we can find a $\mathcal{T}^*$-model N such that $M \subseteq N$, then $N \in \mathcal{F}'$ and since $\mathcal{F}'$ is model complete, $M \prec N$, therefore M is also a $\mathcal{T}^*$-model, i.e. $\mathcal{F}' = \mathcal{F} = \mathrm{Mod}(\mathcal{T}^*)$; this shows $\mathcal{F}$ to be the forcing companion and $\mathcal{T}^* = \mathrm{Th}(\mathcal{F})$. ∎

By Th. 4.3 this also shows that the model companion, with it exists, is unique. A theory with a model companion is also called *companionable.*

To describe the connexion between model completion and model companion we introduce the following notions. A class Σ of structures is said to have the *joint embedding property* if any two Σ-structures have a common extension in Σ. A related property is the *amalgamation property* which is said to hold if any two Σ-structures with a common Σ-substructure M have a common extension over M in Σ (cf. the notion of 'directed class', p. 259). A theory $\mathcal{T}$ is said to have the joint embedding (or amalgamation) property when this is true of $\mathrm{Mod}(\mathcal{T})$.

Theorem 5.2

Let $\mathcal{T}$ be a theory and $\mathcal{T}^*$ an extension theory of $\mathcal{T}$ which is a model companion for $\mathcal{T}$, then

(i) *$\mathcal{T}^*$ is a complete theory if and only if $\mathcal{T}$ has the joint embedding property,*

(ii) *$\mathcal{T}^*$ is a model completion for $\mathcal{T}$ if and only if $\mathcal{T}$ has the amalgamation property.*

Proof:

(i) Assume that $\mathcal{T}^*$ is complete and let M_1, M_2 be $\mathcal{T}$-models. Since

$\mathscr{T}^*$ is model consistent with $\mathscr{T}$, we can find a $\mathscr{T}^*$-model M_i' extending M_i. Now the theories generated by both $\mathscr{T}^* \cup \Delta(M_1', M_1)$ and $\mathscr{T}^* \cup \Delta(M_2', M_2)$ contain the complete theory $\mathscr{T}^*$; thus the theory generated by $\mathscr{T}^* \cup \Delta(M_1', M_1) \cup \Delta(M_2', M_2)$ has a model M say, and since $\mathscr{T}^*$ extends $\mathscr{T}$, M is a $\mathscr{T}$-model and $M \supseteq M_i$.

Conversely, if $\mathscr{T}$ has the joint embedding property, let M_1, M_2 be any $\mathscr{T}^*$-models, then the M_i are also $\mathscr{T}$-models and by the joint embedding property there exists a $\mathscr{T}$-model M extending M_1, M_2. Now $\mathscr{T}^*$ is model consistent with $\mathscr{T}$, hence M is contained in a $\mathscr{T}^*$-model M'. By the model completeness of $\mathscr{T}^*$ we have $M_i \prec M'$, hence M_1 and M_2 are elementarily equivalent, and it follows that $\mathscr{T}^*$ is complete.

To prove (ii) we take a fixed $\mathscr{T}$-model M and apply (i) to the theory generated by $\mathscr{T} \cup \Delta_M$. ∎

The different concepts introduced in this chapter may be combined in a table as follows.

class of structures	axiomatizable case
inductive class Σ	model class defined by a set of $\forall\exists$-sentences
class of existentially closed Σ-structures	model completion
class of infinitely generic Σ-structures (forcing companion)	model companion

6. EXAMPLES

We have seen examples of theories with a model completion. As an example of a theory with a model companion but no model completion we may take the theory of formally real fields. Its model companion is the theory of real closed fields. To show that it lacks a model completion, take the rational numbers $\mathbf{Q}$ and form the quadratic extension $\mathbf{Q}(\gamma)$ generated by a root of the equation $x^2 = 2$. $\mathbf{Q}(\gamma)$ is formally real and we can order γ so that either (i) $\gamma > 0$, or (ii) $\gamma < 0$. Accordingly we obtain a real closed field containing $\mathbf{Q}(\gamma)$ in which γ has a square root or $-\gamma$ has a square root. Hence $\mathbf{Q}(\gamma)$ has two real closures which are not elementarily equivalent, since the sentence $\bigvee_x(x^2 = \gamma)$ holds in (i) but not in (ii).

The theory of groups forms an example of a non-companionable theory. This is most easily seen by showing that the class of all generic groups does not admit ultraproducts (and so cannot be axiomatizable, cf. Prop. 5.1; the model companion would be a model completion, because the amalgamation property holds). Let G_n for $n = 2, 3, \ldots$ be a generic group with elements x_n, y_n of orders n, $2n$ respectively, and take an ultraproduct G of the G_n with a non-principal ultrafilter, then $x = \overline{(x_n)}$ and $y = \overline{(y_n)}$ both have infinite order, hence are conjugate in some extension group (by a theorem of Higman–Neumann–Neumann [49]): $x = z^{-1}yz$ for some z. If G were generic, x and y would be conjugate in G, hence x_n, y_n would be conjugate in G_n for infinitely many n, which is impossible, because they have different orders. We remark that the same proof shows that the class of all existentially closed groups is not axiomatizable.

For many purposes existentially closed groups are adequate, but sometimes generic groups are needed, especially in dealing with logical questions. Thus the finite forcing companion (a more delicate tool than the infinite sort) has been used by A. Macintyre [72] to show that a finitely generated group has a solvable word problem if and only if it can be embedded in every existentially closed group (the necessity of the condition was proved by B. H. Neumann [73]). Thus a precise connexion is established between an algorithmic and a purely algebraic property.

Another example of a non-companionable theory is provided by skew fields. Let $D(k)$ be the theory of skew fields which are k-algebras, for a given field k. We can find generic skew fields K_n with elements x_n, y_n that are algebraic of degrees n, $2n$ respectively over k. Now the same argument as before can be applied to obtain an ultraproduct of the K_n which is not generic (cf. Cohn [71]). Since $D(k)$ is inductive, there is again an infinite forcing companion, but the theory is less well developed than for groups, partly because so far there is no analogue of Higman's theorem (Higman [61], cf. also Macintyre [79]).

The theory of abelian groups has as model completion the theory of divisible abelian groups with infinitely many elements of order p, for each prime p. For any ring R, the theory of right R-modules has a model completion if and only if R is coherent (cf. Eklof–Sabbagh [71]).

Finally we remark that the theory of commutative rings is not companionable. To see this we require the following.

Lemma 6.1

Let R be a commutative ring and let $a \in R$ be non-nilpotent, then R can be embedded in a ring R' with an element b such that ab is a non-zero idempotent.

Proof:

Form the polynomial ring $R[x]$ and write $f = (ax)^2 - ax$; if (f) is the ideal generated by f, then $(f) \cap R = 0$, because the elements of (f) all have zero constant term, hence on writing $R' = R[x]/(f)$, we have an embedding $R \to R'$, and if b is the residue class of x, then ab is an idempotent in R'. Suppose that $ab = 0$, then in $R[x]$ we have an equation

$$ax = ax(1 - ax)(b_0x^n + \cdots + b_n).$$

Multiplying both sides by $1 + ax + a^2x^2 + \cdots + a^Nx^N (N > n)$, we obtain

$$ax(1 + ax + \cdots + a^Nx^N) = ax(1 - a^{N+1}x^{N+1})(b_0x^n + \cdots + b_n).$$

On comparing coefficients of x^{n+2} we find $a^{n+2} = 0$, a contradiction. Hence $ab \neq 0$ as claimed. ∎

Now let R_n be a generic commutative ring with an element a_n such that $a_n^n = 0$, $a_n^{n-1} \neq 0$, and form the ultraproduct R of the R_n with a non-principal ultrafilter. In R we have $a = \overline{(a_n)}$, which is not nilpotent, hence there is an extension R' of R with an element b such that ab is a non-zero idempotent. If R were generic we could find $b_0 \in R$ such that ab_0 is a non-zero idempotent, hence for infinitely many n there exist $b_n \in R_n$ such that a_nb_n is a non-zero idempotent, but $a_n^n = 0$, hence $a_nb_n = (a_nb_n)^n = 0$, a contradiction.

It can be shown that the class of all commutative rings without nilpotent elements $\neq 0$ has a model companion, namely the class of all commutative regular rings whose Boolean algebra of idempotents is atomless and each of whose prime factors is an algebraically closed field (Lipshitz–Saracino [73], Carson [73]).

Chapter X

Miscellaneous Further Results

In this chapter we present a number of brief remarks or new results which have some relevance to the main text.

1. SUBDIRECT PRODUCTS AND PULLBACKS

Let $\mathscr{A}$ be a category; given two $\mathscr{A}$-morphisms f, g with the same target, we consider the different ways of completing them to a commutative

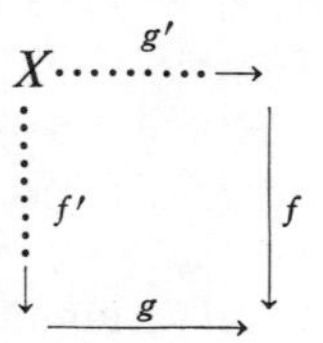

square. By a *pullback* one understands a triple (X, f', g'), where f', g' have the same source X and form with f, g a square as shown: $f'g = g'f$, such that any other square, $f''g = g''f$, has the property that there is a unique morphism h such that $f'' = hf'$, $g'' = hg'$. Frequently we refer simply to X as the pullback, leaving f', g' to be understood. We may regard all squares completing the pair (f, g) as the objects of a category, the maps being $\mathscr{A}$-morphisms between the new vertices so as to obtain a wedge with commutative faces. Then a pullback is just a final object in the category of squares. Dually we can define the *pushout* as a square completing a pair of morphisms with the same source, and initial for all such completions.

To give an example, for Ω-algebras the pushout is the free composition (p. 142), while the pullback is the subdirect product. We consider the latter case in a little more detail.

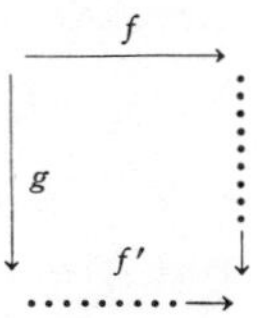

Given three Ω-algebras A, B, C with homomorphisms f: $A \to C$, g: $B \to C$, we can combine f, g to a map(f, g): $A \times B \to C$, and it is easily verified that the pullback of f and g is the subalgebra

$$P = \{(x, y) \in A \times B \mid xf = yg\}.$$

of $A \times B$. Moreover, it is clear that P is a subdirect product of A and B whenever both f and g are surjective; in fact we may intuitively view a pullback of algebras as the part of a direct product in which two quotients (viz. C above) agree. The question now arises to what extent every subdirect product can be described as a pullback. We give a set of conditions, due to Fleischer [55], for this to happen.

Theorem 1.1

Let A be an Ω-algebra with congruences $\mathfrak{q}_1$, $\mathfrak{q}_2$ such that $\mathfrak{q}_1 \cap \mathfrak{q}_2 = \Delta_A$ (the diagonal on A), and write $B_i = A/\mathfrak{q}_i$, so that A is a subdirect product of B_1 and B_2 (p. 99). *Then A with the natural homomorphisms g_i: $A \to B_i$ is a pullback of homomorphisms f_i: $B_i \to C$ for a suitable algebra C if and only if $\mathfrak{q}_1 \circ \mathfrak{q}_2 = \mathfrak{q}_2 \circ \mathfrak{q}_1$.*

Proof:

Suppose first that A is a pullback, thus

$$A \cong A' = \{(u, v) \in B_1 \times B_2 \mid uf_1 = vf_2\}.$$

Let x, $y \in A$ be such that $xg_1f_1 = yg_2f_2$, then $(xg_1, yg_2) \in A'$, but the elements of A' are all of the form (zg_1, zg_2), for some $z \in A$, hence $xg_1 =$

$zg_1, yg_2 = zg_2$, i.e. $(x, y) \in \mathfrak{q}_1 \circ \mathfrak{q}_2$. This shows that ker $g_1f_1 \subseteq \mathfrak{q}_1 \circ \mathfrak{q}_2$ and the reverse inclusion is obvious, hence (by II. 6.6, p. 89), ker $g_1f_1 = \mathfrak{q}_1 \circ \mathfrak{q}_2 = \mathfrak{q}_2 \circ \mathfrak{q}_1$.

Conversely, assume that $\mathfrak{q}_1 \circ \mathfrak{q}_2 = \mathfrak{q}_2 \circ \mathfrak{q}_1$, put $C = A/(\mathfrak{q}_1 \circ \mathfrak{q}_2)$ and let $f_i\colon B_i \to C$ be the natural homomorphisms. Clearly the natural maps $g_i\colon A \to B_i$ are such that $g_1f_1 = g_2f_2$. Suppose now that $uf_1 = vf_2$; since the g_i are surjective, there exist $x, y \in A$ such that $u = xg_1$, $v = yg_2$, hence $(x, y) \in \ker g_1f_1 = \mathfrak{q}_1 \circ \mathfrak{q}_2$. Let z be such that $(x, z) \in \mathfrak{q}_1$, $(z, y) \in \mathfrak{q}_2$, then $(u, v) = (xg_1, yg_2) = (zg_1, zg_2)$, therefore A is the pullback of f_1, f_2. ∎

Since all congruences on a group commute (p. 90), we have

Corollary 1.2.

Every subdirect product of two groups may be represented as a pullback. ∎

2. THE REDUCTION TO BINARY OPERATIONS

The estimates on p. 148 show that the number of n-ary operations on a finite set grows rapidly with n. Nevertheless it is possible to express all finitary operations in terms of binary ones, by allowing sufficiently long concatenations of operators. This fact is due to Sierpiński [45] and we reproduce his proof below.

Let A be a set with more than one element and denote the subclone of $\mathcal{O}(A)$ generated by all the binary operations on A by S. Given $a, b, a_1, \ldots, a_n \in A$, where $a \neq b$, we define the *discriminator* at $(a_1, \ldots, a_n)$ as the function

$$d_n(x_1, \ldots, x_n) = \begin{cases} a & \text{if } \ x_i = a_i \ (i = 1, \ldots, n), \\ b & \text{otherwise.} \end{cases} \tag{1}$$

We claim that $d_n \in S$; for $n \leq 2$ this is clear. When $n > 2$, define f by the equations

$$f(x, y) = \begin{cases} a & \text{if } \ (x, y) = (a, a_n), \\ b & \text{otherwise,} \end{cases}$$

and put $d_n(x_1, \ldots, x_n) = f(d_{n-1}(x_1, \ldots, x_{n-1}), x_n)$. If d_{n-1} satisfies (1), then so does d_n, defined in this way, and now it follows by induction on n that $d_n \in S$.

Theorem 2.1

Let A be any finite set, then every finitary operation on A may be expressed in terms of binary ones.

Proof:

Write again S for the subclone generated by the binary operations. When $|A| \leq 1$, there is nothing to prove, so assume that $|A| \geq 2$ and let f, g be n-ary operations whose values differ for at most one set of arguments $(a_1, \ldots, a_n)$, and denote by d_n the discriminator (1) for this n-tuple $(a_1, \ldots, a_n)$ and any $a \neq b$. Let φ be the binary operation given by

$$\varphi(x, y) = \begin{cases} g(a_1, \ldots, a_n) & \text{if} \quad x = a, \\ y & \text{otherwise.} \end{cases}$$

Clearly $\varphi \in S$ and $g(x_1, \ldots, x_n) = \varphi(d_n, f)$, therefore if f is in S, so is g. Hence every n-ary operation g differing from an n-ary operation f in S in just one point belongs itself to S. By induction, any g differing from $f \in S$ in a finite number of points lies in S. But A^n is itself finite, therefore every n-ary operation lies in S, so $\mathcal{O}(A) = S$, as claimed. ∎

3. INVARIANCE OF THE RANK OF FREE ALGEBRAS

In Theorem III.5.6, p. 140, conditions were given for the free algebras (of a given category of Ω-algebras) of different ranks to be non-isomorphic. Let us briefly consider what happens without such conditions. In any category $\mathscr{K}$ the free algebra of rank n, F_n may be regarded as the free composition, also called coproduct, of n copies of F_1. If $\mathbf{N}$ is the category of natural numbers with the sum of numbers as coproduct (as in VIII.4), then we have a coproduct-preserving functor from $\mathbf{N}$ to $\mathscr{K}$, given by $f\colon n \mapsto F_n$. Essentially this is a semigroup homomorphism and the rank of free algebras is unique precisely when f is injective on objects. In general denote by $\mathfrak{q}$ the kernel of f. These kernels are certain congruences on the free cyclic semigroup, and we begin by describing the latter.

Lemma 3.1

Let $\mathbf{N}$ *be the semigroup of non-negative integers under addition, then any congruence on* $\mathbf{N}$ *is either*

(i) $\Delta_{\mathbf{N}}$, *the diagonal on* $\mathbf{N}$, *or*

(ii) $\mathbf{N}^2$, *the universal congruence on* $\mathbf{N}$, *or*

(iii) *the congruence* $\mathfrak{q}_{r,d}$ *generated by* $(r, r + d)$, *where* $r \geq 0$, $d > 1$. *Moreover,* $I_{r,d} = \mathbf{N}/\mathfrak{q}_{r,d}$ *is the cyclic semigroup defined by*

(1) $m \equiv n$ *if and only if* $m = n$ *or* $m \geq r$, $n \geq r$ *and* $d \mid m - n$.

Proof:

It is clear that each $\mathfrak{q}_{r,d}$ just describes the semigroup $I_{r,d}$ as in (1), and all are distinct from each other and from $\Delta_{\mathbf{N}}$ and $\mathbf{N}^2$ (note that $\mathfrak{q}_{0,1} = \mathbf{N}^2$). Now let $\mathfrak{q}$ be a congruence on $\mathbf{N}$ different from $\mathbf{N}^2$, $\Delta_{\mathbf{N}}$, and denote by $(r, r + d)$ the lexicographically least element of $\mathfrak{q}\backslash\Delta_{\mathbf{N}}$. Clearly $\mathfrak{q} \supseteq \mathfrak{q}_{r,d}$ and if the inequality is strict, we can choose $(r + h, r + h + d') \in \mathfrak{q}\backslash\mathfrak{q}_{r,d}$, where $h \geq 0$, $d \nmid d'$. Let us take the lexicographically least such element; if $d' < d$, take m such that $md \geq h$, then (mod $\mathfrak{q}$) we have $r \equiv r + md \equiv r + h + (md - h) \equiv r + h + d' + (md - h) \equiv r + d'$, and this contradicts the minimality of (r, d). If $d' \geq d$, then (mod $\mathfrak{q}$) $r + h \equiv r + h + d' \equiv (r + d) + h + (d' - d) \equiv r + h + (d' - d)$ and this contradicts the minimality of $(r + h, r + h + d')$. Hence $\mathfrak{q} = \mathfrak{q}_{r,d}$ as claimed. ▮

Returning to the homomorphism $f: \mathbf{N} \rightarrow F$, denote by $\mathfrak{q}$ the kernel of f, then the semigroup of free $\mathscr{K}$-algebras is isomorphic to $\mathbf{N}/\mathfrak{q}$. Hence using Lemma 3.1, we obtain the following result, due to Goetz–Ryll-Nardzewski [60]:

Theorem 3.2

Let $\mathscr{K}$ *be a category of* Ω*-algebras with free algebras, and denote by* F_n *the free* $\mathscr{K}$*-algebra on n free generators, then there are three possibilities:*

(i) $F_m \cong F_n$ *if and only if* $m = n$; F_n *has unique rank,*
(ii) $F_m \cong F_n$ *for all* m, n; $\mathscr{K}$ *is trivial,*
(iii) *there exist* $r \geq 1$ *and* $d > 1$ *such that* $F_m \cong F_n$ *if and only if either* $m = n$ *or* $m, n \geq r$ *and* $d \mid m - n$.

The proof follows by examining the kernel of the map $n \mapsto F_n$; when all the free algebras are isomorphic, then $F_n = F_0$ is trivial, hence every $\mathscr{K}$-algebra is trivial and so $\mathscr{K}$ is then trivial. Conversely, when $\mathscr{K}$ is trivial, (ii) must hold. ▮

Consider again the case (iii); we have $F_r \cong F_s$, where $s = r + d$, and this means that there are r derived s-ary operators $\alpha_1, \ldots, \alpha_r$ and s derived

r-ary operators $\beta_1, \ldots, \beta_s$ such that, on writing $x = (x_1, \ldots, x_r)$, $y = (y_1, \ldots, y_s)$, we have

$$(2) \qquad (x\beta_1, x\beta_2, \ldots, x\beta_s)\,\alpha_i = x_i, \qquad (y\alpha_1, y\alpha_2, \ldots, y\alpha_r)\,\beta_j = y_j,$$

identically in the x's and y's.

By considering algebras with s r-ary operators α_i and r s-ary operators β_j satisfying the laws (2) it can be shown that all the types enumerated in Th. 3.2 are actually realized (Clark [69], cf. also Ex. III. 5.5, p. 142).

4. THE DIAMOND LEMMA FOR RINGS

The diamond lemma (I.4.9, p. 25) is frequently used in normal form proofs, and a general situation in which it can be applied is described in Th. III.9.3, p. 159, but as G. M. Bergman [78] has pointed out, specific instances of the lemma often involve subtleties which are not apparent from this general form. In particular he has worked out the application to rings and linear algebras in some detail. We shall describe the result here and refer for the proof (and many illustrations, motivating remarks and applications) to Bergman's original paper.

Let X be a set, $\langle X \rangle$ the free monoid (= semigroup with 1) on X and $F = k\langle X \rangle$ the free associative algebra on X over a commutative coefficient ring k. Further, let S be an indexed subset of $\langle X \rangle \times F$; S is called a *reduction system*. For each reduction pair $\sigma = (w, f) \in S$ and for any monomials u, v we define a *reduction* $r_{u\sigma v}$ as the k-module endomorphism of F which maps uwv to ufv and leaves all monomials $\neq uwv$ unchanged. An element c of F is called *reduction-finite* if in any infinite chain of successive reductions only finitely many produce a change; if all the irreducible values so obtained by applying chains of reductions to c are the same, c is called *reduction-unique*.

In applying the diamond lemma to rings we have to deal with two kinds of ambiguity: Let u, v, w be monomials, then in reducing uvw, we have an *overlap ambiguity* if we can reduce either uv or vw, while an *inclusion ambiguity* arises if we can reduce either v or uvw. If the two expressions obtained by reductions of uvw can be reduced to the same element of F, the ambiguity is said to be *resolvable*. Now Bergman proves

Theorem 4.1

Let $F = k\langle X \rangle$ *be the free k-algebra on* X *and let* $S = \{(w_\sigma, f_\sigma)\}$ *be*

a reduction system for F, and suppose that the monoid $\langle X\rangle$ *has a (partial) ordering compatible with the monoid structure and satisfying the descending chain condition, and such that* f_σ *is a linear combination of monomials* $< w_\sigma$, *for each* σ. *Then every element of F is reduction-finite and the following conditions are equivalent*:

(a) *All ambiguities of S are resolvable.*

(b) *All elements of F are reduction-unique under S.*

(c) *If I is the ideal of F generated by all* $w_\sigma - f_\sigma$, *then a transversal in F for the algebra F/I is given by the k-submodule spanned by the S-irreducible elements of* $\langle X\rangle$.

This result (whose proof, with the help of the diamond lemma, is quite straightforward), makes explicit the principle used e.g. in proving the Birkhoff–Witt theorem (cf. the proof in Birkhoff [37]).

5. THE EMBEDDING OF RINGS IN SKEW FIELDS

On p. 277 the problem was mentioned of finding a ring which is not embeddable in a skew field, but whose non-zero elements form a monoid embeddable in a group. This problem, which was first raised by Malcev in the 1930's, was solved simultaneously and independently by three people in 1966: L.A. Bokut' [69], A. J. Bowtell [67], and A. A. Klein [67]. A little later a criterion was found for any ring to be embeddable in a skew field (Cohn [71″]) which, together with a method of construction described by Cohn [69] made it relatively straightforward to give examples answering Malcev's problem. We shall give a brief outline which clarifies the difference between 'embeddability in groups' and 'embeddability in skew fields'.

Let R be any ring. A relation

$$x_1 y_1 + \cdots + x_n y_n = 0 \quad (x_i,\, y_i \in R) \tag{1}$$

is called *trivial* if for each $i = 1, \ldots, n$, either $x_i = 0$ or $y_i = 0$. Given a relation (1), which we shall write in vector form as $x \cdot y = 0$, if there exists an invertible $n \times n$ matrix P over R such that the relation $xP \cdot P^{-1}y = 0$ is trivial, then (1) is called *trivializable*. We shall refer to (1) as an *n-term relation*. A non-zero ring in which all m-term relations for $m \leq n$ are trivializable is called an *n-fir*; if *all* relations are trivializable, the ring is called a *semifir* (= semi-free ideal ring, for such a ring is also characterized by the fact that all its finitely generated left, or equivalently, right ideals are free, of unique rank). By an *atom*, or irreducible element, in a ring we understand a non-unit which cannot be written as a product of two non-units, and a ring in which every non-zero element is either a unit or a product of atoms is called *atomic*. Now the following results hold, where $R^* = R\backslash\{0\}$:

F.1. If R is an atomic 2-fir, then R^* is embeddable in a group (Cohn [71]),

F.2. For every $n = 1, 2, \ldots$ there exists an atomic n-fir which is not embeddable in a skew field (Cohn [69]),

F.3. Every semifir is embeddable in a skew field (Cohn [71″]).

If we compare **F.1** with **F.3**, we see that what matters for embeddability in a group is the trivializability of 2-term relations, whereas the embeddability in a skew field involves n-term conditions for every n (although the sufficient condition in **F.3** is of course not necessary).

By **F.2**, we can find an atomic 2-fir not embeddable in a skew field, but its multiplicative monoid R^* is embeddable in a group, by **F.1**, and this provides a solution of Malcev's problem. The examples by Bowtell and Klein are similar in principle to this one. By contrast Bokut's example is quite different; his construction is more complicated, but the example he obtains is of a monoid ring; this could not have been found by the above methods.

It follows from general considerations that the class of rings embeddable in a skew field is axiomatizable, and it may be of interest to make this statement more precise. The class in question is not a quasi-variety, beause it does not admit direct products (cf. VI.4.4, p. 235), but we shall see that it comes quite close to being a quasi-variety. Namely we shall find that there is a quasi-variety $\mathscr{T}$ such that the subclass of $\mathscr{T}$ consisting of the non-zero rings without zero-divisors is precisely the class of rings embeddable in skew fields.

By an *integral domain* we shall understand a non-zero ring without zero-divisors. A ring R is said to be *strongly regular* if for each $a \in R$ there exists $x \in R$ such that $a^2x = a$. The class of strongly regular rings is thus elementary; we denote by $\mathscr{T}$ the class of all subrings of strongly regular rings (by definition, every ring has a unit element and subrings inherit the unit element, hence any subring of a non-zero ring is again non-zero). By Theorem VI.2.1, p. 223 and Theorem VI.2.8, p. 226, $\mathscr{T}$ is a universal class (i.e. it can be defined by a set of universal sentences).

Moreover, $\mathscr{T}$ admits direct products, because the class of strongly regular rings clearly does so, therefore $\mathscr{T}$ is a quasi-variety, by Cor. VI.4.4, p. 235. Further, any skew field is clearly strongly regular, so any ring embeddable in a skew field is in $\mathscr{T}$, and it is also an integral domain. Conversely, any integral domain in $\mathscr{T}$ is embeddable in a skew field; this depends on the fact that a strongly regular ring is a subdirect product of skew fields and will not be proved here (cf. e.g. Cohn [71″ ′]).

Thus we see that the class of rings embeddable in skew fields can be defined by the quasi-identities characterizing subrings of strongly regular rings, together with the conditions for an integral domain:

$$\bigwedge_{x,y} xy = 0 \Rightarrow x = 0 \vee y = 0, \quad 1 \neq 0. \tag{2}$$

There remains the question whether the number of quasi-identities needed can be taken to be finite or is necessarily infinite. We can see as follows that it must be infinite (Cohn [74]). If the number could be taken finite, this would mean that the class of rings embeddable in skew fields is elementary. Let $\mathscr{F}_n$ be the class of all n-firs that are not embeddable in skew fields; this class is axiomatizable and by definition, $\mathscr{F}_1 \supseteq \mathscr{F}_2 \supseteq \cdots$, while $\mathscr{F}_n \neq \emptyset$ for all n, by **F.2**. Hence $\bigcap \mathscr{F}_n \neq \emptyset$ by the compactness theorem (Th. V.6.7, p. 218), but this would mean that there is a semifir not embeddable in a skew field, which contradicts **F.3**. Hence the class of all subrings of skew fields, though axiomatizable, is not elementary.

For a detailed study of the skew fields associated with a semihereditary ring, with the help of dependence relations, see Bergman [81].

Chapter XI

Algebra and Language Theory

1. INTRODUCTION

In the early days of high-speed computers there was a hope that it would be possible to program a computer to translate from one language to another, and this led to an intensive study of language structure. The result has been disappointing in that we are still far from making translations by computer, but that is no cause for despair. In the first place it shows that we have underestimated the richness and intricacy of our natural languages, even when used for quite prosaic ends. Secondly, this study of language has led to mathematical models of languages which admittedly are too simple to reflect all the complexities of a natural language like English, but – quite apart from their intrinsic interest — these models are found to give a good approximation to certain programming languages. And thirdly, there has been a vigorous interaction with some parts of noncommutative algebra, with benefit to 'mathematical linguist' and algebraist alike.

There is an extensive literature in the field, much of it written from the point of view of automata theory, but overlapping with orthodox linguistics, communication theory, probability, logic, and even algebra.

*Reprinted from: *Bull. London Math. Soc.* **7** (1975), 1–29.

Likewise there are several excellent expositions; we mention in particular Booth [67], Arbib [69], Hopcroft–Ullman [69], Gross–Lentin [70], Claus [71], and for an introduction which emphasizes the links with natural languages, Gross [72]. But many of these books are written for non-mathematicians, spending pages of explanation on points that a mathematical audience would take for granted, and nearly all concentrate on a particular aspect, depending on their bias. Since the subject is mathematical at heart, and perhaps not as widely known among mathematicians as it deserves to be, it seemed worthwhile to present a brief survey of the main results. An advantage in writing such a survey is that the author may skip long or technical proofs; but in this field even quite technical results often have an intuitive content, more readily appreciated in a sketch than a rigorous presentation, and I have tried to include such sketch proofs where possible. Like other accounts, this one has a bias, not unnaturally towards algebra. One of my aims has been to make clear the connexion with power series rings, and to include results on the latter which may be relevant to the study of language theory.

The methods used to study languages fall into three classes: (i) grammars, (ii) machines and (iii) algebraic methods. Of these, (i) and (ii) are discussed in §§2 and 3 respectively, and (iii) in §5 (monoids) and §6 (power series). §4 deals with transductions, really a formalization of the idea of translation, and §7 looks briefly at the attempts to relate the existing models to natural languages.

This article is based on a lecture given to the London Mathematical Society, on 15th November, 1973. I am grateful to Professor D. G. Kendall, the President, for encouraging me to write it up. My thanks are also due to Professor G. M. Bergman and Dr. M. Fliess for correcting some inaccuracies and providing additional references.

2. GRAMMARS

(i) Let X be a finite set, our *alphabet*. The set of all *words*, or strings of letters from X, is written X^*; it can be thought of as the free monoid (semigroup with 1) on X, with juxtaposition as multiplication, and including 1 to represent the empty word. The number of letters in a word u is called its *length*, written $l(u)$. By a *language* on X we understand any subset of X^*. We make no distinction between words and sentences; e.g. we could take one symbol x_0 of X as a blank space.

The first method of singling out interesting languages is by means of a grammar. Traditionally, grammar is used to parse a sentence, e.g.

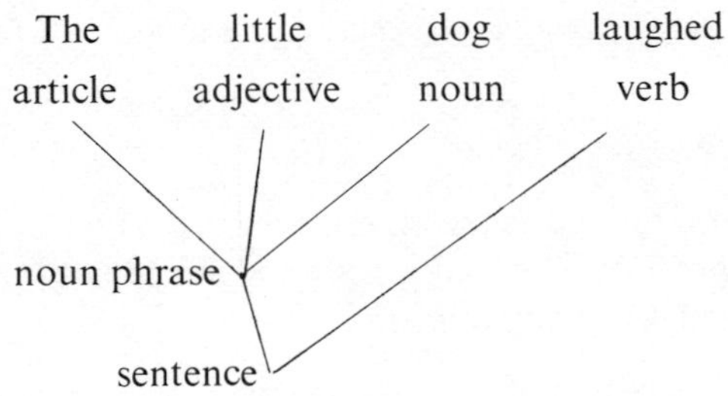

We shall use grammar to build up a sentence, thus a grammar for us is a set of rules of the form: sentence → {noun phrase, verb}, noun → dog etc. which will lead us to the sentences of the language and no others. This amounts to reading the above diagram (also called a *derivation tree*) from the bottom upwards.

Formally we shall define a *phrase structure grammar* G to consist of

(i) an alphabet X; its letters are also called *terminal letters*;

(ii) a set V of *clause-indicators* or *variables*, also called *non-terminal letters*, including a symbol σ for a complete sentence. We write $A = X \cup V$.

(iii) a finite set of *rewriting rules*: $u \to v$, where $u, v \in A^*$ and u contains at least one variable.

A string of letters from A is called *terminal* if it belongs to X^*, *non-terminal* otherwise. We apply the rewriting rule $u \to v$ by replacing a string fug in A by fvg. To get a sentence f of the language we start from σ and apply the rewriting rules until no variables (clause-indicators) are left. The sequence of rules used constitutes a *derivation* of f; we also write $\sigma \to f$. In this way we obtain the language $L(G)$ generated by the given grammar G; it consists of the strings on X that can be reached by a derivation from σ, and no others.

Any language generated by a phrase structure grammar is called a *phrase structure language* or is said to be of *type* 0. It is not hard to see that the languages of type 0 are precisely the recursively enumerable subsets of X^*, so the study of these languages is just a part of recursive function theory, and further restrictions are needed to impart a typical linguistic flavour. This is achieved by imposing restrictions on the rewriting rules. Before giving examples of languages, we list the most widely studied types of grammars.

Type 1. *Context-sensitive* or *-dependent* grammars. All rewriting rules have the form

$$f\alpha g \to fug, \quad \text{where} \quad \alpha \in V,\ u \in A^*,\ u \neq 1.$$

This rule can be interpreted to mean: α is replaced by *u in the context* $f\alpha g$. A grammar of this form is called a *CS-grammar* (or of type 1) and the language generated a *CS-language*.

Type 2. *Context-free* (Chomsky). All rewriting rules have the form

$$\alpha \to u, \quad \text{where} \quad \alpha \in V,\ u \in A^*,\ u \neq 1.$$

This means that α is replaced by u independently of the context. This is called a *CF-grammar* (or of type 2), and it generates a *CF-language*.

Type 3. *Regular* or *finite-state* (Kleene[1]), The rewriting rules have the form

(1) $$\alpha \to \beta x, \quad \text{where } x \in X,\ \alpha \in V,\ \beta \in V \text{ or } \beta = 1.$$

Here the special features is that each variable is replaced either by a letter or by a variable followed by a letter. We speak of a *regular grammar* or a *K-grammar* and *-language*. Instead of writing the variable β on the left of the terminal letter in (1) we could also restrict our rules so as to have β on the right of the terminal letter throughout. It can be shown that this leads to the same class of grammars.

(ii) If $\mathscr{L}_i$ denotes the class of languages of type i ($i = 0, 1, 2, 3$), then it is clear that

$$\mathscr{L}_0 \supseteq \mathscr{L}_1 \supseteq \mathscr{L}_2 \supseteq \mathscr{L}_3,$$

and here all the inclusions can be shown to be strict, but in general it is not easy to tell where a given language belongs, since there will usually be many grammars generating it.

From the above definition it follows that no CS-language can include the empty word. This is often inconvenient; in fact a minor modification

[1]These grammars were studied by Kleene [56] to obtain a model of the neuron.

of the rewriting rules will allow us to include the empty word, if desired. We shall not enter into the details, but in what follows we shall feel free to include rewriting rules that lead to the empty word.

Example 1

$\sigma \to x, \sigma \to \sigma y$. Let us derive a typical sentence:

$$\sigma \to \sigma y \to \sigma y^2 \to xy^2.$$

The complete set of sentences is x, xy, xy^2, ..., or briefly xy^*. Each sentence has just one derivation; we call such a language *unambiguous.* Clearly this is a regular language; here is a K-grammar generating the same language, but with the variables on the right:

$$\sigma \to x\lambda,\ \lambda \to y\lambda,\ \lambda \to 1.$$

Example 2

$\{x^n y^n \mid n \geqslant 0\}$. This is generated by the grammar

$$\sigma \to x\sigma y,\ \sigma \to 1.$$

This is a CF-grammar, so we have a CF-language. As it turns out, the language is not regular, but to show this we have to check, not merely that the above grammar is not regular, but that there exists no regular grammar generating the language. We shall reach this conclusion later by an indirect method.

Example 3

$\{x^n z^m y^n \mid m, n \geqslant 0\}$. This is again context-free but not regular. It is generated by the grammar

$$\sigma \to x\sigma y,\ \sigma \to \lambda,\ \lambda \to z\lambda,\ \lambda \to 1.$$

Here we cannot do without the variable λ; e.g. $\sigma \to x\sigma y$, $\sigma \to z\sigma$, $\sigma \to 1$ would also give zxy.

Example 4

$\{x^m y x^n \mid m \geqslant n \geqslant 0\}$. This is a CF-language, with grammar

(2) $$\sigma \to y, \sigma \to x\sigma, \sigma \to x\sigma x.$$

This grammar is ambiguous, for the sentence $x^m y x^n$ has $\binom{m}{n}$ different derivations.

But the ambiguity is a property of the grammar rather than the language;

an unambiguous grammar is

$$\sigma \to y,\ \lambda \to y,\ \sigma \to x\lambda,\ \lambda \to x\lambda,\ \sigma \to x\sigma x.$$

In this form a definite order for the application of the rules (2) is laid down.

Example 5

The free monoid X^* has an antiautomorphism which leaves X fixed. This is written $f \mapsto \tilde{f}$ and is called *reversal*; e.g. the reverse of $x^2 yzx$ is $xzyx^2$. Now $\{f\tilde{f} \mid f \in X^*\}$ is a *CF*-language, but not regular. Taking $X = \{x, y\}$, we have as a generating grammar:

$$\sigma \to x,\ \sigma \to y,\ \sigma \to x\sigma x,\ \sigma \to y\sigma y,\ \sigma \to 1.$$

By contrast, the language $\{f^2 \mid f \in X^*\}$ is not context-free (it turns out to be context-sensitive).

Example 6

$\{x^n z^n y^n \mid n \geqslant 0\}$ is a *CS*-language, but not *CF*. A generating grammar is

$$\sigma \to x\sigma\lambda\mu,\ \sigma \to xz\mu,\ \mu\lambda \to \lambda\mu,\ z\lambda \to z^2,\ \sigma \to 1,\ \mu \to y.$$

The first two rules generate the words $x^n z\mu(\lambda\mu)^{n-1}$, the next moves the λ's past the μ's next to z and the fourth rule replaces each such λ by z.

Example 7

The universal language X^* is regular: $\sigma \to \sigma x$ $(x \in X)$, $\sigma \to 1$, and so is the empty language: $\sigma \to \sigma x$.

(iii) We add some remarks on the different types of grammar. The regular languages can be characterized explicitly as follows:

Theorem 2.1 (***Kleene*** [*56*])

The regular languages form the smallest class containing all finite languages and closed under the operations of union, product and *. ∎

Here the *-operation replaces a language L by

$$L^* = \{u_1 \dots u_n \mid u_i \in L,\ n = 0, 1, 2, \dots\}.$$

All four types of languages are closed under unions, products, reversal, and all except type 2 (*CF*-languages) are closed under intersection (cf. §4). Further, each type other than 1 (*CS*-languages) is closed under substitution: If L of type v ($v = 0, 2, 3$) has the alphabet $X = \{x_1, \ldots, x_m\}$ and x_1 is replaced by a language of type v, with an alphabet disjoint from X then the resulting language is again of type v. This follows by performing the substitution on the rewriting rules. E.g. xy and z^* are regular languages; replacing y by z^* we obtain the regular language xz^*. Here we can replace x by y^* and obtain the regular language y^*z^*. On the other hand, we can substitute any language L of type v into x^* and obtain L^* which is again of type v.

Let X, Y be any alphabets and L a language on X. If each $x \in X$ is replaced by a single word in Y, we obtain the image of L under a homomorphism $X^* \to Y^*$. Thus the homomorphic image of any language of type $v \neq 1$ is again of type v. For a *CS*-language we have closure under substitution, provided that the languages substituted do not contain the empty word. This follows by looking at the derivations (which in a *CS*-grammar cannot decrease the length).

(iv) Context-sensitive languages are just those languages that can be generated by a grammar with rules of the form $u \to v$, where $l(u) \leqslant l(v)$. For the *CS*-grammars are of this form, and conversely, given $u \to v$, where $u = u_1 \ldots u_r$, $v = v_1 \ldots v_s$ and $r \leqslant s$, we take new variables $w_1, \ldots, w_r$ and rules

$$u_1u_2 \ldots u_r \to w_1u_2 \ldots u_r \to \ldots \to w_1w_2 \ldots w_r \to v_1w_2 \ldots w_r \to$$
$$\to \ldots \to v_1 \ldots v_{r-1}\, w_r \to v_1 \ldots v_{r-1}\, v_r \ldots v_s.$$

By treating all rules in this way we obtain a *CS*-grammar.

It follows from this description that every *CS*-language is a recursive subset of X^*; therefore $\mathscr{L}_1 \subset \mathscr{L}_0$. However, not every recursive subset of X^* is a *CS*-language, for we can enumerate the *CS*-languages: L_1, $L_2, \ldots$, and the strings on X: $u_1, u_2, \ldots$. Now the set L defined by the rule "$u_i \in L$ if and only if $u_i \notin L_i$, for $i = 1, 2, \ldots$" is recursive, but not a *CS*-language (Chomsky [59]).

The extra power of *CS*-languages over *CF*-languages is that rules of the form $uv \to vu$ are admitted in the former; as Chomsky [59] observes, this is really a defect as far as natural languages are concerned. On the other hand, *CF*-languages are not enough to describe natural languages. This can be seen by forming sentences of the following type:

Alex came by aeroplane,

Alex and Bill came by aeroplane and boat, respectively,

Alex, Bill and Colin came by aeroplane, boat and car, respectively. Clearly the series can be continued indefinitely (given sufficient ingenuity), and it corresponds to the sentences

(3) $$u_1v_1,\ u_1u_2v_1v_2,\ u_1u_2u_3v_1v_2v_3, \ldots$$

But as Example 5 above shows, this is not a pattern obtainable by a *CF*-grammar, which would be limited to generating

$$u_1v_1, u_2u_1v_1v_2,\ u_3u_2u_1v_1v_2v_3, \ldots$$

More precisely, any *CF*-grammar capable of producing the series (3) would also produce many other sentences which one would not want to include in the language.

This type of argument suggests that a natural language such as English is not context-free. On the other hand, many programming languages can be described quite accurately by *CF*-grammars.

(v) Any *CF*-language can be generated by a grammar whose rules are all of the form $\alpha \to \beta\gamma$ or $\alpha \to x$ ($\alpha, \beta, \gamma \in V$, $x \in X$). This is called the *Chomsky normal form* (Chomsky [59]); it follows by a double induction (i) on the number of rules not of this form and (ii) the greatest length of any right-hand side. If the grammar contains a rule $\alpha \to a_1 \ldots a_n$ ($a_i \in A$), we replace this by $\alpha \to bc$, $c \to a_2 \ldots a_n$ and $b = a_1$ or $b \to a_1$ according as a_1 is in V or X.

Let G be a grammar of type 0 and $L(G)$ the language generated. Various quite simple questions about $L(G)$ are in general undecidable. E.g., whether a given word w is in $L(G)$ is essentially the halting problem for Turing machines and so is undecidable. For a *CS*-grammar the corresponding problem is decidable, because $L(G)$ is then recursive, but even here the question whether $L(G)$ is empty is undecidable (it can be reduced to Post's correspondence problem, cf. Gross–Lentin [70]). But for a *CF*-grammar this problem can be decided:

Theorem 2.2

There is an algorithm to determine whether the language generated by a given CF-grammar is empty.

For if w is a word in $L(G)$ whose derivation takes more than r steps, where r is the number of variables in G, some variable must occur twice on the left of a rule, say there is a step $\alpha \to u$, and later in the derivation there is a step $\alpha \to v$. Now we can get a word with a shorter derivation by substituting the part of the derivation that starts with $\alpha \to v$ into that starting with $\alpha \to u$. Hence if $L(G) \neq \varnothing$, it contains words whose derivation has at most r steps, and we need only check finitely many derivations to see if this is so. ▮

This result can be used to simplify *CF*-grammars. E.g., given a *CF*-grammar G, for any variable α we can determine whether there is a derivation starting from α, i.e. whether the grammar G_α obtained by taking α as the sentence symbol σ generates the empty language or not. If $L(G_\alpha) = \varnothing$, we can omit α and all rules involving α from G without affecting the language generated. If $L(G_\alpha)$ is finite, $= \{u_1, \ldots, u_n\}$ say, we can again omit α so long as we add the rules $\sigma \to u_i$ $(i = 1, \ldots, n)$ to G. Thus our grammar is now such that there are infinitely many words derivable from each variable α(unless the whole language is finite). We can also omit a variable α if no derivation starting from σ produces a word containing α. Finally if there is a rule $\alpha \to \beta$, where $\alpha, \beta \in V$, we can again omit it, provided that we add, for each rule $\beta \to u$ $(u \in A^*)$ a rule $\alpha \to u$. We have now reduced our grammar to a form where (i) there are infinitely many derivations starting from each variable, (ii) each variable occurs in some derivation and (iii) no variable occurs by itself on the right of a rule.

A grammar is called *self-embedding* if there is a variable α and a derivation $\alpha \to u\alpha v$, where u, v are non-empty words. If we apply the above reduction to a *CF*-grammar which is not self-embedding, we obtain a grammar which is again not self-embedding. Our aim will be to show that the language generated by such a grammar is in fact regular. The variables of G can be partially ordered by writing $\alpha < \beta$ whenever there are derivations $\alpha \to u\beta v$ with $u \neq 1$ and with $v \neq 1$ (not necessarily the same derivation). This relation is irreflexive because G is not self-embedding, so we have indeed a partial ordering. Let γ be maximal in the partial ordering, then the rules with γ on the left all have the form

$$\gamma \to \beta u, \gamma \to u \qquad (u \in X^*),$$

or all have the form

$$\gamma \to u\beta, \gamma \to u \qquad (u \in X^*).$$

In either case the language $L_\gamma = L(G_\gamma)$ is regular. Now replace γ by a new terminal letter z, then we obtain a grammar G' which is again non-self-embedding, but with fewer variables. By induction the language L' it generates is regular. If in L' we substitute L_γ for z we obtain $L(G)$, which is therefore regular. Together with the obvious remark that no regular grammar is self-embedding, this proves

Theorem 2.3

A language L is regular if and only if there exists a CF-grammar G generating L which is not self-embedding. ▮

The same idea is exploited in the following result, sometimes called the pumping lemma (Bar Hillel–Perles–Shamir [61], Arbib [69]):

Theorem 2.4

Let G be a CF-grammar. Then there exist integers p, q such that every word $z \in L(G)$ *of length greater than p can be written as* $z = xuwvy$, *where* $uv \neq 1$, $l(uwv) \leqslant q$ *and* $xu^nwv^ny \in L(G)$ *for all* $n \geqslant 1$.

For we can simplify G as before; then the only rules $\alpha \to u$ with $l(u) = 1$ are those where u is a terminal letter. We can omit such a rule, provided that we add all rules obtained from the existing ones by replacing α by u on the right. The effect on $L(G)$ will be to omit at most one word from the language, namely u. Now let r be the number of non-terminal letters in G, then any derivation with more than r steps has a variable occurring twice on the left. If the longest word obtainable by r derivations has p letters, then any word z longer than p has in its chain of derivations the subchain $\alpha \to u'\alpha v' \to uwv$, where $u' \to u$, $v' \to v$, $\alpha \to w$ are derivations and $uv \neq 1$. Thus z has the form $z = xuwvy$, and by repeating the stage $\alpha \to u'\alpha v'$ of the derivation n times we get xu^nwv^ny. Moreover, the derivation $\alpha \to uwv$ can be taken to have at most $r + 1$ steps, so uwv has bounded length. ▮

In the case of a regular grammar the terminal letters all occur on the same side of the variables, and the conclusion of Theorem 2.4 simplifies accordingly:

Corollary

Let G be a regular grammar. Then there exists r such that every word w of L(G) of length greater than r has the form $w = xyz$, *where* $y \neq 1$, $l(yz) \leqslant r$ *and* $xy^nz \in L(G)$ *for all* $n \geqslant 1$. ▮

Here r can again be chosen to be the number of variables in G.

(vi) A great deal has been written on ambiguity. From the definition it is not at all easy to tell whether a given grammar is ambiguous (i.e. whether some word has more than one derivation), in fact the problem is insoluble for general *CF*-grammars (it can be reduced to Post's correspondence problem, cf. Cantor [62], Arbib [69]). A *CF*-language is *unambiguous* if there is an unambiguous *CF*-grammar generating it, otherwise it is *inherently ambiguous*. An example of the latter is

$$L = \{x^m y^m z^n,\ x^m y^n z^n \mid m, n \geqslant 0\};$$

this is the union of two *CF*-grammars and $x^n y^n z^n$ is generated twice, cf. Parikh [61], Arbib [69].

The following instructive example of an ambiguous *CF*-grammar is taken from Gross-Lentin [70]: The alphabet is $\{e, +, \times\}$ and σ is the only variable. The rules are

$$\sigma \to \sigma + \sigma,\ \sigma \to \sigma \times \sigma,\ \sigma \to e\sigma,\ \sigma \to e.$$

The sentences are just the "calculations" in elementary arithmetic; if we abbreviate *e*, *ee*, *eee*, ... as 1, 2, 3, ..., a typical sentence is $3 \times 5 + 4$, and the ambiguity of this grammar is reflected in the fact that this expression has different meanings according to its derivation. The ambiguity can be resolved in several ways; (i) by priority rules ($\times$ precedes $+$) and using brackets: $(3 \times 5) + 4$ or $3 \times (5 + 4)$; or (ii) by writing the operation signs on one side of the variables: $35 \times 4 +$ or $354 + \times$.

The second method is often used in abstract algebra (cf. Chapter III) and it leads to the Łukasiewicz grammar: this has the alphabet $x_0, x_1, \ldots, x_m$, variable σ and rules $\sigma \to \sigma^i x_i$ $(i = 0, 1, \ldots, m)$. Here x_μ is to be interpreted as a μ-ary operator. Let us define a homomorphism $X^* \to \mathbf{Z}$ by setting $f(x_i) \to 1 - i$, then the words of the language Ł are just the strings u such that $f(u) = 1$ and $f(u') > 0$ for every non-trivial left factor u' of u. More generally, every string u such that $f(u') > 0$ for each non-trivial left factor u' of u, can be factorized uniquely as the product of $f(u)$ words in Ł (Th. III. 2.3, p. 118).

3. Machines

(i) Logical machines form a convenient means of studying recursive functions. In particular, the Turing machines (Davis [58], Booth [67]) lead precisely to recursively enumerable sets; thus any phrase structure grammar can be obtained by a Turing machine and we would expect the more restricted types 1–3 of grammars to correspond to more special machines. The basic notion is that of a sequential machine:

A *sequential machine* (*Mealy machine*, Hartmanis–Stearns [66], Booth [67]) or also *deterministic automaton* M is given by three non-empty sets (usually finite in what follows): X = *input*, Y = *output* and S = *set of states*, with an *initial state* s_0 say. The action of M is described by two functions, the *transition function* $\delta : S \times X \to S$ (giving the next state) and the *output function* $\lambda : S \times X \to Y$.

M acts *sequentially*, i.e. it reads each word letter by letter, and *synchronously*, i.e. for each $x \in X$ it gives out a $y \in Y$. Generally the input will be a word on X; the output is a word on Y, of the same length, obtained as follows. Define mappings δ', λ' by

$$\delta'(s, 1) = s,\ \delta'(s, ux) = \delta(\delta'(s, u), x) \qquad s \in S,\ x \in X,\ u \in X^*,$$

$$\lambda'(s, 1) = 1,\ \lambda'(s, ux) = \lambda'(s, u)\lambda(\delta'(s, u), x).$$

This defines a mapping $\delta' : S \times X^* \to S$ which extends δ and defines a representation of the free monoid X^* by mappings of S. Likewise λ': $S \times X^* \to Y^*$ is a mapping extending λ. We may therefore without risk of confusion omit the dashes from δ', λ'.

To operate the machine we start with an input $u \in X^*$ and with the machine in the initial state s_0. The machine then moves into the state $\delta(s_0, u)$ and produces the output $\lambda(s_0, u)$. We see that an automaton differs from a Turing machine chiefly in having no specific memory (tape). If S, λ are absent from M and instead we have a set $F \subseteq S$ of *final states*, we speak of an *acceptor*. Now there is no output, but merely a set of final states; the set of words in X which take us from s_0 to a state in F is the set *accepted* by M.

Algebraically an acceptor may be described as a set S with a representation of the free monoid X^* by mappings of S into itself. An acceptor may also be represented as a graph: we take the set of states as nodes, with a segment $s_i \to s_j$ for each letter x such that $\delta(s_i, x) = s_j$. Similarly an automaton may be represented by a graph, with a segment $s_i \to s_j$ labelled by x/y if $\delta(s_i, x) = s_j$ and $\lambda(s_i, x) = y$. The importance of acceptors for us rests in

Theorem 3.1

A language is regular if and only if it is the precise set accepted by some (finite) acceptor. ∎

This follows easily using Theorem 2.1 (Arbib [69]). As an example, here is an acceptor for the language $\{xy^n | n \geqslant 0\}$ (Example 1). There are three states: s_0 (initial), s_1, s_2 and only s_1 is final. The transition function δ is given by the table

δ	x	y
s_0	s_1	s_2
s_1	s_2	s_1
s_2	s_2	s_2

Let M, M' be two machines with inputs X, X', outputs Y, Y' and set of states S, S'. A *homomorphism* f: $M \to M'$ is a triple of mappings $X \to X'$, $Y \to Y'$, $S \to S'$ all denoted by f for short, such that the following diagrams commute:

$$\begin{array}{ccc} S \times X & \xrightarrow{\delta_M} & S \\ \downarrow f & & \downarrow f \\ S' \times X' & \xrightarrow{\delta_{M'}} & S' \end{array} \qquad\qquad \begin{array}{ccc} S \times X & \xrightarrow{\lambda_M} & Y \\ \downarrow f & & \downarrow f \\ S' \times X' & \xrightarrow{\lambda_{M'}} & Y' \end{array}$$

In the special case when $X = X'$, $Y = Y'$ and f is the identity of X and Y, f is called a *state homomorphism*.

Given any machine, we can often find a simpler machine doing the same work by taking a state homomorphism to a machine with fewer states. If the number of states cannot be diminished in this way, the machine is said to be *reduced*. Every machine has a reduced homomorphic image, which can be constructed by identifying states having the same effect.

Instead of using the functions δ, λ to describe M we can also take the set of all quadruples (s_i, x, y, s_j) such that $\delta(s_i, x) = s_j$ and $\lambda(s_i, x) = y$. In this form it is evident how to define the *adjoint* M^* of M: it has input Y, output X, the same set of states S as M and quadruples (s_j, y, x, s_i) corresponding to each (s_i, x, y, s_j) for M. The functions δ, λ defining M^* will in general be many-valued, i.e. M^* will usually be a *non-deterministic* automaton; however, it is deterministic whenever M was reduced (Arbib [69]). Using adjoints it is easy to show that the reverse of any regular language is again regular.

(ii) Acceptors can be used to estimate the number of words in a language. To see this we note the following description of automata by matrices.

Let M be an automaton with N states $s_1, \ldots, s_N$, then M can be described by a set of $N \times N$ matrices $P(x|y)(x \in X, y \in Y)$, where

$$P(x|y) = p_{ij}(x|y)),\ p_{ij}(x|y) = \begin{cases} 1 \text{ if } \delta(s_i, x) = s_j \text{ and } \lambda(s_i, x) = y, \\ 0 \text{ otherwise.} \end{cases}$$

The $P(x|y)$, called the *transition matrices*, describe the action of M completely. More generally, we can define $P(u|v)$ recursively for any $u \in X^*$, $v \in Y^*$ by the equations

$$P(ux|vy) = P(u|v)P(x|y), \quad P(u|v) = 0 \quad \text{if} \quad l(u) \neq l(v)$$

It is easily verified that for any $u, u' \in X^*$, $v, v' \in Y^*$ we have

$$(1) \qquad P(uu'|vv') = P(u|v)P(u'|v').$$

For each $x \in X$, let us write

$$P(x) = \sum_y P(x|y),$$

where y ranges over Y, then $P(x)$ is a matrix with precisely one 1 in each row and the rest 0's. In particular, for an acceptor we have only the matrices $P(x)$, and (1) takes the form

$$P(uu') = P(u)P(u').$$

Thus P is the matrix representation of X^* acting on S. Let π be the row vector describing the initial state: $\pi_i = 1$ if s_i is the initial state and $\pi_i = 0$ otherwise, and let f be the column vector with entry 1 for final and 0 for non-final states, then for any $u \in X^*$

$$\pi P(u)f = \begin{cases} 1 & \text{if } u \text{ is accepted,} \\ 0 & \text{if not.} \end{cases}$$

Put $A = \sum_x P(x)$, then $a_{ij}(n)$, the (i, j)-entry of A^n, represents the number of words of length n which send s_i to s_j. In terms of the graph of M it is the number of paths of length n from s_i to s_j; in particular, $a_{ii}(n)$

is the number of loops (closed circuits) of length n based at s_i. Let us write $\lambda(n)$ for the number of words of length n in the language, and write

$$f(t) = \sum_0^\infty \lambda(n)t^n$$

for the length generating function, then

$$f(t) = \sum \pi\, A^n t^n f = \pi\,(I - At)^{-1}f.$$

To get an asymptotic value for $\lambda(n)$ we take the characteristic equation of A in the form

$$A^N + c_1A^{N-1} + \cdots + c_N = 0,$$

then for $n \geqslant N$,

$$a_{ij}(n + 1) + c_1a_{ij}(n) + \cdots + c_Na_{ij}(n - N + 1) = 0.$$

Taking s_1 to be the initial state, we have $\lambda(n) = \sum a_{1j}(n)$, where the sum is over all j such that s_j is a final state; hence we obtain the following recursion relations for the λ's:

$$\lambda(n + 1) + c_1\lambda(n) + \cdots + c_N\lambda(n - N + 1) = 0.$$

The characteristic values are just the eigenvalues of A: $\lambda_1, \ldots, \lambda_N$ say. Assuming them to have distinct moduli for the moment, we obtain as the general solution

(2) $$\lambda(n) = b_1\,\lambda_1^n + \cdots + b_N\,\lambda_N^n,$$

where the b_i are constants. If the λ_i are so numbered that $|\lambda_1| > |\lambda_i|$ for all i, then $\lambda(n)/\lambda_1^n \to b_1$ as $n \to \infty$, and

(3) $$\lim \frac{\log \lambda(n)}{n} = \log |\lambda_1|.$$

In general λ_1 is an r-fold eigenvalue, and b_1 in (2) is now a polynomial of degree r in n. Thus $\log \lambda(n) \sim n \log |\lambda_1| + r \log n + c$, and (3) still holds.

The limit on the left of (3) is known as the *channel capacity* in information theory; for this reason $\log|\lambda_1|$ may be regarded as the information capacity of the language. If the logarithms are taken to base 2, this gives the capacity of the language in bits per letter. As an example, suppose that the graph of the acceptor has just one loop, of length r say. The number of loops of length n is $tr\ A^n$, hence $\sum \lambda_i^v = \delta_{vr}$ and it follows that A has the characteristic equation $A^N - A^{N-r} = 0$, and $\log |\lambda_1| = 0$.

There is a curious property of regular languages. Let m be the number of letters in the alphabet X, then the universal language X^* has $\mu(n) = m^n$ words of length n, so $\lambda_1 = m$ in this case. Given any regular language L, its complement L' is again regular, and if $\lambda(n)$, $\lambda'(n)$ are the number of words of length n in L, L' respectively, then

$$\lambda(n) + \lambda'(n) = m^n. \tag{4}$$

Hence

$$\frac{\lambda(n)}{\mu(n)} \sim c(\lambda_1/m)^n,$$

where c is at most a polynomial in n. Now if $|\lambda_1| < m$, then $c(\lambda_1/m)^n \to 0$ as $n \to \infty$, hence by (4), $\lambda'(n)/\mu(n) \to 1$. This means that L' has a greatest eigenvalue of absolute value m. We obtain

Theorem 3.2

Let L be a regular language on m letters; either L has capacity log *m, or the probability that a sufficiently long string of letters forms a word in L is arbitrarily close to zero.* ∎

(iii) We now turn to context-free languages. If we take a typical *CF*-language such as $\{x^n y^n | n \geqslant 0\}$ and try to build an acceptor for it we find that we are hampered by the fact that the acceptor has no memory. Here we see again that English is not a regular language, since we clearly need a memory to parse (or form) such pairs of sentences as

Those		die	
	whom the gods love		young.
He		dies	

The memory to be described is of a rather simple sort, a 'first-in, last-out' store, where we only have access to the last item in the store.

Formally, a *pushdown acceptor* (PDA for short) is an acceptor which in addition to its set S of states and input alphabet X has a set Σ of *store symbols*, with initial symbol λ_0, and a transition function δ on $S \times X \times \Sigma$ with values in $S \times \Sigma^*$, but for a given triple of arguments there may be several or no values (this machine is nondeterministic). At any stage the machine is described by a triple (s_i, w, α), where $s_i \in S$, $w \in X^*$, $\alpha \in \Sigma^*$. We apply δ to the triple consisting of s_i, the first letter of w and the last letter of α. If $w = xw'$, $\alpha = \alpha'\lambda$ say, and (s_j, β) is a value of

$\delta(s_i, x, \lambda)$, then

$$(s_i, xw', \alpha'\lambda) \to (s_j, w', \alpha'\beta)$$

is a possible move. We say that $w \in X^*$ is *accepted* by the machine if starting from (s_0, w, λ_0) there is a series of moves to take us to $(s_r, 1, \gamma)$, where s_r is a final state. Now we have

Theorem 3.3

The context-free languages constitute the precise class of sets accepted by pushdown acceptors. ▮

For a proof we refer to Arbib [69]. As an example we describe a PDA for $\{x^n y^n | n \geqslant 0\}$. Its states are s_0 (initial state), s_1, s_2 where s_2 is final. The store symbols are λ (initial symbol), μ, ν. We give the values for $(s_i, ., .)$ in the form of a table for each s_i:

s_0	λ	μ	ν
x	$s_0\mu$	$s_0\mu\nu$	$s_0\nu^2$
y		$s_2\mu$	s_1

s_1	λ	μ	ν
x			
y		$s_2\mu$	s_1

Blanks and the remaining values (for s_2) are undefined. It is easily seen that the strings accepted are just $x^n y^n$; note how the store acts as a memory, remembering how many factors x have been taken off. If we think of the store arranged vertically, at each stage we remove the topmost symbol and add a number of symbols to the top, just like a stack of plates in a cafeteria; this explains the name.

Another example of a *CF*-language is the set L_1 of all strings u in x, y such that in any left-hand factor of u the degree in x is at least equal to the degree in y. It is easy to construct a PDA for L_1 and the reader may like to try this for himself. A somewhat harder example is the set L_2 of strings in x, y whose degree in x is at least equal to the degree in y. For L_2 the probability that a string of given even length is in L_2 is 1/2. By Theorem 2.4 Corollary, L_2 is not regular; likewise L_1 is not regular.

(iv). We can think of a PDA as a Turing machine, by regarding the store as part of the tape[1]. An estimate of the length of tape needed may be given as follows. Let n be the length of the input word, a the greatest

[1] A PDA is most naturally regarded as a 2-tape Turing machine, but of course this is equivalent to the usual (1-tape) sort.

length of any word added to the store at any stage and b the length of tape taken up with instructions, then the total length of tape needed is $an + b$, a linear function of n. We shall find that a more general type of acceptor, with a similar limitation on the length of tapes, can be used for general CS-languages.

A *linear-bounded automaton* is an automaton in which the input and output alphabet are the same, X say, with an initial state s_0 and final states, and with a tape on which the input word is written. The configuration of the machine at any stage is described by a state s_i and a square of the tape being scanned, and in addition to the transition function δ and output function λ there is a transport function

$$\tau: S \times X \to \{L, R, O\},$$

where L, R, O denotes motion of the tape by one square to the left, right or not at all. As input we take the symbol on the square being scanned and replace it by the output. In general these functions will again be many-valued partial functions. A word w is *accepted* if there is at least one computation starting with w printed on the tape, the leftmost letter being scanned, and in the initial state s_0, and terminating with the last letter being read as the machine enters a final state. Such a machine differs from the general Turing machine in having a tape length which is bounded by a linear function of the input word.

***Theorem 3.4* (*Landweber* [63], *Kuroda* [64]).**
The context-sensitive languages constitute the precise class of languages accepted by linear-bounded automata. ▮

E.g. it is not hard to construct a linear-bounded automaton to accept

$$\{x^n y^n z^n \mid n \geqslant 0\}.$$

Intuitively we first move across the word to check that it is a block of x's followed by a block of y's followed by a block of z's. Now erase (i.e. replace by a further, neutral symbol) one z, one y, one x and recommence. The formal description is rather longer and will be omitted.

(v) The range of languages can be extended in another direction by using stochastic automata. We recall that any automaton can be described by a set of transition matrices $P(x|y)$; here the $P(x|y)$ are non-negative (i.e. with non-negative entries) and such that $P(x) = \sum_y P(x|y)$

has one entry 1 in each row and the rest 0's. If, more generally we take $P(x|y)$ to be a non-negative matrix such that $P(x)$ is *stochastic* i.e. has row sums 1, we have a *stochastic automaton* or *acceptor* as the case may be.

In a stochastic acceptor A it is natural to replace the initial state by an *initial distribution* π, i.e. a row vector whose components are non-negative and add up to 1; by f we denote again the column vector with 1 for final and 0 for non-final states. With each word $u \in X^*$ we can associate a real number between 0 and 1:

$$\lambda_\mu = \pi P(u) f,$$

representing the 'strength of acceptance' of the word u by A. We say that u is *λ-accepted* by A if $\pi P(u) f > \lambda$.

Given $0 \leqslant \lambda \leqslant 1$ and $L \subseteq X^*$, we say that L is *λ-stochastic* if there is a stochastic acceptor A such that L is the language λ-accepted by A, i.e.

$$L = L(A, \lambda) = \{u \in X^* \mid \pi P(u) f > \lambda\}.$$

Now a *stochastic language* is a language that is λ-stochastic for some λ. Every 0-stochastic language or more generally, every λ-stochastic language with a deterministic acceptor is regular; but in general a stochastic language need not even be of type 0.

As an example of a stochastic acceptor let us take the alphabet $X = \{0, 1, \ldots, m-1\}$, where $m \geqslant 2$, and define an acceptor B with two states s_1 (initial state), s_2 (final state) and

$$P(x) = \begin{pmatrix} 1 - x/m & x/m \\ 1 - (x+1)/m & (x+1)/m \end{pmatrix} \quad x = 0, 1, \ldots, m-1.$$

Each word $u = x_1 \ldots x_r$ may be represented in m-adic form as $0.x_r x_{r-1} \ldots x_1$. Then the language λ-accepted by B is

$$L(B, \lambda) = \{x_1 \ldots x_r \in X^* \mid 0.x_r \ldots x_1 > \lambda\}.$$

It is not hard to verify that $L(B, \lambda)$ is regular if and only if λ is rational (Paz [66], Claus [71]), so that we obtain languages that are not regular by choosing λ irrational. Moreover, since $L(B, \lambda) \neq L(B, \lambda')$ for $\lambda \neq \lambda'$, we have in fact uncountably many stochastic languages. In particular this shows that not every stochastic language is of type 0. An explicit example of a stochastic language that is not regular is

$$\{x^m y^n \mid n - 2 \geqslant m \geqslant 0\} \text{ (cf. Claus [71]).}$$

Stochastic automata are of interest because real machines are never fully determinate, but in fact show stochastic behaviour. Moreover, if we drop 'stochastic', i.e. allow arbitrary real matrices $P(x)$ in the above definition, the remarkable fact emerges that the languages so obtained are nevertheless stochastic. This result is due to Turakainen [69] (cf. also Claus [71]).

4. TRANSDUCTIONS

(i) To formalize the notion of translation from one language to another one introduces transductions. A *transduction* from X^* to Y^* is just a subset T of $X^* \times Y^*$. Given a transduction T from X^* to Y^* we can apply T to any language L on X by forming

$$T(L) = \{v \in Y^* \mid (u, v) \in T \text{ for some } u \in L\}.$$

This is called the *image* of L under T. It is clear how transductions $T \subseteq X^* \times Y^*$ and $U \subseteq Y^* \times Z^*$ can be composed (in the usual manner of correspondences, cf. p.10) to give a transduction $TU \subseteq X^* \times Z^*$, and the inverse $T^{-1} \subseteq Y^* \times X^*$ can be formed.

A transduction T is said to be *regular* or a *K-transduction* if there is an automaton A with a pair of states s_1, s_2 (not necessarily distinct) such that T consists precisely of those pairs (u, v) for which the input u in state s_1 leads to output v in state s_2; A is also called a *transducer* for T. Thus an automaton with N states defines N^2 transductions; a K-transduction by a machine with a single state is just a homomorphism $X^* \to Y^*$. It is clear that any product of regular transductions is again regular.

What we have just defined should more precisely be called a *right regular* transduction, because the action of X^* on A is a right action. Left regular transductions may be defined similarly. To be explicit, let $\sim$ be the reversal on

$$X^* \times Y^* : (u, v) \mapsto (\tilde{u}, \tilde{v}),$$

then R is left regular if and only if $\tilde{R}$ is right regular. It can be shown that every regular language in Y^* can be obtained from X^* be applying first a right and then a left regular transduction (cf. Schützenberger [61]).

(ii) Let us consider the effect of a transduction T on a CF-language L. The operation of T may be described in terms of quadruples (s_i, x, y, s_j), as we have seen; we shall again denote the initial state by s_0. The rules of L can be written in Chomsky normal form as $\alpha \to \beta\gamma$ or $\alpha \to x(\alpha, \beta, \gamma \in V, x \in X)$. The image L' of L has again a CF-grammar, with non-terminal variables of the form (s_i, x, s_j) and (s_i, α, s_j) and rules

(i) $\sigma \to (s_0, \sigma, s_i)$ for all i such that s_i is final,

(ii) if $\alpha \to \beta\gamma$ in L, then $(s_i, \alpha, s_j) \to (s_i, \beta, s_k)(s_k, \gamma, s_j)$ for all i, j, k,

(iii) if $\alpha \to x$ in L, then $(s_i, \alpha, s_j) \to (s_i, x, s_j)$ for all i, j,

(iv) if (s_i, x, y, s_j) is in T, then $(s_i, x, s_j) \to y$.

It may be verified that $T^{-1}(L') = L \cap \Lambda$, where Λ is the regular language accepted by T. Thus T accepts the intersection of L with the maximal regular language it accepts.

Now given any regular language Λ, there is an acceptor for Λ, which can be turned into a automaton by recopying the input word. Thus T is a subset of the diagonal and $L' = L \cap \Lambda$. This proves

Theorem 4.1 **(*Bar Hillel-Perles-Shamir* [*61*]).**

The intersection of a CF-language and a regular language is a CF-language. ∎

In the same way one can show that the image of a regular language by a transducer is regular and hence the intersection of two regular languages is regular. By contrast the intersection of two CF-languages need not be CF, e.g.

$$\{x^m y^n z^n \mid m, n \geqslant 0\} \cap \{x^m y^m z^n \mid m, n \geqslant 0\} = \{x^n y^n z^n \mid n \geqslant 0\}.$$

5. MONOIDS[1]

(i) Let M be a monoid and S a finite set with a right M-*action*, i.e. to each pair $(s, x) \in S \times M$ there corresponds an element $sx \in S$ such that $s(xy) = (sx)y$ and $s1 = s$. We fix $s \in S$ and define a relation $\mathfrak{r}_s$ on M as follows:

$$x\mathfrak{r}_s y \text{ if and only if } sx = sy.$$

Clearly this is an equivalence on M, which moreover is *right admissible*, i.e.

$$x\mathfrak{r}_s y \Rightarrow xz\mathfrak{r}_s yz \text{ for all } x, y, z \in M.$$

[1] All monoids in this section are understood to be finitely generated.

The classes of this equivalence correspond just to the images of s under the action of M and, since S is finite, the *index* of $\mathfrak{r}_s$, i.e. the number of $\mathfrak{r}_s$-classes, is also finite.

If we do this for each $s \in S$ and put

$$\mathfrak{q} = \bigcap \mathfrak{r}_s, \tag{1}$$

we obtain another equivalence which is left as well as right admissible: $x\mathfrak{q}y$ means that $sx = sy$ for all $s \in S$, hence $szx = szy$ for all $z \in M$. Thus $\mathfrak{q}$ is a *congruence* on M and we can form the quotient monoid $M/\mathfrak{q}$. To obtain an interpretation for $\mathfrak{q}$, let us write $Map(S)$ for the monoid of all mappings of the set S into itself, then the right M-action on S defines a monoid homomorphism

$$\phi \colon M \to Map(S),$$

and now the congruence $\mathfrak{q}$ defined in (1) is just the kernel of ϕ, while $M/\mathfrak{q}$ is the image (up to isomorphism). Since $Map(S)$ is finite, this shows that $\mathfrak{q}$ has again finite index.

We saw that each $\mathfrak{r}_s$ is a right admissible equivalence of finite index. Conversely, let $\mathfrak{r}$ be any right admissible equivalence on M of finite index and write $S = M/\mathfrak{r}$ for the set of $\mathfrak{r}$-classes. By definition of $\mathfrak{r}$, right multiplication by $x \in M$ maps each $\mathfrak{r}$-class again into an $\mathfrak{r}$-class, thus we have a right M-action on S, and this shows that the M-actions on finite sets correspond to right admissible equivalences on M of finite index.

(ii) These remarks can be applied to regular languages, bearing in mind that the acceptors for these languages are just finite sets with an X^*-action. Thus let A be an acceptor with initial state 0 and set of final states F, and let L be the language accepted by A. Then L is the union of all the $\mathfrak{r}_0$-classes mapping 0 to some state in F; hence L is the union of certain $\mathfrak{r}_0$-classes, where $\mathfrak{r}_0$ is a right admissible equivalence of finite index. Conversely, if L is the union of a finite number of classes of a right admissible equivalence of finite index, we can form an acceptor for the language by taking the quotient with respect to this equivalence. In this way we obtain another criterion for the regularity of a language:

Given $L \subseteq X^*$, we form the least right admissible equivalence $\mathfrak{r}(L)$ on X^* for which L is a union of $\mathfrak{r}(L)$-classes; $\mathfrak{r}(L)$ may be defined explicitly by the rule

$$u\mathfrak{r}(L)v \text{ if and only if } uw \in L \Leftrightarrow vw \in L \text{ for all } w \in X^*.$$

It is called the *Nerode equivalence* associated with L on X^*. Now it is clear that we have (Nerode [58], Teissier [51])

Theorem 5.1

A language L in X is regular if and only if the Nerode equivalence $\mathfrak{r}(L)$ on X^* is of finite index. ∎

A similar discussion can be carried through for the least congruence $\mathfrak{q}$ such that L is a union of $\mathfrak{q}$-classes. This is the *Myhill congruence* $\mathfrak{q}(L)$ associated with L and may be defined by the rule

$$u\mathfrak{q}(L)v \text{ if and only if } yuz \in L \Leftrightarrow yvz \in L \text{ for all } y, z \in X^*.$$

As before we obtain (Myhill [57], Teissier [51]):

Theorem 5.2

A language L is regular if and only if the Myhill congruence $\mathfrak{q}(L)$ on X^* is of finite index. ∎

A slight reformulation leads to the

Corollary

A language L is regular if and only if there is a homomorphism $f: X^* \to M$ to a finite monoid M such that $L = f^{-1}(f(L))$. ∎

E.g. the regular languages on one letter are all of the form x^{an+b} $(n = 0, 1, 2, \ldots)$, for fixed a, b. The language $\{x^m y^n \mid n \geqslant m \geqslant 0\}$ is easily seen not to be regular, for if $i < j$, we have $x^i y^{j-1} \in L$ but $x^i y^{i-1} \notin L$, hence x^i, x^j lie in different classes of the Nerode equivalence, which therefore has infinite index.

(iii) For some purposes (e.g. transductions) it is useful to consider direct products of free monoids; regularity corresponds here to several inequivalent notions.

Given any monoid M, a subset L of M is said to be *rational* if it can be obtained from finite subsets of M by the operations of union, product and * (i.e. forming from P the submonoid P^* generated by P). When $M = X^* \times Y^*$, this can be shown to agree with the notion of a regular transduction defined in §4.

A subset L of a monoid M is said to be *recognizable* if it is the union of congruence classes of a congruence of finite index. Let us call a topological monoid *coarse* if the topology is induced by a homomorphism to a finite monoid (with the discrete topology) then a recognizable subset of M is a subset which is closed in some coarse topology on M.

It is easily seen that if $f: M \to N$ is a homomorphism of monoids, then f preserves rational sets, while f^{-1} preserves recognizable sets. Moreover, by Theorems 2.1 and 5.2, the notions "rational" and "recognizable" coincide for free monoids. It follows that any recognizable set is rational. For if L is recognizable in M, and $f: F \to M$ is a surjective homomorphism from a free monoid F (which always exists), then $f^{-1}(L)$ is recognizable in F, hence rational, and so $L = f(f^{-1}(L))$ is rational. But in general these notions are distinct, e.g. on a direct product of free monoids we have the following criterion (cf. Fliess [70] where this result is ascribed to Mezei):

Theorem 5.3

A subset of $M_1 \times M_2$ is recognizable if and only if it is a finite union of products $R_1 \times R_2$ of recognizable subsets of M_1 and M_2.

This follows because any coarse topologies on M_1 and M_2 (always compatible with the monoid structure) lead to a coarse topology on $M_1 \times M_2$, via the projections $M_1 \times M_2 \to M_i$ and conversely, via the injections $M_i \to M_1 \times M_2$. ∎

(iv) Each CF-language may be written as a monoid equation, by replacing the non-terminal variable α by L_α, the set of words derivable from α. Thus each nonterminal variable is replaced by a subset of X^* and σ becomes the language itself. It will be convenient to use + as union sign and to identify 1-element sets with their members. Let us take e.g. $\{x^m y x^n | m \geqslant n \geqslant 0\}$ (Example 4); replacing σ, λ by L, A respectively, we have

$$L = y + xA + xLx, \quad A = y + xA.$$

We shall apply this description to a particularly simple type of regular language. Let L be given by a grammar with alphabet $X = \{x_1, \ldots, x_m\}$, non-terminal variables $\sigma, \alpha_1, \ldots, \alpha_m$ and rules such that σ occurs in just one rule on the left, while the right-hand side of any rule is of the form x_i or $x_i\alpha_i$. In equational form (writing A_i for α_i) we have

$$L = x_{i_1}A_{i_1} + \cdots + x_{i_r}A_{i_r} + x_{k_1} + \cdots + x_{k_t},$$

$$A_i = \delta_{ij_1}x_{j_1} + \cdots + \delta_{ij_s}x_{j_s} + \delta_{i1}x_1A_1 + \cdots + \delta_{im}x_mA_m.$$

Here $\delta_{ij} = 1$ or 0 according as A_i contains a word beginning with x_j or not,

$$I = \{x_{i_1}, \ldots, x_{i_r}\}$$

are the initial letters of words in L, $J = \{x_{j_1}, \ldots, x_{j_s}\}$ are the final letters of words in L and $I \cap J = \{x_{k_1}, \ldots, x_{k_t}\}$. Thus L may also be described as

$$L = \{IX^* \cap X^*J\} \backslash X^*DX^*,$$

where $D \subseteq (X^*)^2$ is the set of pairs $x_i x_j$ such that $\delta_{ij} = 0$. A language of this form is called a *standard K-language* or also a *local language*; to test if a word belongs to it we need not look at the word as a whole but only at the first and last letter and at pairs of adjacent letters. Thus if we form a graph on X as set of nodes, with a segment from x to y whenever $xy \notin D$, then the words of the language are represented by all the paths beginning in I and ending in J.

The importance of K-languages stems from the following result (Myhill [57]):

Theorem 5.4

Given any regular language L, there exists a standard K-language L_0 and a homomorphism ϕ such that $\phi(L_0) = L$.

The proof proceeds by taking a regular grammar for L and associating with each letter x_i occurring on the right of a rule the symbol (x_i, σ), with each pair $x_i\ \alpha_j$ the symbol (x_i, α_j) $[x_i, \alpha_j]$ and with σ the symbol $[\sigma]$. We then obtain a standard K-language with alphabet (x_i, α_j) and non-terminal variables $[x_i, \alpha_j]$, $[\sigma]$, while $(x_i, \alpha_j) \mapsto x_i$ is the required homomorphism. ∎

There is an analogous characterization of regular transductions (Nivat [68]):

Theorem 5.5

Any homomorphism of a standard K-language into $X^* \times Y^*$ is a regular transduction, and conversely, every regular transduction may be obtained in this way, as homomorphic image of a standard K-language. ∎

This result shows that the inverse of a regular transduction is again regular, a fact which was not obvious from the definition.

(v) There is a similar analysis of CF-languages, with the help of an important special class, the Dyck languages. We begin with some general remarks on homomorphisms of monoids.

Given any homomorphism of monoids, $f\colon M \to N$, the inverse image of 1 under f is called the *kernel* of f, $\ker f$. This kernel is always a submonoid, and if M is free then so is $\ker f$. To see this we recall the criterion (Cohn [62, 71″]):

A submonoid T of a free monoid M is free if and only if for any a, b, a', $b' \in T$ satisfying $ab' = ba'$ there exists $c \in T$ such that $a = bc$ or $b = ac$.

Given $f\colon M \to N$, where M is free, and $ab' = ba'$, where $a, b, a', b' \in \ker f$, we have $a = bc$ or $b = ac$, say the former, for some $c \in M$, because M is free. Now $1 = f(a) = f(b)f(c) = f(c)$, hence $c \in \ker f$.

We shall take M to be free on $X \cup X'$, where X' is a set bijective with X by the rule $x \leftrightarrow x'$, and take $f\colon M \to F_X$ to be the homomorphism from M to F_X, the free group on X, in which $x \mapsto x$ and $x' \mapsto x^{-1} (x \in X)$. The kernel is written D^* and is called the *Dyck language* on $X \cup X'$. By what has been said, D^* is a free monoid; its elements are all words in $X \cup X'$ which reduce to 1 on omitting all occurrences of xx' and $x'x$. The free generating set of D^* is the set D of words of the form xfx' or $x'fx$, where $f \in D^*$. This justifies the notation D^*.

We obtain a submonoid D_r of D^* by allowing only those words which reduce to 1 on omitting occurrences of xx'; this is called a *restricted Dyck language*. It may be obtained as the kernel of a homomorphism to a certain monoid (the bicyclic monoid and its polycyclic generalizations, cf. Clifford-Preston [61] and Nivat [68]), and so is again free. An example of a restricted Dyck language is the set of bracket formations obtainable from a given set of pairs of brackets $(_i,)_i$ (cf. p. 268).

The Dyck language is context-free, a *CF*-grammar being

$$\sigma \to \sigma\sigma, \ \sigma \to x\sigma x', \ \sigma \to x'\sigma x (x \in X), \ \sigma \to 1.$$

This grammar turns out to be ambiguous, but an unambiguous grammar is easily constructed (Gross-Lentin [70], p. 187).

Let C be a standard K-language, then the intersection $C \cap D^*$ with a Dyck language is necessarily a *CF*-language, as we saw in §4. Any language of the form $C \cap D^*$, where C is a standard K-language and D^* is a Dyck language, possibly restricted, is called a *standard CF-language*. Thus a standard *CF*-language is subject to two kinds of restriction, those due to C, which are "local", while those due to D^*, though more subtle, are the same in all cases. What gives standard CF-languages their importance is the following result (Chomsky-Schützenberger [63]):

Theorem 5.6

Every CF-language L can be obtained as a homomorphic image of a standard CF-language. Thus there is a Dyck language D^* *(possibly restricted) and a standard K-language C, with a homomorphism* ϕ *such that* $L = \phi(C \cap D^*)$.

The proof consists essentially in enclosing all occurrences of variables on the right of the rewriting rules in brackets, serially numbered: $(_1,)_1$, $(_2,)_2$, ..., and replacing each terminal letter x by xx'. We thus obtain a Dyck language which is also regular (in fact it is a standard K-language): each word is bracketed so that it can be tested by a finite acceptor. Now the homomorphism consists in setting the added symbols equal to 1. (cf. Arbib [69] for a detailed proof). E.g., to obtain the mirror language $L = \{f\tilde{f} \mid f \in X^*\}$, let C be the standard K-language on $X \cup X'$ with initials $x \in X$ and forbidden pairs $x'y (x, y \in X)$, then $C \cap D^* = \{f\bar{f} \mid f \in X^*\}$, where $f \mapsto \bar{f}$ is the anti-isomorphism $X \to X'$ defined by $\bar{x} = x'$. Now the homomorphism $x \mapsto x$, $x' \mapsto x$ maps $C \cap D^*$ to L. ∎

This theorem throws some light on the structure of CF-languages: while the class of standard CF-languages is closed under intersections, the class of all CF-languages is not, so the irregularities are in some sense due to the homomorphism. This is borne out by the fact that, given two homomorphisms $f, g: X^* \to Y^*$ between free monoids, there is no general procedure for determining whether a non-empty word u in X exists such that $f(u) = g(u)$; in essence this is again Post's correspondence problem (Gross-Lentin [70]).

6. POWER SERIES

(i) The description of languages by formal power series in non-commuting indeterminates, in many ways the simplest and most natural, is due to Schützenberger [61]. Starting from any commutative ring K, we can form the free K-algebra on X, $K\langle X\rangle$ as the monoid algebra of X^* over K. Its power series completion is denoted by $K\langle\langle X\rangle\rangle$; since X is finite, no ambiguity arises (for infinite X the structure of $K\langle\langle X\rangle\rangle$ depends on whether the degrees of the elements of X are bounded).

An element of $K\langle\langle X\rangle\rangle$ is said to be *rational* if it can be obtained from the elements of $K\langle X\rangle$ by repeated division (by elements with constant term 1) together with the other ring operations. As Schützenberger [62] has observed, this is so if and only if the element can be obtained as a component of the solution of a matrix equation

$$Au = a, \tag{1}$$

where a is a column vector in $K\langle X\rangle$ and A a matrix which is invertible over $K\langle\langle X\rangle\rangle$. This observation, suitably reformulated, turns out to be valid for entirely arbitrary rings, and it lies at the basis of the author's study of homomorphisms of rings into skew fields (Cohn [71″], Chapter 7).

Let $\varepsilon: K\langle\langle X\rangle\rangle \to K$ be the augmentation mapping, i.e. the K-algebra homomorphism defined by $\varepsilon(X) = 0$. We observe that a square matrix A over $K\langle X\rangle$ is invertible over $K\langle\langle X\rangle\rangle$ precisely when its image $\varepsilon(A)$ is invertible over K. In that case we can multiply by $\varepsilon(A)^{-1}$ to make a reduction to the case where $\varepsilon(A) = I$. Thus we may in (1) take $A = I - B$, where B has entries with zero constant terms; (1) now becomes

$$u = Bu + b, \tag{2}$$

or in components: $u_i = \sum b_{ij}u_j + b_i$, where b_i, $b_{ij} \in K\langle X\rangle$, $\varepsilon(b_{ij}) = 0$. The rational power series form a subalgebra of $K\langle\langle X\rangle\rangle$, denoted by $K_{\text{rat}}\langle X\rangle$.

An element $f \in K\langle\langle X\rangle\rangle$ is said to be *algebraic*, if it is of the form $f = u_1 + \gamma$, where $\gamma \in K$ and u_1 is a component of the solution of a system of equations

$$u_i = \phi_i(u, x), \tag{3}$$

where ϕ_i is a (non-commutative!) polynomial in the u's and x's without constant term or linear term in the u's. It is easily seen that we can solve for u_i by successive substitution and obtain a unique solution in $K\langle\langle X\rangle\rangle$. The algebraic power series form again a subalgebra $K_{\text{alg}}\langle X\rangle$, and we have the inclusions:

$$K\langle X\rangle \subseteq K_{\text{rat}}\langle X\rangle \subseteq K_{\text{alg}}\langle X\rangle \subseteq K\langle\langle X\rangle\rangle.$$

(ii) The rules of a given grammar, which were interpreted as monoid equations in §5, can now be interpreted as equations in the power series ring. Thus let G be a grammar on X; with each word w on X we associate a non-negative integer a_w, the number of ways of deriving w in G. The power series

(4) $$f = \sum a_w w \text{ in } Z\langle\langle X\rangle\rangle$$

describes the resulting language completely, in fact it takes multiplicities into account. The language $L(G)$ is just the *support* of f. In (4) all the coefficients are non-negative, but we can interpret a series with coefficients of both signs as the *difference* of two languages. The operations of union, product and * occurring in Theorem 2.1 correspond to sum, product and quasi-inverse: $x^* = 1 + x + x^2 + \cdots = (1 - x)^{-1}$.

To give an example, the language xy^* has rules $\sigma \to x$, $\sigma \to \sigma y$. These correspond to the monoid equation $L = x + Ly$; the power series equation is obtained by replacing L by a letter such as u:

$$u = x + uy,$$

and it has the solution $u = x(1 - y)^{-1} = \sum xy^n$. Clearly this series is rational and it is not hard to see that every regular grammar gives rise to a rational series with non-negative coefficients (cf. Theorem 2.1).

To state the converse we need the notion of a *semiring*. This is a monoid under addition and multiplication, linked by the distributive laws, thus it only fails to be a ring because subtraction is missing. E.g. the non-negative integers **N** form a semiring, and so does the set $\mathbf{N}\langle\langle X\rangle\rangle$ of formal power series over **N**. Rational and algebraic power series are defined as before. Now we have (Kleene [56]):

Theorem 6.1

A language L is rational if and only if it is the support of a rational power series in $\mathbf{N}\langle\langle X\rangle\rangle$. ∎

Similarly the algebraic power series corrpesond to *CF*-languages (Schützenberger [62]):

Theorem 6.2

A language L is context-free if and only if it is the support of an algebraic power series in $\mathbf{N}\langle\langle X\rangle\rangle$. ∎

Thus the series $u = \sum x^n y^n$ is algebraic, for it is of the form $u = v + 1$, where

$$v = xvy + xy.$$

That the series $\sum x^n y^n$ is not rational can be seen by inspecting its Hankel matrix, soon to be defined.

(iii) Let M be a direct product of free monoids; we can again form the monoid algebra $R = K[M]$ of M over K, and its completion by formal power series, $\hat{R}$. Rational power series may be defined as before, as components of solutions of equations of the type (1). A series f is said to be *recognizable*, if there is a matrix representation $\mu: M \to K_n$ and a matrix $P \in K_n$ such that

$$f = \sum tr(P\mu(w))w, \tag{5}$$

or equivalently, if there is a matrix representation $\mu: M \to K_n$ and $\gamma \in K$ such that $f - \gamma$ is the $(1, n)$-entry of $\sum \mu(w)w$ (Schützenberger [62], Fliess [70]). It can then be shown that a subset of M is recognizable (or rational) if and only if it is the support of a recognizable (or rational) series with coefficients 0 or 1.

To obtain a connexion between rational and recognizable series, corresponding to Theorem 5.3, suppose first that M is free on $x_1, \ldots, x_m$. Then

$$\begin{aligned}\sum \mu(w)w &= I + \sum \mu(x_i)x_i + \sum \mu(x_ix_j)x_ix_j + \cdots \\ &= I + A + A^2 + \cdots = (I - A)^{-1},\end{aligned}$$

where $A = \sum \mu(x_i)x_i$. Thus a recognizable series in $K\langle\langle X\rangle\rangle$ is rational, and conversely, because A in (1) can always be taken to be linear in the x_i. This is the process of linearization by enlargement ('Higman's trick'): to get rid of a product ab, in the (n, n)-position say, we take the diagonal sum with 1 and apply elementary transformations:

$$\begin{pmatrix} c + ab & 0 \\ 0 & 1 \end{pmatrix} \to \begin{pmatrix} c + ab & a \\ 0 & 1 \end{pmatrix} \to \begin{pmatrix} c & a \\ -b & 1 \end{pmatrix}$$

(where only the bottom right-hand corner is shown).

Next let $M = M' \times M''$, where M' is free on $x_1, \ldots, x_r$ and M'' free on $y_1, \ldots, y_s$. Then

$$\begin{aligned}\sum \mu(w)w &= I + \sum \mu(x_i)x_i + \sum \mu(y_p)y_p + \sum \mu(x_ix_j)x_ix_j \\ &\qquad + \sum \mu(x_iy_p)x_iy_p + \sum \mu(y_py_q)y_py_q + \cdots \\ &= (I + \sum \mu(x_i)x_i + (\sum \mu(x_i)x_i)^2 + \cdots)(I + \sum \mu(y_p)y_p \\ &\qquad + (\sum \mu(y_p)y_p)^2 \cdots) = (I - A)^{-1}(I - B)^{-1},\end{aligned}$$

where $A = \sum \mu(x_i)x_i$, $B = \sum \mu(y_p)y_p$. These equations show very clearly the crucial effect of commutativity on the form the series $\sum \mu(w)w$ can take. The result (cf. Fliess [70″]) may be summed up as

Theorem 6.3

Let $X = \{x_i\}$ and $Y = \{y_p\}$ be two sets of non-commuting indeterminates, where the x's commute with the y's. Then the recognizable series in the x's and y's constitute the tensor product $K_{\text{rat}}\langle X\rangle \otimes K_{\text{rat}}\langle Y\rangle$. ▮

Thus a recognizable series in two commuting variables x, y has the form

$$f(x, y)/g(x)h(y),$$

where f, g, h are polynomials and $g(0) \neq 0$, $h(0) \neq 0$.

Every recognizable series is rational, but not conversely, e.g. if x, y commute, $(1 - xy)^{-1}$ is rational but not recognizable. To obtain a general criterion we introduce the Hankel matrix. With any series $f = \sum a_w w$ we associate the infinite matrix whose rows and columns are indexed by the monoid M:

$$H(f) = (a_{uv}).$$

This is the *Hankel matrix* of f; its rank is defined in the obvious way as the supremum of the ranks of its finite submatrices. Now one can prove (Schützenberger [61′], Carlyle-Paz [71], Fliess [74]):

Theorem 6.4

Let R be the formal power series ring of a direct product M of free monoids, with coefficients in a field. Then $f \in R$ is recognizable if and only if its Hankel matrix $H(f)$ has finite rank. If the rank is N, then the matrices in (5) *may be taken to be $N \times N$.* ▮

There is also an analogue of the pumping lemma for recognizable series (Fliess [71]):

Theorem 6.5

Let M, R be as in Theorem 6.4 *and let $f = \sum a_w w \in R$ be recognizable but not a polynomial. Then there exist words u, v, $w \in M$, $v \neq 1$ such that the series in the central indeterminate t:*

$$\sum a_{uv^n w}\, t^n$$

is rational (*i.e. in $K(t)$*) *but not a polynomial in t.* ▮

As we have seen, when M is free, rational and recognizable series are the same, and either of Theorems 6.4, 6.5 may be used to show that $\sum x^n y^n$ is not rational.

With every subset P of a monoid we can define a Hankel matrix $H(f)$, where f is the characteristic series of P: $f = \sum a_w w$, where $a_w = 1$ or 0 according as w is in P or not. Then $H(f)$ has finite rank if and only if P is recognizable (Fliess [72]).

(iv) As an example of an algebraic series consider the restricted Dyck language on $x_1, \ldots, x_m, x_1', \ldots, x_m'$ (with $x_i x_i' \mapsto 1$). This has the equation

$$u = 1 + \textstyle\sum_i u x_i u x_i'.$$

Similarly, the Łukasiewicz language on $x_0, \ldots, x_m$ is defined by

$$u = x_0 + u x_1 + \cdots + u^m x_m.$$

The following necessary condition for an algebraic series, based on a theorem of Eisenstein (Pólya-Szegö [25]) is often useful (Fliess [72]):

Theorem 6.6

Let A be an integral domain with field of fractions K. Given any algebraic series f in $K\langle\langle X\rangle\rangle$, there exists $m \in A$, $m \neq 0$ such that on substituting mx in f for each $x \in X$ we obtain an algebraic series in $A\langle\langle X\rangle\rangle$.

The proof is as in the classical case: in the equations for f we can make the substitution $x \mapsto mx$ so that all coefficients come to lie in A. E.g. the criterion shows that $\sum x^n/2^{n^2}$ is not algebraic. ∎

(v) In order to describe the intersection of languages we introduce the following operation on power series. With $f = \sum a_u u$, $g = \sum b_u u$ we associate the series

$$f \odot g = \textstyle\sum c_u u, \text{ where } c_u = a_u b_u.$$

This operation is of course well known in analysis, where it is called the *Hadamard product*, and this name is used here too. If f, g have supports L, M respectively, then $f \odot g$ has support $L \cap M$. Clearly the Hadamard product is commutative, associative and distributive over addition. E. Borel observed that the Hadamard product of two rational series in one variable is again rational, but this does not extend to two commuting indeterminates, as the following example due to Hurwitz [99] shows;

$$f = \sum x^n y^n = (1 - xy)^{-1},$$

$$g = \sum \binom{m+n}{n} x^m y^n = (1 - x - y)^{-1},$$

$$f \odot g = \sum \binom{2n}{n} x^n y^n = (1 - 4xy)^{-1/2}.$$

For any ring R, denote the $n \times n$ matrix ring over R by $\mathfrak{M}_n(R)$ or also R_n, then it is not hard to see that

$$\mathfrak{M}_n(K_x([M]) = \mathfrak{M}_n(K)_x[M],$$

where M is a direct product of free monoids, and x = rat, alg or rec (= recognizable) If $\mu: X^* \to K_n$ is a representation and we substitute in any element of $K_x[M]$, $\mu(x)x$ for x, we obtain a series in $\mathfrak{M}_n(K_x[M])$. In this way we find (Schützenberger [62])

Theorem 6.7

In the power series ring over a direct product of free monoids, the Hadamard product of a rational (*or recognizable or algebraic*) *series and a recognizable series is again rational* (*or recognizable or algebraic*). ∎

In particular, applied to $\mathbf{Z}\langle\langle X\rangle\rangle$ this yields another proof that the intersection of a rational and a *CF*-language is *CF*. For series in one variable Theorem 6.7 was first proved by Jungen [31], who also observed that the Hadamard product of two algebraic series need not be algebraic:

$$f = \sum \binom{2n}{n} x^n = (1 - 4x)^{-1/2},$$

$$f \odot f = \sum \binom{2n}{n}^2 x^n = \frac{2}{\pi}\int_0^{\pi/2} \frac{d\theta}{\sqrt{(1-4x.\sin^2\theta)}} = \frac{2}{\pi}\int_0^1 \frac{dt}{\sqrt{((1-t^2)(1-4xt^2))}}.$$

However, over a perfect field of characteristic $p \neq 0$, the Hadamard product of two algebraic series is algebraic (Furstenberg [67], Fliess [72]).

(vi) The free algebra and the free power series ring over a field have been studied from various aspects (Cohn [71″]), and it is natural to examine their relation to $K_{\text{rat}}\langle X\rangle$ and $K_{\text{alg}}\langle X\rangle$ more closely. In $K\langle\langle X\rangle\rangle$ we can equate cofactors of the generators x_i, or more generally, if the elements $u_i \in X^*$ are such that none is a right multiple of any other, then $\sum u_i f_i = \sum u_i g_i$ implies $f_i = g_i$. Hence the operation $\sum u_i f_i \mapsto f_\nu$ (for any ν) is well-defined; it is called a *right transduction*, and left transductions are defined similarly. These operations correspond to the transductions defined in §4. A transduction maps each rational or algebraic series to a series of the same type (Fliess [72]), and this has many interesting consequences. To describe them we need some definitions.

A ring R is called a *free right ideal ring* or *right fir* for short, if every right ideal of R is free of unique rank. *Left firs* are defined similarly and a left and right fir is called a *fir*. For commutative rings a fir is just a principal ideal domain, and in fact any principal ideal domain, commutative or not, is a fir, but the class of firs is much wider than this, e.g. any free algebra over a field is a fir. By a *semifir* one understands a ring in which any finitely generated right ideal (or equivalently, every finitely generated left ideal) is free, of unique rank. Commutative semifirs are just Bezout rings; to give a general example, the free power series ring over a field, $K\langle\langle X\rangle\rangle$ is a semifir, but not a fir, unless X reduces to a single element.

A ring extension $R \subseteq S$ is said to be *inert* if any factorization of an element of R in S can be reduced to one in R, i.e. given $c \in R$, if $c = ab$ $(a, b \in S)$, then there exists a unit u in S such that $au, u^{-1}b \in R$. Let M be a right R-module; a family (u_i) of elements of M is said to be (linearly) *independent* over R if $\sum u_i a_i = 0 (a_i \in R)$ implies $a_i = 0$ for all i. The family is called *weakly independent* over R if no u_i lies in the R-module generated by the others. Clearly any independent family is weakly independent, but the converse need not hold. Now if S is a semifir and R is a subring such that any weakly independent family in S, qua right R-module, is independent, then it is easily shown that R is again a semifir and is inert in S. This condition is easily verified for the pair $K_{\text{rat}}\langle X\rangle$, $K\langle\langle X\rangle\rangle$, where K is a field, using left transductions, hence $K_{\text{rat}}\langle X\rangle$ is a semifir, inert in $K\langle\langle X\rangle\rangle$. Corresponding results hold for $K_{\text{alg}}\langle X\rangle$. The ring $K_{\text{rat}}\langle X\rangle$ is a fir (Cohn-Dicks [76]), but $K_{\text{alg}}\langle X\rangle$ is not a fir (Cohn [71″], Exercise 5.8.6, p. 209).

The ring $K\langle\langle X\rangle\rangle$ has rather simple factorization properties, when K is a field: it is a *rigid UFD* (Cohn [62′], [71″], Chapter 3). This means that every element not 0 or unit can be written as a product of atoms (i.e. unfactorable elements): $a = p_1 \dots p_r$ and if $a = q_1 \dots q_s$ is another such factorization, then $r = s$ and $q_i u_i = u_{i-1} p_i (i = 1, \ldots, r)$, where the u's are units and $u_0 = u_r = 1$. Let S be a rigid *UFD* and R an inert subring such that any non-unit of R is a non-unit in S; then R is again a rigid *UFD*. It follows that both $K_{\text{rat}}\langle X\rangle$ and $K_{\text{alg}}\langle X\rangle$ are rigid *UFDs* (Fliess [72]).

7. TRANSFORMATIONAL GRAMMARS

We have already seen that the classes of formal languages described earlier fail to do justice to natural languages: context-sensitive languages are too wide, while context-free languages are too restrictive. Here is another very simple illustration of the same fact. Many natural languages —including English—use two alphabets, capitals and lower case letters, and it is a rule that the first letter in each sentence is a capital. It is not at all obvious how to accommodate this type of rule in a formal grammar. To give a further example, many sentences occur in several different forms, such as active and passive; e.g. consider the sentences:

1. The cat ate the mouse.
2. The mouse ate the cat.
3. The mouse was eaten by the cat.
4. The cat was eaten by the mouse.

A moment's glance shows that 1 and 3 mean the same thing, as do 2 and 4, but *not* 1 and 2, nor 1 and 4. Again there is no mechanism (within the frame-work of phrase-structure grammars) to take account of these facts. This difficulty was recognized by Chomsky, who proposed the notion of transformation rules. Roughly speaking, instead of merely providing rules of sentence formation, one also has rules of sentence transformation. Without entering into the details, we can easily see how this might apply to the above examples. A language might have the transformation rule which replaces the first letter of a sentence by the corresponding capital, and we only accept sentences where this rule has been applied. Similarly one can formulate rules for turning an active form into the passive, and so obtain from each sentence (if expressed in active form) another one (in passive form).

The study of transformations (transformational grammar) is less developed than that of phrase-structure grammars. It also has less of an algebraic character, which is why we do not discuss it in greater detail.

A serious difficulty is that transformation rules applied to context-free grammars can lead to all languages of type 0 (Peters and Ritchie [71]). This means that the transformation rules are not sufficiently restrictive. However, one would hope that a comparison with the rules of natural grammars will eventually lead to a more accurate picture; for an attempt in this direction see Ginsburg-Partee [69].

Bibliography and Name Index

The page references at the end of each entry indicate the places in the text where the entry is quoted; other references to an author are listed after his name.

Albert, A. A. 278

[34] On a certain algebra of quantum mechanics, *Ann. of Math.* **35** (1934) 65–73. 298

[47] A structure theory for Jordan algebras, *Ann. of Math.* **48** (1947) 546–567. 305

[50] A note on the exceptional Jordan algebra, *Proc. Nat. Acad. Sci. USA* **36** (1950) 372–374. 298

Albert, A. A. and Paige, L. J.

[59] On a homomorphism property of certain Jordan algebras, *Trans. Amer. Math. Soc.* **93** (1959) 20–29. 299

Amitsur, S. A. 278

Arbib, M. A. *see also* **Kalman, R. E.**

[69] *Theories of Abstract Automata*, Prentice-Hall (Englewood Cliffs, 1969). 346, 354f., 357, 361, 371

Artin, E., Nesbitt, C. J., and Thrall, R. M.

[44] *Rings with Minimum Condition*, Univ. of Michigan Press (Ann Arbor, 1944). 44

Bar-Hillel, Y., Perles, M., and Shamir, E.

[61] On formal properties of simple phrase structure grammars, *Z. Phonetik*,

Sprachwiss. Kommunikat. **14** (1961) 143–172. Reprinted in *Handbook of Math. Psychology*, Vol. 3, J. Wiley (New York, 1965). 354, 365

Barwise, J.

[77] *Handbook of Logic* (ed.), North-Holland (Amsterdam, 1977). 251, 329

Barwise J. and Robinson, A.

[70] Completing theories by forcing, *Ann. Math. Logic* **2** (1970) 119–142.

Bates, Grace E.

[47] Free nets and loops and their generalizations, *Amer. J. Math.* **69** (1947) 499–550. 279

Beck, J. 318

Bell, J. L. and Slomson, A. B.

[69] *Models and Ultraproducts*, North-Holland (Amsterdam, 1969).

Bendix, P. B., *see* **Knuth, D. E.**

Bergman, G. M. 160

[75] Some category-theoretic ideas in algebra, *Proc. Internat. Cong. Math. Vancouver* 1974 (Vancouver, 1975) 285–296.

[77] On the existence of subalgebras of direct products with prescribed *d*-fold projections, *Alg. Universalis* **7** (1977) 341–356.

[78] The diamond lemma in ring theory, *Adv. in Math.* **29** (1978) 178–218. 160, 341

[81] Dependence relations and rank functions on free modules, to appear. 344

Bergman, G. M., and Clark, W. E.

[73] On the automorphism class group of the category of rings, *J. Algebra* **24** (1973) 80–99.

Berstel, J.

[79] *Transductions and Context-Free Languages*, Teubner (Stuttgart, 1979).

Bing, K.

[55] On arithmetical classes not closed under direct union, *Proc. Amer. Math. Soc.* **6** (1955) 836–846.

Birkhoff, G.

[33] On the combination of subalgebras, *Proc. Cambridge Phil. Soc.* **29** (1933) 441–464.

[35] On the structure of abstract algebras, *Proc. Cambridge Phil. Soc.* **31** (1935) 433–454. 55, 169

[37] Representability of Lie algebras and Lie groups by matrices, *Ann. of Math.* **38** (1937) 526–532. 294

[44] Subdirect unions in universal algebras, *Bull. Amer. Math. Soc.* **50** (1944) 764–768. 100

[46] Universal algebra, *Proc. Canad. Math. Cong.* (Montreal, 1946) 310–326.

[48] *Lattice Theory*, rev. ed., Amer. Math. Soc. Coll. Publ. Vol. 25 (New York, 1948). 77

Birkhoff, G. and Frink, O.

[48] Representation of lattices by sets, *Trans. Amer. Math. Soc.* **64** (1948) 299–316. 85

Birkhoff, G. and Whitman, P. M.

[49] Representation of Jordan and Lie algebras, *Trans. Amer. Math. Soc.* **65** (1949) 116–136.

Bokut', L. A.

[69] On Malcev's problem (Russian), *Sibirsk. Mat. Ž.* **10** (1969) 965–1005. 342

Boole, G. 4, 191

Boone, W. W.

[57] Certain simple unsolvable problems of group theory, *Indag. Math.* **19** [= *Proc. Kon. Ned. Akad.* (A) **60**] (1957) 22–27, 227–232. 155

Booth, T. L.

[67] *Sequential Machines and Automata Theory*, J. Wiley (New York, 1967). 346, 355f.

Borel, E. 376

Bourbaki, N.

[51] *Topologie générale*, Hermann (Paris, 1951) Ch. III-IV. 60

[54] *Théorie des ensembles*, Hermann (Paris, 1954) Ch. I-II. 1, 3

Bowtell, A. J.

[67] On a question of Malcev, *J. Algebra* **6** (1967) 126–139. 342

Britton, J. L.

[58] The word problem for groups, *Proc. London Math. Soc.* (3) **8** (1958) 493–506. 155

Burgess, J. P. 329

Cantor, D. G.

[62] On the ambiguity problem of Backus systems, *J. Assoc. Comput. Mach.* **9** (1962) 477–479. 355

Cantor, G. 239

Carlyle, J. W. and Paz, A.

[71] Realizations by stochastic finite automata, *J. Comput. System Sci.* **5** (1971) 26–40. 375

Carmichael, R.

[37] *Introduction to the Theory of Groups of Finite Order*, Ginn (Boston, 1937). 146

Carson, A. B.

[73] The model completion of the theory of commutative regular rings, *J. Algebra* **27** (1973) 136–146. 335

Cartan, H. and Eilenberg, S.

[56] *Homological Algebra*, Princeton Univ. Press (Princeton, 1956). 78, 296

Cartier, P.

[58] Remarques sur le théorème de Birkhoff-Witt, *Ann. scuola norm. sup. Pisa, Sci. fis. mat.* III Ser. **12** (1958) 1–4. 296

Chang, C. C.

[59] On unions of chains of models, *Proc. Amer. Math. Soc.* **10** (1959) 120–127. 222, 322

Chang, C. C. and Keisler, H. J.

[73] *Model Theory*, North-Holland (Amsterdam, 1973). 321

Chang, C. C. and Morel, A. C.

[58] On closure under direct product, *J. symb. Logic* **23** (1958) 149–154. 236

Cherlin, G.

[72] The model companion of a class of structures, *J. symb. Logic* **37** (1972) 546-556.

[73] Algebraically closed commutative rings, *J. symb. Logic* **38** (1973) 493–499.

Chomsky, N. 348, 379

[59] On certain formal properties of grammars, *Information and Control* **2** (1959) 137–167. 351f.

[59′] A note on phrase structure grammars, *Information and Control* **2** (1959) 393–395.

[64] *Aspects of the Theory of Syntax*, MIT Press (Cambridge, Mass., 1964).

Chomsky, N. and Miller, G. A.

[58] Finite state languages, *Information and Control* **1** (1958) 91–112.

Chomsky, N. and Schützenberger, M.-P.

[63] The algebraic theory of context-free languages, in *Computer Programming and Formal Systems* (ed. P. Braffort and D. Hirschberg), North-Holland (Amsterdam, 1963). 371

Church, A.

[56] *Introduction to Mathematical Logic,* Vol. I, Princeton Univ. Press (Princeton, 1956). 203, 206, 213

Clark, D. M.

[69] Varieties with isomorphic free algebras, *Colloq. Math.* **20** (1969) 181–187. 341

Clark, W. E. *see* **Bergman, G. M.**

Claus, V.

[71] *Stochastische Automaten*, Teubner (Stuttgart, 1971). 346, 363f.

Clifford, A. H. and Preston, G. B.

[61] *The Algebraic Theory of Semigroups*, Vol. 1, Amer. Math. Soc. (Providence, 1961). 370

Cohen, P. J.

[63] The independence of the continuum hypothesis, *Proc. Nat. Acad. Sci. USA* **50** (1963) 1143–1148, **51** (1964) 105–110. 239, 329

[66] *Set Theory and the Continuum Hypothesis*, Benjamin (New York, 1966). 329

Cohn, P. M.

[52] A theorem on the structure of tensor spaces, *Ann. of Math.* **56** (1952) 254–268. 284

[54] On homomorphic images of special Jordan algebras, *Canad. J. Math.* **6** (1954) 253–264. 299, 306

[56] Embeddings in semigroups with one-sided division, *J. London Math. Soc.* **31** (1956) 169–181. 282

[59] On the free product of associative rings, *Math. Zeits.* **71** (1959) 380–398. 144

[61] On the embedding of rings in skew fields, *Proc. London Math. Soc.* (3) **11** (1961) 511–530. 276

[62] On subsemigroups of free semigroups, *Proc. Amer. Math. Soc.* **13** (1962) 347–351. 370

[62′] Factorization in non-commutative power series rings, *Proc. Cambridge Phil.*

Soc. **58** (1962) 452–464. 378

[63] A remark on the Birkhoff-Witt theorem, *J. London Math. Soc.* **38** (1963) 197–203. 296

[64] Subalgebras of free associative algebras, *Proc. London Math. Soc.* (3) **14** (1964) 618–632. 255

[69] Dependence in rings II. The dependence number, *Trans. Amer. Math. Soc.* **135** (1969) 267–279. 342f.

[70] *Pure and Applied Algebra*, Inaugural lecture, Bedford College (London, 1970).

[71] The embedding of firs in skew fields, *Proc. London Math. Soc.* (3) **23** (1971) 193–213. 334, 343

[71′] Un critère d'immersibilité d'un anneau dans un corps gauche, *C. R. Acad. Sci. Paris*, Sér. A **272** (1971) 1442–1444.

[71″] *Free Rings and their Relations*, LMS monographs No. 2, Academic Press (London, New York, 1971) 342f., 370, 372, 377f.

[71‴] Rings of fractions, *Amer. Math. Monthly* **78** (1971) 596–615. 344

[74] The class of rings embeddable in skew fields, *Bull. London Math. Soc.* **6** (1974) 147–148. 344

[74′] *Universal Algebra*, Section 5 of 'Algebraic Structures' in *Encyclopedia Britannica* 15th ed. (Chicago, 1974).

[77] *Algebra*, Vol. 2, J. Wiley (Chichester, 1977). 313f.

[77′] *Skew Field Constructions*, LMS Lecture Notes No. 27, Cambridge Univ. Press (Cambridge, 1977). 261

Cohn, P. M. and Dicks, W.

[76] Localization in semifirs II, *J. London Math. Soc.* (2) **13** (1976) 411–418. 378

Conway, J. H.

[71] *Regular Algebra and Finite Machines*, Chapman and Hall (London, 1971).

Davis, Anne C. *see also* **Morel**

[55] A characterization of complete lattices, *Pacif. J. Math.* **5** (1955) 311–319.

Davis, M.

[58] *Computability and Unsolvability*, McGraw-Hill (New York, 1958). 355

Dedekind, J. W. R.

[00] Über die von drei Moduln erzeugte Dualgruppe, *Math. Ann.* **53** (1900) 371–403 [Ges. Werke II, 236–271]. 65

Diamond, A. H. and McKinsey, J. C. C.

[47] Algebras and their subalgebras, *Bull. Amer. Math. Soc.* **53** (1947) 959–962. 286

Dicker, R. M.

[63] The substitutive law, *Proc. London Math. Soc.* (3) **13** (1963) 493–510.

Dicks, W. *see* **Cohn, P. M.**

Dieudonné, J. 103

Dilworth, A. P. and Gleason, A. M.

[62] A generalized Cantor theorem, *Proc. Amer. Math. Soc.* **13** (1962) 704–705. 24

Douady, A. 312

Dubreil, P.

[43] Sur les problèmes d'immersion et la théorie des modules, *C. R. Acad. Sci. Paris* **216** (1943) 625–627. 273

Dwinger, Ph.

[61] *Introduction to Boolean Algebras*, Physica-Verlag (Würzburg, 1961). 198

Dwinger, Ph. and Yaqub, F. M.

[63] Generalized free products of Boolean algebras with an amalgamated subalgebra, *Indag. Math.* **25** (1963) 225–231. 199

Dyck, W. 370

[1882] Gruppentheoretische Studien, *Math. Ann.* **20** (1882) 1–44. 152

Ehrenfeucht, A.

[61] An application of games to the completeness problem for formalized theories, *Fund. Math.* **49** (1961) 129–141.

Ehrig, H. and Pfender, M.

[72] *Kategorien und Automaten*, W. de Gruyter (Berlin, 1972).

Eilenberg, S. *see* **Cartan, H.**

Eilenberg, S.

[74] *Automata, Languages and Machines*, vol. A, Academic Press (New York, 1974).

Eilenberg, S. and Mac Lane, S.

[45] General theory of natural equivalences, *Trans. Amer. Math. Soc.* **58** (1945) 231–294.

Eilenberg, S. and Moore, J. C.

[65] Adjoint functors and triples, *Ill. J. Math.* **9** (1965) 381–398.

Eilenberg, S. and Steenrod, N.

[52] *Foundations of Algebraic Topology*, Princeton Univ. Press (Princeton, 1952). 312

Eilenberg, S. and Wright, J. B.

[67] Automata in general algebras, *Information and Control* **11** (1967) 452–470.

Eisenstein, F. G. M. 376

Eklof, P. C.

[76] Whitehead's problem is undecidable. *Amer. Math. Monthly* **83** (1976) 775–788.

Eklof, P. C. and Sabbagh, G.

[71] Model completions and modules, *Ann. Math. Logic* **2** (1971) 251–295. 334

Erdös, P. 125

Eršov, Yu. L.

[62] On axiomatizable model classes with infinite signature (Russian), *Algebra i Logika Sem.* **1**, No. 4 (1962) 32–44. 224

Evans, T.

[51] On multiplicative systems defined by generators and relations, I. Normal form theorems, *Proc. Cambridge Phil. Soc.* **47** (1951) 637–649. 279

[53] Embeddability and the word problem. *J. London Math. Soc.* **28** (1953) 76–80.

[67] The spectrum of a variety, *Z. Math. Logik Grundl. Math.* **13** (1967) 213–218.

[78] Word problems, *Bull. Amer. Math. Soc.* **84** (1978) 789–802.

Falb, P. L. *see* **Kalman, R. E.**

Feigelstock, S.

[65] A universal subalgebra theorem, *Amer. Math. Monthly* **72** (1965) 884–888.

Felscher, W.

[65] Zur Algebra unendlich langer Zeichenreihen, *Z. Math. Logik Grundl. Math.* **11** (1965) 5–16.

[65′] Adjungierte Funktoren und primitive Klassen, *Sitzber. Heidelberg. Akad. d. Wiss. Math.-Naturw.* Kl. (1965) 447–509.

Fleischer, I.

[55] A note on subdirect products, *Acta Math. Acad. Sci. Hungar.* **6** (1955) 463–465. 337

[68] On Cohn's theorem, *J. London Math. Soc.* **43** (1968) 237–238.

Fliess, M.

[70] Inertie et rigidité des séries rationnelles et algébriques, *C. R. Acad. Sci. Paris*, Sér. A, **270** (1970) 221–223. 368, 374

[70′] Transductions algébriques, *RIRO* R-1 (1970) 109–125.

[70″] Séries reconnaissables, rationnelles et algébriques, *Bull. Sci. Math.* **94** (1970) 231–239. 375

[71] Deux applications de la representation matricielle d'une série rationnelle non-commutative, *J. Algebra* **19** (1971) 344–353. 375

[72] *Sur certaines familles de séries formelles*, Thèse, Paris (Université de Paris VII, 1972) 376ff.

[74] Matrices de Hankel, *J. Math. pures et appl.* **53** (1974) 197–224. 375

Foster, A. L. 179f., 199

[55] The identities of – and unique subdirect factorization within – classes of universal algebras, *Math. Zeits.* **62** (1955) 171–188.

[59] An existence theorem for functionally complete universal algebras, *Math. Zeits.* **71** (1959) 69–82.

Frayne, T., Morel, A., and Scott, D.

[62] Reduced direct products, *Fund. Math.* **51** (1962) 195–228. 210, 244

Frege, G. 247

Freyd, P. 311

[66] Algebra valued functors in general and tensor products in particular, *Colloq. Math.* **24** (1966) 89–106.

Friedman, H. 279

Frink, O. *see* **Birkhoff, G.**

Funayama, N. 98

Furstenberg, H.

[67] Algebraic function fields over finite fields, *J. Algebra* **7** (1967) 271–277. 377

Gabriel, P.

[62] Des catégories abeliennes, *Bull. Soc. Math. France* **90** (1962) 323–448. 6, 311

Galler, B. A.

[57] Cylindric and polyadic algebras, *Proc. Amer. Math. Soc.* **8** (1957) 176–183. 208

Ginsburg, S.

[66] *The Mathematical Theory of Context-Free Languages*, McGraw-Hill (New York, 1966).

Ginsburg, S. and Partee, B.
[69] A mathematical model of transformational grammars, *Information and Control* **15** (1969) 297–334. 380

Gleason, A. M. *see* **Dilworth, R. P.**

Glennie, C. M.
[63] *Identities in Jordan Algebras*, Ph.D. thesis (Yale University, 1963). 299
[66] Some identities valid in special Jordan algebras but not valid in all Jordan algebras, *Pacif. J. Math.* **16** (1966) 47–59.

Gödel, K. 206f., 213, 322
[40] *The Consistency of the Axiom of Choice and of the Generalized Continuum Hypothesis with the Axioms of Set Theory*, Princeton Univ. Press (Princeton, 1940). 1, 3, 239
[47] What is Cantor's continuum problem? *Amer. Math. Monthly* **44** (1947) 515–525. 240

Goetz, A. and Ryll-Nardzewski, C.
[60] On bases of abstract algebras, *Bull. Acad. Polon. d. Sci., Ser. Math. Astr. et Phys.* **8** (1960) 157–162. 340

Goldie, A. W.
[52] The scope of the Jordan-Hölder theorem in abstract algebra, *Proc. London Math. Soc.* (3) **2** (1952) 349–368. 94

Grätzer, G.
[78] *General Lattice Theory*, Birkhäuser (Basel, 1978).
[79] *Universal Algebra* (2nd ed.), Springer (Berlin, 1979).

Green, J. A.
[52] A duality in abstract algebra, *J. London Math. Soc.* **27** (1952) 64–73.

Gross, M.
[72] *Mathematical Models in Linguistics*, Prentice-Hall (Englewood Cliffs, 1972). 346

Gross, M. and Lentin, J.
[70] *Introduction to Formal Grammars*, Springer (Berlin, 1970). 346, 352, 355, 370f.

Grothendieck, A. 311

Hadamard, J. 376

Haigh, G.T. 179

Hales, A. W.
[64] On the non-existence of free complete Boolean algebras, *Fund. Math.* **54** (1964) 45–66. 318

Hall, M. Jr.
[50] A basis for free Lie rings and higher commutators in free groups, *Proc. Amer. Math. Soc.* **1** (1950) 575–581. 292
[59] *The Theory of Groups*, Macmillan (New York, 1959). 54, 255

Hall, P. 81, 105, 125, 127, 132, 171
[58] Some word problems, *J. London Math. Soc.* **33** (1958) 482–496. 118, 292

Halmos, P. R. 244
[61] *Naive Set Theory*, v. Nostrand (Princeton, 1961). 20

[62] *Algebraic Logic*, Chelsea (New York, 1962) [This is a collection of the author's papers on monadic and polyadic algebras, over the years 1954–1959]. 198, 208

Hankel, H. 375

Harris, Z. S.

[51] *Structural Linguistics* (formerly entitled: *Methods in Structural Linguistics*), Univ. of Chicago Press (Chicago, 1951, 1960).

Hartmanis, J. and Stearns, R. E.

[66] *Algebraic Structure Theory of Sequential Machines*, Prentice-Hall (Englewood Cliffs, 1966). 356

Harzheim, E.

[66] Über die Grundlagen der universellen Algebra, *Math. Nachr.* **31** (1966) 39–52.

Hausdorff, F. 27

[14] *Mengenlehre*, W. de Gruyter (Berlin, 1914), translated as: *Theory of Sets*, Chelsea (New York, 1957).

Henkin, L. 252

[49] The completeness of the first order functional calculus, *J. symb. Logic* **14** (1949) 159–166.

[50] Completeness in the theory of types, *J. symb. Logic* **15** (1950) 81–91. 207

[53] Some interconnections between modern algebra and mathematical logic, *Trans. Amer. Math. Soc.* **74** (1953) 410–427. 223

[60] On mathematical induction, *Amer. Math. Monthly* **67** (1960) 323–338. 247

Henkin, L. and Tarski, A.

[61] *Cylindric Algebras*, Proc. Symp. on pure Math. II: Lattice theory, Amer. Math. Soc. (1961) 81–113. 208

Herrlich, H. and Strecker, G. E.

[73] *Category Theory*, Allyn and Bacon (Boston, 1973).

Higgins, P. J. 47

[56] Groups with multiple operators, *Proc. London Math. Soc.* (3) **6** (1956) 366–416. 52

[63] Algebras with a scheme of operators, *Math. Nachr.* **27** (1963) 115–132. 127

Higman, G. 46, 374

[52] Ordering by divisibility in abstract algebras, *Proc. London Math. Soc.* (3) **2** (1952) 326–336. 122, 276

[59] Some remarks on varieties of groups, *Quarterly J. Math.* (2) **10** (1959) 165–178. 173

[61] Subgroups of finitely presented groups, *Proc. Roy. Soc.* **A262** (1961) 455–475. 155, 334

Higman, G. and Neumann, B. H.

[52] Groups as groupoids with one law, *Publ. Math. Debrecen* **2** (1952) 215–221. 165

Higman, G., Neumann, B. H., and Neumann, H.

[49] Embedding theorems for groups, *J. London Math. Soc.* **24** (1949) 247–254. 334

Higman, G. and Stone, A. H.

[54] On inverse systems with trivial limits, *J. London Math. Soc.* **29** (1954) 233–236. 312

Hirschfeld, J. and Wheeler, W. H.

[75] *Forcing, Arithmetic, Division Rings*, Lecture Notes in Math. No. 454, Springer (Berlin, 1975).

Hopcroft, J. E. and Ullman, J. D.

[69] *Formal Languages and their Relation to Automata*, Addison-Wesley (Reading, Mass., 1969). 346

Horn, A.

[51] On sentences which are true of direct unions of algebras, *J. symb. Logic* **16** (1951) 14–21. 235

Howie, J. M. and Isbell, J. R.

[67] Epimorphisms and dominions II, *J. Algebra* **6** (1967) 7–21.

Hurwitz, A.

[1899] Sur un théorème de M. Hadamard, *C. R. Acad. Sci. Paris* **128** (1899) 350–353, reprinted in *Ges. Werke* I, 482–484. 377

Isbell, J. R. *see* **Howie, J. M.**

Jacobson, N.

[54] Structure of alternative and Jordan bimodules, *Osaka Math. J.* **6** (1954) 1–71. 302ff.

[56] *Structure of Rings*, Amer. Math. Soc. Coll. Publ. Vol. 37 (Providence, 1956). 97, 132

[62] Macdonald's theorem on Jordan algebras, *Arch. Math.* **13** (1962) 241–250. 308

[68] *Structure and Representations of Jordan Algebras*, Amer. Math. Soc. Coll. Publ. Vol. 39 (Providence, 1968).

Jónsson, B. 99

[57] On direct decompositions of torsionfree abelian groups, *Math. Scand.* **5** (1957) 230–235. 97

[62] Algebraic extensions of relational systems, *Math. Scand.* **11** (1962) 197–205. 262

[66] The unique factorization problem for finite relational structures, *Colloq. Math.* **14** (1966) 1–32.

[68] Equational classes of lattices, *Math. Scand.* **22** (1968) 187–196.

[72] *Topics in Universal Algebra*, Lecture Notes in Mathematics, No. 250, Springer (Berlin, 1972).

Jónsson, B. and Tarski, A.

[47] *Direct Decompositions of Finite Algebraic Systems*, Univ. of Notre Dame Press (Notre Dame, 1947). 56, 97

[61] On two properties of free algebras, *Math. Scand.* **9** (1961) 95–101. 140ff.

Jungen, R.

[31] Sur les séries de Taylor n'ayant que des singularités algébrico-logarithmiques sur leur cercle de convergence, *Comment. Math. Helv.* **3** (1931) 266–306. 377

Kaiser, K.
[69] Über eine Verallgemeinerung der Robinsonschen Modellvervollständigung, *Z. Math. Logik Grundl. Math.* **15** (1969) 37–48.
Kalicki, J. and Scott, D. S.
[55] Equational completeness in abstract algebras, *Indag. Math.* **17** (1955) 650–659. 180
Kalman, R. E., Falb, P. L., and Arbib, M. A.
[69] *Topics in Mathematical Systems Theory*, McGraw-Hill (New York, 1969).
Kaplansky, I. 288
Kargapolov, M. I. 228
Keisler, H. J. *see also* **Chang, C. C.** 329
[61] Ultraproducts and elementary classes, *Indag. Math.* **23** (1961) 477–495. 220, 239, 241, 243
[61′] On some results of Jónsson and Tarski concerning free algebras, *Math. Scand.* **9** (1961) 102–106.
Kelley, J. L.
[55] *General Topology*, v. Nostrand (Princeton, 1955). 1, 20, 28, 31, 136
Kertész, A. 262
[60] On independent sets of elements in an algebra, *Acta Sci. Math. Szeged* **21** (1960) 260–269.
Kleene, S. C.
[52] *Introduction to Metamathematics*, North-Holland (Amsterdam, 1952). 203, 207
[56] Representation of events in nerve sets, in *Automata Studies*, (ed. C. E. Shannon and J. McCarthy), Princeton Univ. Press (Princeton, 1956), 3–40. 348, 350, 373
Klein, A. A.
[67] Rings nonembeddable in fields with multiplicative semigroups embeddable in groups, *J. Algebra* **7** (1967) 100–125. 342
[69] Necessary conditions for embedding rings into fields, *Trans. Amer. Math. Soc.* **137** (1969) 141–151. 269
Kleisli, H. 317
Knuth, D. E. and Bendix, P. B.
[70] Simple word problems in universal algebras, in *Computational Problems in Abstract Algebra*, (ed. J. Leech), Pergamon Press (Oxford, 1970) 263–297.
Kochen, S. B.
[61] Ultraproducts in the theory of models, *Ann. of Math.* **74** (1961) 221–261. 210, 244
Kogalovskii, S. R.
[59] On universal classes of algebras which are closed with respect to direct products (Russian), *Izvestiya vysš. učebn. Zaved. mat.* (1959) 88–96. 235
Krull, W. 196
Kruse, A. H.
[67] An abstract property P for groupoids such that locally locally P is weaker than locally P, *J. London Math. Soc.* **42** (1967) 81–85. 106

Kruskal, J. B.

[60] Well-quasi-ordering, the tree theorem and Vazsonyi's conjecture, *Trans. Amer. Math. Soc.* **95** (1960) 210–225. 124

Kuroda, S. Y.

[64] Classes of languages and linear-bounded automata, *Information and Control* **7** (1964) 207– 233. 362

Kuroš, A. G.

[35] Durchschnittsdarstellungen mit irreduziblen Komponenten in Ringen und sogenannten Dualgruppen, *Mat. Sbornik* **42** (1935) 613–616. 77

[47] Non-associative free algebras and free products of algebras (Russian, English summary), *Mat. Sbornik* **20** (62) (1947) 239–262. 287

[56] *Theory of Groups*, Chelsea (New York, 1956). 228

[63] *Lectures on General Algebra*, Chelsea (New York, 1963). 20, 51, 174

Lallement, G.

[79] *Semigroups and Combinatorial Applications*, J. Wiley (New York, Chichester, 1979).

Lambek, J.

[66] *Completions of Categories*, Lecture Notes in Math. No. 24, Springer (Berlin, 1966).

Landweber, P. S.

[63] Three theorems on phrase structure grammars of type 1, *Information and Control* **6** (1963) 131–136. 362

Lang, S.

[58] *Introduction to Algebraic Geometry*, Interscience (New York, 1958). 39, 274

Lawvere, F. W.

[63] Functorial semantics of algebraic theories, *Proc. Nat. Acad. Sci. USA* **50** (1963) 869–872. 318

[63′] Algebraic theories, algebraic categories and algebraic functors, in *The Theory of Models*, 1963 Internat. Symp. Berkeley, North-Holland (Amsterdam, 1965) 413–418.

Lazard, M.

[54] Sur les algèbres enveloppantes universelles de certaines algèbres de Lie, *Publ. Sci. Univ. Alger*, Sér. A, **1** (1954) 281–294. 296

Lewin, J.

[68] On Schreier varieties of linear algebras, *Trans. Amer. Math. Soc.* **132** (1968) 553–562.

Lipshitz, L. and Saracino, D.

[73] The model companion of the theory of commutative rings without nilpotent elements, *Proc. Amer. Math. Soc.* **37** (1973) 381–387. 335

Lorenzen, P.

[53] Eine Bemerkung über die Abzählbarkeitsvoraussetzung in der Algebra, *Math. Zeits.* **57** (1953) 241–243.

Łos, J. 322

[55] On the extending of models I, *Fund. Math.* **42** (1955) 38–54. 226

[55′] *Quelques remarques, théorèmes et problèmes sur les classes définissables d'algèbres*, Math. Interpretation of formal systems, North-Holland (Amsterdam, 1955) 98–113. 210

Łukasiewicz, J. 120, 355, 376

Lyndon, R. C.

[54] Identities in finite algebras, *Proc. Amer. Math. Soc.* **5** (1954) 8–9. 173

[59] Properties preserved under algebraic constructions, *Bull. Amer. Math. Soc.* **65** (1959) 287–299.

[59′] Properties preserved under homomorphism, *Pacif. J. Math.* **9** (1959) 143–154. 237

[59″] Properties preserved in subdirect products, *Pacif. J. Math.* **9** (1959) 155–164.

Macdonald, I. G.

[60] Jordan algebras with three generators, *Proc. London Math. Soc.* (3) **10** (1960) 395–408. 308

Macintyre, A.

[72] On algebraically closed groups, *Ann. of Math.* **96** (1972) 53–97. 334

[79] Combinatorial problems for skew fields. I. Analogue of Britton's lemma and results of Adyan-Rabin type, *Proc. London Math. Soc.* (3) **39** (1979) 211–236.

Mac Lane, S. *see also* **Eilenberg, S.**

[63] *Homology*, Springer (Berlin, 1963) 60, 108, 141, 162

[65] Categorical algebra, *Bull. Amer. Math. Soc.* **71** (1965) 40–106.

[71] *Categories for the Working Mathematician*, Springer (Berlin, 1971). 316, 318

Malcev, A. I. 98, 142, 149, 162, 213, 228, 277, 322, 343

[37] On the immersion of an algebraic ring into a field, *Math. Ann.* **113** (1937) 686–691. 269

[39] Über die Einbettung von assoziativen Systemen in Gruppen I (Russian, German summary), *Mat. Sbornik* **6** (48) (1939) 331–336. 228, 265, 268

[40] Über die Einbettung von assoziativen Systemen in Gruppen II (Russian, German summary), *Mat. Sbornik* **8** (50) (1940) 251–264. 228, 269

[48] On the embedding of group algebras in division algebras (Russian), *Dokl. Akad. Nauk SSSR* **60** (1948) 1499–1501. 276

[50] On algebras with identical defining relations (Russian), *Mat. Sbornik* **26** (68) (1950) 19–33. 284

[52] On a representation of nonassociative rings (Russian), *Uspekhi Mat. Nauk* **7** (1952) 181–185. 184

[54] On the general theory of algebraic systems (Russian), *Mat. Sbornik* **35** (77) (1954) 3–20. 148, 165

[56] Subdirect product of models (Russian), *Dokl. Akad. Nauk SSSR* **109** (1956) 264–266. 236

[58] The defining relations in categories (Russian), *Dokl. Akad. Nauk SSSR* **119** (1958) 1095–1098.

[58′] The structural characteristic of some classes of algebras (Russian), *Dokl. Akad. Nauk SSSR* **120** (1958) 29–32.

[58″] On certain classes of models (Russian), *Dokl. Akad. Nauk SSSR* **120** (1958) 245–248. 236

[62] Axiomatizable classes of locally free algebras of certain types (Russian), *Sibirsk. Mat. Ž.* **3** (1962) 729–743.

[67] Multiplication of classes of algebraic systems (Russian), *Sibirsk. Mat. Ž.* **8** (1967) 346–365.

[73] *Algebraic Systems*, Springer (Berlin, 1973).

Manes, E. G.

[76] *Algebraic Theories*, Springer (Berlin, 1976). 315, 318

Marczewski, E.

[51] Sur les congruences et les propriétés positives d'algèbres abstraites, *Colloq. Math.* **2** (1951) 220–228.

[66] Independence in abstract algebras, results and problems, *Colloq. Math.* **14** (1966) 169–188.

McKinsey, J. C. C., *see also* **Diamond, A. H.**

[43] The decision problem for some classes of sentences without quantifiers, *J. symb. Logic* **8** (1943) 61–76. 235

McNaughton, R., *see* **Wang, H.**

Mezei, N. 368

Miller, G. A., *see* **Chomsky, N.**

Mitchell, B.

[65] *Theory of Categories*, Academic Press (New York, 1965) 311

Mlitz, R.

[77] *Jacobson's Density Theorem in Universal Algebra.* ***Contributions to Universal*** *Algebra*, North-Holland (Amsterdam, 1977) 331–340.

Moore, J. C., *see* **Eilenberg, S.**

Morel, A. C., *see* **Chang, C. C.,** *also* **Frayne, T.**

Morimoto, A. 169

Morley, M.

[65] Categoricity in power, *Trans. Amer. Math. Soc.* **114** (1965) 513–518.

[73] *Studies in Model Theory* (ed.), Math. Assn. America (Buffalo, 1973). 329

Mostowski, A.

[55] The present state of investigations on the foundations of mathematics, *Rozprawy Mat.* **IX** (Warsaw, 1955).

[57] A generalization of quantifiers, *Fund. Math.* **44** (1957) 12–36.

Moufang, Ruth

[37] Einige Untersuchungen über angeordnete Schiefkörper. *J. reine angew. Math.* **176** (1937) 203–223. 277

Myhill, J.

[57] *Finite Automata and the Representation of Events*, Wright Air Development Command Tech. Report No. 57–624 (1957) 112–137. 367, 369

[60] *Linear Bounded Automata*, Wright Air Development Command Tech. Report No. 60–165 (1960).

Nakayama, T. 98

Nash-Williams, C. St. J. A.

[63] On well-quasi-ordering finite trees, *Proc. Cambridge Phil. Soc.* **59** (1963) 833–835. 124

Nerode, A.
[58] Linear automaton transformations, *Proc. Amer. Math. Soc.* **9** (1958) 541–544. 367
Nesbitt, C. J., *see* **Artin, E.**
Neumann, B. H., *see also* **Higman, G.**
[37] Identical relations in groups I, *Math. Ann.* **114** (1937) 506–526. 177
[37′] Some remarks on infinite groups, *J. London Math. Soc.* **12** (1937) 120–127. 81
[49] On ordered division rings, *Trans. Amer. Math. Soc.* **66** (1949) 202–252. 276
[49′] On ordered groups, *Amer. Math. J.* **71** (1949) 1–18. 276
[51] Embedding non-associative rings in division rings, *Proc. London Math. Soc.* (3) **1** (1951) 241–256. 282
[54] An essay on free products of groups with amalgamations, *Phil. Trans. Roy. Soc.*, Ser. A **246** (1954) 503–554.
[73] The isomorphism problem for algebraically closed groups, in *Word problems* (ed. W. W. Boone *et al.*), North-Holland (Amsterdam, 1973) 553–562. 334
Neumann, B. H., Neumann, H., and Neumann, P. M.
[62] Wreath products and varieties of groups, *Math. Zeits.* **80** (1962) 44–62. 177
Neumann, H. *see also* **Higman, H. and Neumann, B. H.**
[56] On varieties of groups and their associated near-rings, *Math. Zeits.* **65** (1956) 36–69. 175
[67] *Varieties of Groups*, Springer (Berlin, 1967).
Neumann, P. M., *see* **Neumann, B. H.**
Newman, M. H. A.
[42] On theories with a combinatorial definition of "equivalence", *Ann. of Math.* **43** (1942) 223–243. 25f.
Nielsen, J. 255
Nivat, M.
[68] Transduction des langages de Chomsky, *Ann. Inst. Fourier* (Grenoble) 18, **1** (1968) 339–346. 369f.
[70] *Séries rationnelles et algébriques en variables non-commutatives*, Cours du DEA 1969–70.
Noether, E. 59
[21] Idealtheorie in Ringbereichen, *Math. Ann.* **83** (1921) 24–66. 77
Novikov, P. S.
[55] The algorithmic insolubility of the word problem in group theory, *Trudy Mat. Inst. Steklov, Akad. Nauk SSSR* No. 44 (1955); *AMS Translations* Ser. 2, **19** (1958) 1–122. 155
Ore, O.
[31] Linear equations in non-commutative fields, *Ann. of Math.* **32** (1931) 463–477. 275
[35] On the foundation of abstract algebra I, *Ann. of Math.* **36** (1935) 406–437.
[36] On the foundation of abstract algebra II, *Ann. of Math.* **37** (1936) 265–292. 77
[42] Theory of equivalence relations, *Duke Math. J.* **9** .(1942) 573–627.
[44] Galois connexions, *Trans. Amer. Math. Soc.* **55** (1944) 493–513. 44

Paige, L. J., *see* **Albert, A. A.**

Pareigis, B.

[69] *Kategorien und Funktoren*, Teubner (Stuttgart, 1969). 316, 318, 320

Parikh, R.

[61] Language generating devices, *MIT RLE Quarterly Progress Report* **60** (1961) 199–212. Reprinted as "On context-free languages", *J. Assoc. Comput. Mach.* **13** (1966) 570–581. 355

Paz, A. *see also* **Carlyle, J. W.**

[66] Some aspects of probabilistic automata, *Information and Control* **9** (1966) 26–60. 363

Peano, G. 7, 247

Perles, M. *see* **Bar-Hillel, Y.**

Peters, P. S. and Ritchie, R. W.

[71] On restricting the base component of transformational grammars, *Information and Control* **18** (1971) 483–501. 380

Pfender, M. *see* **Ehrig, H.**

Pierce, R. S. 142

[68] *Introduction to the Theory of Abstract Algebras*, Holt, Rinehart and Winston (New York, 1968).

Pixley, A. F.

[63] Distributivity and permutability of congruence relations in equational classes of algebras, *Proc. Amer. Math. Soc.* **14** (1963) 105–109.

Pólya, G. and Szegö, G.

[25] *Aufgaben und Lehrsätze aus der Analysis II*, Springer (Berlin, 1925). 376

Preston, G. B., *see* **Clifford, A. H.**

Rasiowa, H.

[52] A proof of the compactness theorem for arithmetical classes, *Fund. Math.* **39** (1952) 8–14.

[55] Algebraic models of axiomatic theories, *Fund. Math.* **41** (1955) 291–310.

Rasiowa, H. and Sikorski, R.

[50] A proof of the completeness theorem of Gödel, *Fund. Math.* **37** (1950) 193–200.

[51] A proof of the Skolem-Löwenheim theorem, *Fund. Math.* **38** (1951) 230–232.

[63] *The Mathematics of Metamathematics*, Pan. Wyd. Nauk (Warsaw, 1963).

Reid, J. D. 145

Ritchie, R. W., *see* **Peters, P. S.**

Robinson, A. 324, 329

see also **Barwise, J.**

[63] *Introduction to Model Theory and the Metamathematics of Algebra*, North-Holland (Amsterdam, 1963). 218, 223, 226

[71] On the notion of algebraic closedness for non-commutative groups and fields, *J. symb. Logic* **36** (1971) 441–444.

Rosenbloom, P. C.

[50] *Elements of Mathematical Logic*, Dover (New York, 1950). 118

Russell, Bertrand. 1f.

[19] *Introduction to Mathematical Philosophy*, Allen and Unwin (London, 1919).

Ryll-Nardzewski, C. *see* **Goetz, A.**
Sabbagh, G. *see* **Eklof, P. C.**
Sacks, G.
[72] *Saturated Model Theory*, Benjamin (New York, 1972).
Salomaa, A. and Soittola, M.
[78] *Automata-Theoretic Aspects of Formal Power Series*, Springer (Berlin, 1978).
Samuel, P. *see also* **Zariski, O.**
[48] On universal mappings and free topological groups, *Bull. Amer. Math. Soc.* **54** (1948) 591–598. 108, 136
Saracino, D. *see* **Lipshitz, L.**
Schmidt, J.
[52] Über die Rolle der transfiniten Schlussweisen in einer allgemeinen Idealtheorie, *Math. Nachr.* **7** (1952) 165–182. 45, 81
[55] Eine verallgemeinerte Wohlordnung und die Endlichkeitsbedingungen der Wohlordnungstheorie, *Arch. Math.* **6** (1955) 374–381.
Schubert, H.
[72] *Categories*, Springer (Berlin, 1972).
Schützenberger, M.-P. 180
see also **Chomsky, N.**
[61] A remark on finite transducers, *Information and Control* **4** (1961) 185–196. 364, 371
[61′] On the definition of a family of automata, *Information and Control* **4** (1961) 245–270. 375
[62] On a theorem of Jungen, *Proc. Amer. Math. Soc.* **13** (1962) 885–890. 372ff., 377
Scott, D. S. *see also* **Frayne, T. and Kalicki, J.**
[56] Equationally complete extensions of finite algebras, *Indag. Math.* **18** (1956) 35–38. 177f.
Scott, W. R.
[51] Algebraically closed groups, *Proc. Amer. Math. Soc.* **2** (1951) 118–121. 262
Šestakov, I. P. *see* **Ževlakov, K. A.**
Shamir, E. *see also* **Bar-Hillel, Y.**
[67] A representation theorem for algebraic and context-free power series in non-commuting variables, *Information and Control* **11** (1967) 239–254.
Shelah, S.
[71] Every two elementarily equivalent models have isomorphic ultrapowers, *Isr. J. Math.* **10** (1971) 224–233. 243
[71′] Stability, the finite cover property and superstability, *Ann. Math. Logic* **3** (1971) 271–371. 321
[74] Infinite abelian groups, – Whitehead's problem and some constructions, *Isr. J. Math.* **18** (1974) 243–256.
Shoda, K.
[49] Allgemeine Algebra, *Osaka Math. J.* **1** (1949) 182–225. 153
Sierpiński, W.
[45] Sur les fonctions de plusieurs variables, *Fund. Math.* **33** (1945) 169–173. 338

[58] *Cardinal and Ordinal Numbers*, Pan. Wyd. Nauk (Warsaw, 1958). 28, 32

Sikorski, R. *see also* **Rasiowa, H.**

[53] Products of abstract algebras, *Fund. Math.* **39** (1953) 211–228. 144

[60] *Boolean Algebras*, Springer (Berlin, 1960). 198

Širšov, A. I., *see also* **Ževlakov, K. A.**

[53] On a representation of Lie rings in associative rings (Russian), *Uspekhi Mat. Nauk* **5** (1953) 173–175. 296

[54] The subalgebras of free commutative and free anticommutative algebras (Russian), *Mat. Sbornik* **34** (76) (1954) 81–88. 287

[56] On special *J*-rings (Russian), *Mat. Sbornik* **38** (80) (1956) 149–166. 307

[58] On free Lie rings (Russian), *Mat. Sbornik* **45** (87) (1958) 113–122. 292

Skolem, Th. 213

Skornyakov, L. A.

[57] *T*-homomorphisms of rings (Russian), *Mat. Sbornik* **42** (84) (1957) 425–440. 282

Slinko, A. M. *see* **Ževlakov, K. A.**

Słominski, J.

[59] The theory of abstract algebras with infinitary operations, *Rozprawy Mat.* **XVIII** (Warsaw, 1959). 55

Slomson, A. B. *see* **Bell, J. L.**

Soittola, M. *see* **Salomaa, A.**

Sonner, J.

[62] The formal definition of categories, *Math. Zeits.* **80** (1962) 163–176. 6

Specht, W.

[50] Gesetze in Ringen, I, *Math. Zeits.* **52** (1950) 557–589. 284

Stanley, M. G.

[66] Generation of full varieties, *Michigan Math. J.* **13** (1966) 127–128. 172

Stearns, R. E. *see* **Hartmanis, J.**

Steenrod, N. *see* **Eilenberg, S.**

Stevens, M. L.

[69] *On Certain Varieties of Universal Algebras*, Ph.D. thesis (London University, 1969).

Stone, A. H. *see* **Higman, G.**

Strecker, G. E. *see* **Herrlich, H.**

Suszko, R. 322

Suzuki, M.

[56] *Structure of a Group and the Structure of Its Lattice of Subgroups*, Springer (Berlin, 1956). 71

Szabo, I.

[78] Concrete representations of related structures of universal algebras I, *Acta Sci. Math.* **40** (1978) 175–184.

Szegö, G. *see* **Pólya, G.**

Taimanov, A. D.

[59] A class of models closed with respect to direct union (Russian), *Dokl. Akad. Nauk SSSR* **127** (1959) 1173–1175.

[62] Characterization of axiomatizable model classes (Russian) *Algebra i Logika Sem.* **1**, No. 4 (1962) 5–31. 215

Tamari, D.

[62] The algebra of bracketings, and their enumeration, *Nieuw Arch. v. Wiskunde* (3) **10** (1962) 131–146. 268, 289

Tarski, A. 162, 180, 195, 197

see also **Henkin, L. and Jónsson, B.**

[36] Der Wahrheitsbegriff in den formalisierten Sprachen, *Studia Philos.* **1** (1936) 261–405. Translated in *Logic, Semantics, Metamathematics*, Oxford Univ. Press (Oxford, 1956).

[46] A remark on functionally free algebras, *Ann. of Math.* **47** (1946) 163–165.

[50] Some notions and methods on the borderline of algebra and metamathematics, *Proc. Int. Cong. Math.* Vol. 1 (Cambridge, Mass., 1950), 705–720.

[54] Contributions to the theory of models, *Indag. Math.* **16** (1954) 572–588; **17** (1955) 56–64.

[55] A lattice-theoretic fixpoint theorem and its applications, *Pacif. J. Math.* **5** (1955) 285–309. 22

[56] Equationally complete rings and relation algebras, *Indag. Math.* **18** (1956) 39–46.

[58] Remarks on predicate logic with infinitely long expressions, *Colloq. Math.* **6** (1958) 171–176.

Tarski, A. and Vaught, R. L.

[57] Arithmetical extensions of relational systems, *Comp. Math.* **13** (1957) 81–102. 229f.

Teissier, M.

[51] Sur les équivalences régulières dans les demi-groupes, *C. R. Paris Acad. Sci.* **232** (1951) 1987–1989. 367

Thrall, R. M. *see* **Artin, E.**

Tietze, H. 153

Turakainen, P.

[69] Generalized automata and stochastic languages, *Proc. Amer. Math. Soc.* **21** (1969) 303–309. 364

Ullman, J. D. *see* **Hopcroft, J. E.**

Vaught, R. L. *see also* **Tarski, A.**

[54] Applications of the Löwenheim-Skolem-Tarski theorem to problems of completeness and decidability, *Indag. Math.* **16** (1954) 467–472. 232

[54′] Remarks on universal classes of relational systems, *Indag. Math.* **16** (1954) 589–591.

[54″] On sentences holding in direct products or relational systems, *Proc. Int. Cong. Math.* Vol. 2 (Amsterdam, 1954) 409–410. 235

[63] Models of complete theories, *Bull. Amer. Math. Soc.* **69** (1963) 299–313.

Volger, H.

[68] Über die Existenz von freien Algebren, *Math. Zeits.* **106** (1968) 312–320.

Vopenka, P.

[62] A method of constructing a non-standard model of the Bernays-Gödel axio-

matic theory of sets, *Dokl. Akad. Nauk SSSR* **143** (1962) 11–12. Translated in *Soviet Math. Doklady* 3 (1962) 309–310. 251

v. d. Waerden, B. L.

[37] *Moderne Algebra* I, Springer (Leipzig, 1937). 59, 218, 252

Wang. H. and McNaughton, R.

[53] *Les systèmes axiomatiques de la théorie des ensembles*, Hermann (Paris, 1953). 1

Welsh, D. J. A.

[76] *Matroid Theory*, LMS Monographs No. 8, Academic Press (London, New York, 1976). 262

Wheeler, W. H. *see* **Hirschfeld, J.**

Whitehead, A. N.

[1898] *A Treatise on Universal Algebra, with Applications* I Cambridge Univ. Press (Cambridge, 1898), reprinted New York, 1960.

Whitman, P. M. *see* **Birkhoff, G.**

Whitney, H. 262

Witt, E.

[37] Treue Darstellung Liescher Ringe, *J. reine angew. Math.* **177** (1937) 152–160. 294–296

[53] Über freie Ringe und ihre Unterringe, *Math. Zeits.* **58** (1953) 113–114. 287

Wood, C.

[76] The model theory of differential fields revisited, *Isr. J. Math.* **25** (1976) 331–352. 326

Wright, J. B. *see* **Eilenberg, S.**

Yaqub, F. M. *see* **Dwinger, P.**

Zariski, O. and Samuel, P.

[58] *Commutative Algebra* I, v. Nostrand (Princeton, 1958). 44, 78, 252

Ževlakov, K. A., Slinko, A. M., Šestakov, I. P., and Širšov, A. I.

[78] *Kol'tsa, blizkie k assotsiativnym* (Rings close to associative ones), Nauka (Moscow, 1978).

List of Special Symbols

Number refers to the page on which the symbol is defined or first used.

$\mathscr{L}(\Omega)$	Lindenbaum algebra over Ω, 207
N	Set of natural numbers (= nonnegative integers), 6 f.
nat $\mathfrak{q}$	Natural mapping associated with the equivalence $\mathfrak{q}$, 15
nat $\mathscr{D}$	Natural mapping of a product associated with a filter $\mathscr{D}$ on the index set, 210
$o(A)$	Order-type of an ordered set A (or ordinal number of a well-ordered set A), 28
$\mathcal{O}(A)$, $\mathcal{O}_n(A)$	Set of all finitary, n-ary operations on a set A, 126
Ob($\mathscr{K}$)	Class of objects of $\mathscr{K}$, 36
P$\mathscr{K}$	Category of products of $\mathscr{K}$-algebras, 105
P$\mathscr{C}$	Class of products of $\mathscr{C}$-models, 221
Q$\mathscr{K}$	Category of quotients of $\mathscr{K}$-algebras, 105
r_a, r_b	Axiom rank, base rank, 173
R$\mathscr{K}$	Category of residually $\mathscr{K}$-algebras, 106
Rg	Category of associative rings, 52
S$\mathscr{K}$	Category of subalgebras of $\mathscr{K}$-algebras, 105
S$\mathscr{C}$	Class of submodels of $\mathscr{C}$-models, 221
$\mathscr{S}(A)$	Set of left segments of an ordered set A, 21
Sg	Category of semigroups, 165
St	Category of sets and mappings, 37
$\mathscr{T}[\Omega]$	Model space over Ω, 208
$U(X)$	Universal functor, applied to X, 110
$V(\Gamma)$	Set of vertices of a graph Γ, 25
$V_\Omega(\Sigma)$	Variety of Ω-algebras defined by a set of laws Σ, 162
$W_\Omega(X)$	Ω-word algebra on X, 116
$\Gamma(\Omega)$	Lattice of categories of Ω-algebras, 105
Δ_A	Diagonal on A, 10
Ω	Operator or predicate domain, 48, 189

$\Omega(n)$	Domain of n-ary operators or $(n+1)$-ary predicates, 48, 189
(Ω)	Category of Ω-algebras and homomorphisms, 50, 104
$[\Omega]$	Class of all Ω-structures, 205
Φ^{-1}	Inverse of a correspondence Φ, 10
$\Phi \circ \Psi$	Composite of correspondences Φ, Ψ, 10
$x^{\mathfrak{q}}$	$\mathfrak{q}$-class containing x, 14
$A/\mathfrak{q}$	Set of $\mathfrak{q}$-classes on A (quotient set), 15
$\lvert A \rvert$	Cardinal number of the set A, 28
A/C	Algebra extension, 54
$[a, b]$	Interval in a lattice, 64
$\prec$	Is subordinate to (for categories), 110
$\coprod, \prod$	Free, direct composition, 113
$\mathscr{K}\{X \mid \Phi\}$	Presentation of a $\mathscr{K}$-algebra with generating set X and defining relations Φ, 150
$\hat{A}$	Variety generated by the algebra A, 172
$M \models \omega(a)$	The relation $\omega(a)$ holds in M, 189
$M \vdash \omega$	ω is valid in M, 189
$\bigvee, \bigwedge$	Existential, universal quantifier, 201
$\Pi M_\lambda / \mathscr{D}$	Reduced product, 210
$\langle M, X \rangle$	Unary enlargement of M by the constants X, 229

Subject Index

Mathematics and Its Applications

1. Willem Kuyk, *Complementarity in Mathematics, A First Introduction to the Foundations of Mathematics and Its History.* 1977.
2. Peter H. Sellers, *Combinatorial Complexes, A Mathematical Theory of Algorithms.* 1979.
3. Jacques Chaillou, *Hyperbolic Differential Polynomials and their Singular Perturbations.* 1979.
4. Svtopluk Fučik, *Solvability of Nonlinear Equations and Boundary Value Problems.* 1980.
5. Willard L. Miranker, *Numerical Methods for Stiff Equations and Singular Perturbation Problems.* 1980.
6. P. M. Cohn, *Universal Algebra.* 1981.
7. Vasile I. Istrăţescu, *Fixed Point Theory, An Introduction.* Forthcoming.

2